Methodologies and Applications of Intelligent Motion Control Systems

R. Vettriselvan
https://orcid.org/0009-0005-1097-0806
AMET University, India

N.V. Suresh
https://orcid.org/0000-0002-0393-6037
ASET College, India

IGI Global
Scientific Publishing
Publishing Tomorrow's Research Today

Vice President of Editorial Melissa Wagner
Director of Acquisitions Mikaela Felty
Director of Book Development Jocelynn Hessler
Production Manager Mike Brehm
Cover Design Jose Rosado

Published in the United States of America by
IGI Global Scientific Publishing
701 East Chocolate Avenue
Hershey, PA, 17033, USA
Tel: 717-533-8845 | Fax: 717-533-7115
Website: https://www.igi-global.com E-mail: cust@igi-global.com

Library of Congress Cataloging-in-Publication Data

Names: Vettriselvan, R. editor | Suresh, N. V., 1992- editor
Title: Methodologies and applications of intelligent motion control systems / edited by: R. Vettriselvan, N.V. Suresh.
Description: Hershey PA : IGI Global Scientific Publishing, [2026] | Includes bibliographical references. | Summary: "This book aims to understand the technologies involved in product design and development, with the focus on Motion Control Systems for increased production"-- Provided by publisher.
Identifiers: LCCN 2025034453 (print) | LCCN 2025034454 (ebook) | ISBN 9798369336373 hardcover | ISBN 9798369351697 paperback | ISBN 9798369336380 ebook
Subjects: LCSH: Motion control devices | Artificial intelligence--Industrial applications
Classification: LCC TJ214.5e .M48 2025 (print) | LCC TJ214.5e (ebook)
LC record available at https://lccn.loc.gov/2025034453
LC ebook record available at https://lccn.loc.gov/2025034454

British Cataloguing in Publication Data
A Cataloguing in Publication record for this book is available from the British Library.

Table of Contents

Detailed Table of Contents

stabilization, vibration correction and thermal offsetting. The complex techniques such as model predictive control (MPC) and machine learning are argued and practical reason demonstrated with the help of Arduino, Marlin, Raspberry Pi, and BeagleBone. Reliability is focused on by calibration, diagnostics and predictive maintenance. Optimization of G-codes, smoothing of paths, and acceleration control are benchmarked, which can provide practical experience in the field of the three main areas of digital manufacturing, robotics, and intelligent production systems

This chapter discusses the combination of collective intelligence and swarm robotics as decentralised methods of motion control. The research is based on an example of biological systems that exhibit strong, scalable, and adaptive global dynamics by the simple rules or programmes applied locally. Reinforcement learning and evolutionary optimization have been demonstrated to be more efficient, resilient, and have better energy performance through simulations, physical verification, and statistical analysis. There is a proposed motion control design that is layered and has a balance between autonomy and hybrid oversight, energy optimization, and ethical governance. The results leave decentralized swarm robotics as the paradigm of resilient, adaptive and ethical motion control in a variety of applications.

In this chapter, the researcher studies how motion analytics enabled by smart sensing and data modeling provides an effective way of predictive maintenance and managing assets within industrial systems. It talks about how time based diagnostics and anomaly detection algorithms in the past time along with the history trends could be used to predict machine failures even before it has happened. The combination of machine learning and Internet of Things (IoT) platforms is stressed as one of the enablers of real-time decision making and low rates of downtimes. Such implementation issues as data quality, sensor calibration, system scalability are also tackled in this chapter. It lays stress on the importance of motion analytics with respect to reliable and efficient planning of lifecycle management and key assets.

This chapter explores how intelligent motion control systems can be utilised to drive sustainable civil engineering through greater efficiency, precision, and safety. Based on a mixed-method study of 650 respondents and an expert interviews, the results show that if motion controlled equipment is implemented on a large scale, it could significantly increase energy efficiency, decrease material waste and emissions, and improve employees' safety. Evidence to substantiate strong relationships between adoption and sustainability outcomes came from ANOVA, regression, and SEM. It also highlights the financial, cultural and technical challenges facing the sustainable development of infrastructures in the Industry 5.0 era and proposes practical solutions and governance strategies, as well as directions for future research

Geopolitical, environmental, and technological uncertainty is becoming a growing disruptive factor within global supply chains, and resilience beyond conventional efficiency models is increasingly important. This chapter discusses the roles of

Preface

The edited volume *"Methodologies and Applications of Intelligent Motion Control Systems"* embodies a concerted global academic effort to advance the understanding and application of Artificial Intelligence (AI) and automation in motion control engineering. This compendium is the culmination of multidisciplinary research contributions from scholars and professionals across seven nations India, Malaysia, Germany, Uzbekistan, Zambia, Oman, and England reflecting the diverse, interconnected, and evolving landscape of intelligent systems research.

In the 21st-century technological paradigm, motion control systems have transcended traditional automation boundaries to become intelligent, adaptive, and predictive. The convergence of AI, machine learning, robotics, and control theory has not only revolutionized industrial processes but has also transformed healthcare, logistics, energy, and infrastructure sectors. This book captures this dynamic transformation through thirty-two rigorously reviewed chapters that explore both foundational theories and real-world applications of intelligent motion control systems.

The chapters collectively address key themes, including AI-optimized controller design, adaptive algorithms for renewable power systems, human–AI collaboration in robotics, smart warehousing, predictive maintenance, cyber-physical systems, and ethical frameworks in automation. Each contribution integrates cutting-edge research with practical insights, presenting comprehensive methodologies, simulations, and case analyses that bridge the gap between academic inquiry and industrial deployment. By incorporating Industry 5.0 principles, the book also underscores the human-centric dimension of technology emphasizing inclusivity, sustainability, and ethical responsibility in intelligent system design.

CHAPTER OVERVIEW

Chapter 1: AI-Based Optimization of Controllers and Converters in Motion Control Systems

This chapter discusses optimization of motion control controllers and power converters using artificial intelligence, whose aim is to combine conventional approaches and artificial intelligence in realizing low-cost adaptive implementations and fault-tolerant behavior. Methods, such as neural networks, Genetic algorithms, fuzzy logic and reinforcement learning are used to optimize motor drive control, torque ripple and converter switching. Emphasis on real-time implementation through MATLAB/Simulink, FPGA, and hardware-in-the-loop is put forward. The chapter includes such points as hybrid intelligent frameworks, energy efficiency, and compatibility with Industry 4.0, IIoT, and digital twins. Of value to engineers and researchers, it deals with the major issues of safety, scalability, and data limitation in the development of intelligent industrial control systems.

Chapter 2: Motion Control Systems for 3D Printing Design, Integration, and Optimization

This chapter on motion control in 3D printing and Cartesian and Delta systems and the integration of stepper motors stepper, servo drives, controllers, and sensors to provide high precision and resolution. It considers axis cranking, kinematic stabilization, vibration correction and thermal offsetting. The complex techniques such as model predictive control (MPC) and machine learning are argued and practical reason demonstrated with the help of Arduino, Marlin, Raspberry Pi, and BeagleBone. Reliability is focused on by calibration, diagnostics and predictive maintenance. Optimization of G-codes, smoothing of paths, and acceleration control are benchmarked, which can provide practical experience in the field of the three main areas of digital manufacturing, robotics, and intelligent production systems

Chapter 3: Collective Intelligence and Swarm Robotics: Decentralized Strategies for Adaptive and Resilient Motion Control

This chapter discusses the combination of collective intelligence and swarm robotics as decentralised methods of motion control. The research is based on an example of biological systems that exhibit strong, scalable, and adaptive global dynamics by the simple rules or programmes applied locally. Reinforcement learning and evolutionary optimization have been demonstrated to be more efficient, resilient, and have better energy performance through simulations, physical verification, and statistical analysis. There is a proposed motion control design that is layered and has a balance between autonomy and hybrid oversight, energy optimization, and ethical governance. The results leave decentralized swarm robotics as the paradigm of resilient, adaptive and ethical motion control in a variety of applications.

Chapter 4: Predictive Maintenance and Asset Management Using Motion Analytics

In this chapter, the researcher studies how motion analytics enabled by smart sensing and data modeling provides an effective way of predictive maintenance and managing assets within industrial systems. It talks about how time based diagnostics and anomaly detection algorithms in the past time along with the history trends could be used to predict machine failures even before it has happened. The combination of machine learning and Internet of Things (IoT) platforms is stressed as one of the enablers of real-time decision making and low rates of downtimes. Such implementation issues as data quality, sensor calibration, system scalability are also tackled in this chapter. It lays stress on the importance of motion analytics with respect to reliable and efficient planning of lifecycle management and key assets.

Chapter 5: Advancing Sustainable Civil Engineering through Intelligent Motion Control: Construction Equipment and Infrastructure Development

This chapter explores how intelligent motion control systems can be utilised to drive sustainable civil engineering through greater efficiency, precision, and safety. Based on a mixed-method study of 650 respondents and an expert interviews, the results show that if motion controlled equipment is implemented on a large scale, it could significantly increase energy efficiency, decrease material waste and emissions, and improve employees' safety. Evidence to substantiate strong relationships between adoption and sustainability outcomes came from ANOVA, regression, and SEM. It also highlights the financial, cultural and technical challenges facing the sustainable development of infrastructures in the Industry 5.0 era and proposes practical solutions and governance strategies, as well as directions for future research

Chapter 6: Cyber-Physical Systems and Cloud Robotics for Global Supply Chain Resilience

Geopolitical, environmental, and technological uncertainty is becoming a growing disruptive factor within global supply chains, and resilience beyond conventional efficiency models is increasingly important. This chapter discusses the roles of Cyber-Physical Systems (CPS) and cloud robotics to achieve resilience through greater visibility, flexibilities, redundancy, and recovery. Descriptive statistics, ANOVA and structural modelling are employed to interpret the findings of adoption using collected empirical data of 650 respondents by industry. Findings confirm that CPS is superior to cloud robotics in monitoring and recovery, and adaptability and redundancy. Recommendations on technological integration, organisational preparedness, regulatory frameworks and future research directions are given towards the end of the chapter

Chapter 7: Decision-Making Frameworks for Integrating Motion Control in Business Operations

This chapter includes some strategic plans of how to incorporate motions control technologies in the business to promote efficiency, agility, and to make data-driven decisions. It cogitates about the real-time motion tracking and the focus on optimization in the logistics sphere, production and asset management through automation and smart analytics. The chapter explains how AI, IoT, control algorithms help harmonize the processes of motion with the goals of organizations. They lay emphasis on decision theory, performance measures and on the flexibility of motion systems in dynamic environments. The chapter has shed light on the ability of motion control to become the engine of innovation and competition in the operation of the contemporary enterprise by offering scalable and practical integration models.

Chapter 8: Digital Twin Integration for Predictive and Real-Time Motion Control in Infrastructure Engineering

Digital twin implementation is a innovative concept of predictive and real-time motion control of infrastructure engineering. Digital twins allow foretelling the future by aligning virtual models with real-life assets using ongoing data communications to provide quick reactivity and improved system efficiency. The chapter examines the conceptual, theoretical, and methodological basis of digital twin-based motion control, compares results, solutions, and directions of future research. The focus is on technical innovation, governance, human capacity, and sustainability and how digital twins can be used to support resilient, adaptive, and sustainable infrastructure systems in line with the global development objectives.

Chapter 9: Governance, Security, and Ethical Considerations in AI-Driven Motion Control Systems

Motion control systems based on AI are revolutionising industries with their accuracy, flexibility, and efficiency but pose serious governance, security and ethical issues. The chapter focuses on the intersection of institutional frameworks, cybersecurity practises and ethical principles in the development of responsible adoption. It presents the connections between governance, resilience, and ethics as interdependent based on interdisciplinary theories, mixed-method analysis, and case studies. Among the crucial findings, there is the necessity to have adaptive regulations, strong security-by-design, and ethical alignment to provide transparency, accountability, and social justice. The chapter suggests the combined solutions and describes the future research directions to ensure the sustainable and reliable innovation in AI-driven motion control.

Chapter 10: AI-Powered Motion Control for Lean Manufacturing and Workflow Automation

In this chapter, this is how computer intelligence is put to use in improving motion control schemes in an environment of lean manufacturing such that automation is achieved in real-time, precision, and flexibility. It evaluates how AI can be connected with robotics, machine vision, and feedback control to automate work processes, reduce wastes, and increase the efficiency of production. Appropriate applications include predictive load balancing, dynamic path optimization and adaptive response to operational variability, which are emphasized in the chapter. It also speaks of the difficulties of AI implementation such as the quality of data, system integration, and labor preparedness. The research highlights the emission ability of AI-based indication control to realize flexible, cost-effective and clever manufacture systems.

Chapter 11: Mathematical Perspectives on Advanced Motion Control Techniques for Modern Mechatronic Applications

The field of technology is advancing, and as much as there is a need to make use of the labor force more competently, innovations of motion control have been invented in order to make the mechatronic system more effective and safer. Advanced control schemes have enabled flawless tracking of motion in mechatronic systems, which also require less energy and less servicing of mechanical parts. Advanced controllers have an excellent performance over traditional controllers where the environment is probabilistic, as such controllers can be trained to adapt to changes in the load, temperature, or dynamics of any system. Hence, every motion control method possesses its controlling nature, and that is used on various applications depending on the situation and requirement. The chapter will now look at the mathematical approach of various advanced techniques of motion control for mechatronic systems. The techniques must be able to alter the position, velocity, acceleration, and deceleration in a required way within a desired span of time, parallelly.

Chapter 12: Optimization of Robotic Material Handling Systems in Industrial Supply Chains

This chapter focuses on how the use of robotic systems is being used to optimize industrial supply chains in material handling. It addresses how doing so using intelligent motion planning, integrating real-time sensor information and applying variable control can result into a higher level of efficiency, accuracy, and throughput. Automated guided vehicles (AGVs), machine vision, and robotic arms are the main technologies which are assessed with reference to their contribution to the decrease in cycle time and operation costs. Scalability, synchronization, and interoperability in a system also come into the picture in the chapter. In pointing out that data-driven logistics is vital to the study, it is important to note the catalyzing aspect of robotics in the construction of resilient, responsive, and intelligent industrial work of the supply chain

The editorial process has adhered to rigorous academic standards. All submissions underwent a double-blind peer review by multiple independent experts, ensuring the highest levels of quality, originality, and scholarly integrity. The volume's thematic coherence and methodological depth were refined through extensive revisions and editorial guidance, supported by a strong international network of reviewers and contributors.

This book is intended to serve as a comprehensive reference for researchers, engineers, academicians, and policy professionals seeking to deepen their understanding of AI-driven control systems. It also provides pedagogical value to postgraduate students and industry practitioners aiming to integrate intelligent motion control technologies into real-world systems. The diversity of its content ensures that readers gain both conceptual clarity and application-oriented perspectives.

We believe that *"Methodologies and Applications of Intelligent Motion Control Systems"* will contribute meaningfully to the global discourse on automation and intelligent control, stimulating further research and collaboration toward sustainable, human-centered innovation.

R. Vettriselvan
Academy of Maritime Education and Training (AMET) University, Chennai, India

N. V. Suresh
ASET College of Science and Technology, Chennai, India

Chapter 1
AI-Based Optimization of Controllers and Converters in Motion Control Systems

Palanivel Rathinasabapathi Velmurugan
https://orcid.org/0000-0002-0395-9060
Berlin School of Business and Innovation, Germany

Mamidala Vijay Karthik
https://orcid.org/0000-0001-7149-7868
CMR Engineering College, India

Savitha Venkatesan
https://orcid.org/0000-0002-3284-9610
Saveetha Engineering College, Chennai, India

A. Sadeesh kumar
https://orcid.org/0000-0001-7173-3605
St. Joseph's College of Engineering, Chennai, India

S. Madhurikkha
Jaya Sakthi Engineering College, Thiruninravur, India

ABSTRACT

This chapter discusses optimization of motion control controllers and power converters using artificial intelligence, whose aim is to combine conventional approaches and artificial intelligence in realizing low-cost adaptive implementations and fault-tolerant behavior. Methods, such as neural networks, Genetic algorithms, fuzzy logic

DOI: 10.4018/979-8-3693-3637-3.ch001

and reinforcement learning are used to optimize motor drive control, torque ripple and converter switching. Emphasis on real-time implementation through MATLAB/Simulink, FPGA, and hardware-in-the-loop is put forward. The chapter includes such points as hybrid intelligent frameworks, energy efficiency, and compatibility with Industry 4.0, IIoT, and digital twins. Of value to engineers and researchers, it deals with the major issues of safety, scalability, and data limitation in the development of intelligent industrial control systems.

INTRODUCTION

The swiftly changing environment of industrial automation, robotics and smart manufacturing has resulted in a dire need of efficient and improved motion control systems with resulting higher levels of precision, flexibility and energy efficiency. There are certain limitations to the traditionally established control structures though these are still applied everywhere and lie at the foundation of the control. The latter is specifically obvious in their failure to cope with nonlinearities of a system, system uncertainty associated with parameters, and real-time fault conditions. To counter this complexity shift, the ability to use Artificial Intelligence (AI) in designing and optimizing controller and converter systems has provided an innovative paradigm shift to create intelligent, more robust and strong motion control systems. The most important aspects of the motion systems are the controllers and converters that direct the flow of electrical energy to electromechanical devices like motors, and hence that determine how the devices react to dynamic signals and to varying loads. Although classical control techniques, including the Proportional-Integral-Derivative (PID) controllers and pulse-width modulation (PWM) strategies have been widely used because they are easy to implement and effective in well-defined operating regimes, they are usually ineffective when simultaneous high speed and high accuracy demands are imposed (Gao et al, 2023). Such is especially the case in the domain of advanced manufacturing where performance demands have gone beyond the static calibration to instead necessitate real-time learning, fault prediction and optimal energy management systems. Using differing approaches, Artificial Intelligence proposes the possibility of modeling those complex systems in those nonlinear systems, which is not presented by specific mathematical formulations, such artificial neural networks (ANNs), genetic algorithms (GAs), fuzzy logic systems, reinforcement learning (RL). These models are trained on data, they can generalize to changes and can improve performance measures according to specific objectives. Another area in optimization through AI is the field of motor controllers (including brushless DC motors (BLDC), permanent magnet synchronous motors (PMSMs), inverter-fed induction machines). The implementation of AI has proven to be very effective in eliminating torque ripple and increases in the promptness of dynamic responses, as well as the development of higher levels of fault tolerance. Converters are used as the interface between power sources and electromechanical loads and AI offers advantage through adaptive switching strategies and predictive modulation techniques that better serve energy efficiency and minimize thermal tasks and electromagnetic noise.

Additionally, the development of Industry 4.0 has introduced greater demand on real-time responsiveness, connectedness and intelligence via data. The motion control systems are anticipated to be a part of the integrated cyber-physical ecosystem and use Industrial Internet of Things (IIoT) architecture and digital twin frameworks. In this respect, intelligent computer-integrated control systems turn not only desirable but mandatory (Li et al, 2021). The unique propensity to learn and adapt through the feedback provided by the system guarantees their stability of operations despite external circumstances, changes in workloads, ageing or deterioration of components. Such dynamic flexibility is critical to the minimization of unplanned downtimes and maximization of equipment life, as well as gaining predictive maintenance levels, which are characteristic of contemporary smart manufacturing facilities. The other major breakthrough has been the innovation and implementation of hybrid control topologies that cleverly combine traditional existing control strategies with compute-based optimization strategies informed by AI. These AI modules do not replace classical controllers, they complement the classical controller, getting its parameters right, anticipating perturbations, producing effective fault solving that can detect and recover faults. There is the added benefit with such hybrid systems, that you have got the stability that has been proven with classical control and then the flexibility that the machine learning can provide. Such combination opens up to the aspects of self-optimization and self-healing which is important in Industrial environments with the importance of mission-critical applications as well. A big section of this chapter will focus on the elaboration of these AI-enhanced architectures and the plan of how they will be implemented in real life (Khan et al, 2022). Particular focus is provided to the platforms that are widely used in the research and prototyping: MATLAB/ Simulink to run simulation and FPGA based systems to run real-time realization of hardware. Furthermore, the open-source platforms like Arduino Mega, Raspberry Pi, and BeagleBone are also taken under consideration because of their role in embedding low cost prototyping and embedded control systems development. These platforms allow both researchers and engineers to ensure that the simulation to deployment gap is rounded out so that theoretical models can be tested and corrected under as-designed situations.

Using this development, the chapter further explains the significance of the AI-founded tuning mechanisms in enhancing major performance gauges which are system stability, response time, energy consumption, and heat management. The application of AI techniques is demonstrated to make the switching behavior of converters smarter achieving the reduction of stress on the electronic components and increasing the long-term reliability. Learning capabilities help AI to realize systems that can predict a future breakdown, and have fail-proof negation due to predictive maintenance. Such systems are able to make intelligent decisions that are not capable of pre-programmed logic or fixed thresholds with the help of historical and real-time data. Also, the issues and drawbacks of the implementation of AI in the motion control systems are discussed. They are shortage of training data, real-time inference latency, AI-driven control action safety certification, and scalability of models on a wide range of hardware. The chapter recognizes these areas of concern and suggests viable mitigation steps like simplification of models, on-device training and using digital twin simulations to validate control strategies in a risk-free simulator. In addition to that, the economic and sustainable sides of AI-enhanced motion control are discussed. Energy-efficient controls are of the essence as industries seek to minimize energy footprint and shift to green mode of manufacturing (Lalitha et al, 2024). Monitoring, predicting and optimising energy usage patterns is the ability which AI can comprises, and which contributes to sustainability goals, as well as reduction in the operational costs. As an example, smart algorithms may modify the switching frequency in converters according to the demand on the load thus cutting power losses and heat. These savings translate into sizeable economic and environmental payoffs in large-scope activities.

This chapter is a two-fold purpose. It is, first of all, a thorough technical introduction into the area of motion control systems development and implementation by engineers, researchers, and graduate students. Second, it provides a future-oriented vision since it places these technologies in the context of the further development of intelligent automation systems. The merger of AI, mechatronics and control theory is no longer a gradual step, but the fundamental paradigm in a manner of conception, operation and maintenance of motion systems. AI will have a fundamental effect by allowing an escalation of real-time, data-driven, and self-adaptive controls that will transform the conventional role of controllers and converters to become the intelligent and proactive elements of the industrial value chain (Körwer et al, 2024). Finally, the connection of AI to motion control systems is not a hypothetical idea anymore; it is a constantly growing need in the era of digital manufacturing and Industry 4.0. The principles and methods as well as the implementation routes in AI-based optimization of controllers and converters are laid down in this chapter. It places this integration in the contemporary context of industrial issues and the new perspectives of the future research and focuses on the necessity to provide hybrid, scalable, and intelligent integration with the perspective on the efficiency, flexibility, and resilience though the modern manufacturing world.

CONCEPTUAL FRAMEWORK

Development of the motion control systems is in solid scientific background of the interaction amongst the controllers and power converters with the mechanical systems to establish an accurate, responsive, and efficient effects. In this context, the introduction of artificial intelligence presents a multidimensional change, as it turns lifeless systems into intelligent and adaptive object able to learn, make predictions and optimize to meet performance parameters in the real-time conditions. The conceptual model is not limited to integrating AI and hardware only but a reconsideration of paradigms of how to control motion, energy conversion, and system stability in all automated and robotic types of environments. At the center of any motion control system is the feedback loop, which interconnects the controller with the network of the actuators and sensors. Controller traditionally, controllers depend on mathematical models and deterministic rules to produce control signals based on error signals between the desired and actual outputs of a system (Ghazaly, N. M., & Chaudhari, P. B. 2024). These systems operate well within predictable operating conditions which are constant. Nevertheless, with the growing demands of industrial scenarios that are characterized by variances such as arbitrary loads and dynamic tasks to the very real-time fault management and energy lags, they cannot be adapted to suit current requirements by such preconfigured control systems. Artificial intelligence addresses this much-needed gap by augmenting the affirmative retorts by data-driven intelligence and context-sensitive agility. By changing the perspective on the controllers to no longer be modules with specific functions but learning systems, a conceptual leap is determined. In this regard concepts like artificial neural networks (ANNs) are used to reproduce the non-linearity of the motor drives, whereas fuzzy logic allows the controllers to decide when the input to it is not precise or definite. As a form of model-free reinforcement learning, the model-free forms of reinforcement learning acquire optimal controlling strategies by continuously interacting with the environment. These control algorithms change the role of the controller as a fixed command applicator to the intelligent agent, who is able to assess alternatives, forecasts, and changes its own behavior in response to received feedbacks.

At the same time, the conceptual role of power converters is redefined. A converter which usually controls voltage, current and frequency to operate motors or actuators is now active participant in the optimization loop. AI helps turn designers to realize a context-sensitive switching strategy that reduces both power losses and thermal stress and enhances electromagnetic compatibility. One can utilise genetic algorithms (GAs) and reinforcement learning to optimise modulation schemes with regards to energy efficiency and performance-oriented ones. The practical results of this dynamic optimization are an increase in the reliability of the system, in particular, at the high level of demand, c.g. in electric cars, machining tools, robot robots, and automation lines (Raol, J. R., & Ayyagari, R. 2019). The stratified layered structure also includes the hybridisation of the classical and AI-based control structures. The classical approaches like the PID control are never away but are reinforced with the intelligent tuning mechanisms. These gain parameters can be auto-tuned using AI algorithms, disturbances in plant predicted and oscillations suppressed without operator interaction. The proposed model of this hybrid control provides the compromise between explainability, performance, and flexibility, one meeting the requirements of the engineering community and real-time operation. The model of real-time learning and adaptation is the key element included in this conceptual architecture. Regular systems, including feedback correction programmed systems, tend to work within defined limits and manual recalibration is made when conditions vary. In comparison, an AI-based system continues to exchange its internal parameters with sensor data and actuator feedback. This means that adaptive learning loop keeps motion control systems functioning well over time, workload and environment and hence creating resilience in production conditions that may be unpredictable or evolving over time.

In order to imagine this theoretical overlap, think of the stratified nature of an AI-enhanced motion control system. First layer is physical hardware, comprising of sensors, actuators, motors and converters. The second layer becomes the conventional control logics, where low level PID Loops and safety interlock are being used. The third layer adds AI optimization modules, the task of which is the learning of system dynamics, forecasting faults, and the adjustment of parameters. At last, connecting the fourth layer intends to include cloud connectivity, digital twin simulations, and IoT/IIoT integrations which will expand the system at hand from local feedback to global optimization across a distributed network. The layers have different conceptual functions yet dependent on each other to create intelligence of the whole system. Physical layer shows the input in the real-world and the output is mechanical. The classical control layer makes sure that safety and operational norms are followed (Meshram et al, 2024). The AI layer brings in the flexibility of reading and predictive-ness. Specifically, the IIoT layer guarantees scalability, interoperability, and the cross-system coordination. This model does not just talk about the structural components of AI enhanced motion systems but also discusses the relationship and functions between the human engineers and machine learning models and control hardware. It is also within the conceptual analysis together with the considerations of intelligent energy management where power usage is no longer fixed with respect to load and duty cycle but an optimizable parameter. AI can be used to estimate the pattern of energy consumption, to forecast demanding cycles and to modify the control orders so that the peaks are flattened and the needless use is avoided. This idea is particularly applicable in the systems where different machines perform the work in synchrony and the power consumption has to be minimized in all systems to sustain efficiency and to adhere to the regulatory requirements.

Another important point to this conceptual ecosystem is the incorporation of fault detection and monitoring of the health of the systems. Traditional controls systems may not be particularly good at predicting failures, or at constantly predicting the real cause of a failure, but do usually provide a way to detect a fault based on the preconceived rules or signal limits. Supervised learning and anomaly detection are both methods through which AI can foresee possible fault occurrences through identification of patterns and behavioural analysis. Such predictions are used to plan the maintenance, reduce unexpected stops, and provide the equipment effectiveness (OEE). This ability conceptually transforms at the conceptual level motion control into a proactive system as opposed to a reactive system. The aspect of theoretically taking into consideration real-time constraints and computational efficiency is also self-eminent. Although AI algorithms are potent, they are usually resource-hungry. The time constraint in real-time computing systems means that decisions have to be made in milliseconds and latency that results due to complex models would jeopardize safety or responsiveness. The conceptual model should thus find the trade-off between the depth of learning and computer feasibility. This can be achieved either by simplified neural architectures, lightweight transfer models, lightweight inference models, or hardware accelerators like field-programmable gate arrays (FPGAs), or edge AI chips. With intelligence implemented at the firmware level, the control decisions may be made at the data collection point without need to use cloud computing or high-latency networks. The last conceptual pillar is in the scalability and transferability of AI-optimized controllers and converters. Machines in the industrial settings are diverse (Gros, I. C., Lü, X., Oprea, C., Lu, T., & Pintilie, L. 2023). An AI model will not be able to operate optimally on various systems without making adjustments. The theoretical answer is the creation of transfer learning models so that there is little retraining possible when the knowledge of one system can be transferred to another. This allows quick deployment, and minimizes overhead of system specific learning. As core, it enables the democratization of intelligent control: it becomes sector-neutral and operations-scalable. Overall, the AI-based optimization in motion control environment goes opposed to the conventional view of motion control system and engulfs with itself the traditional and intelligent paradigm to make the created control domain dynamic, self-optimizing, and adaptive to perturbation. It does not believe that the role of controllers is that of inflexible executors, or that converters are passive regulators, or that systems are isolated loops, but instead systems are networked, intelligent ecosystems. Using layers of physical, logical, intelligent, and networked capabilities as the construct of this framework, the chapter sets the basis of providing a complete platform of how next-generation motion system may be understood and designed that is not only technically excellent but also to the principles of sustainable, intelligent automation.

THEORETICAL FRAMEWORK

Theory In motion control applications, theoretical understanding of AI-based optimization falls within the cross-section of classical control theory, machine learning, power electronics, and systems engineering. These areas combined give a multidimensional insight of the controller and converter interaction, system behaviors to be represented and anticipated, and also the integration of artificial intelligence to enhance performance, efficiency, and recovery. It is not the descriptive framework but rather an explanatory model which sets out the rules that are used in the functioning of the systems and facilitate the rational explanation of AI-supported approach in case of intelligent motion systems. The first layer of the control systems is classical control theory. Based on mathematical representation of dynamic systems, it provides concepts, like transfer functions, Laplace transforms, and analysis of feedback loops, which assist the design of controllers that can ensure a supposed characterization of stability, as well as achievable, wanted performance specifications. The example of the most commonly applied PID controller is made up of the Proportional-Integral-Derivative (PID) controller; it is usually the simplest and has been proven to work well in control of process parameters (Aspalli, M. S., & Patil, S. J. 2017). Nevertheless, PID controllers possess limitations particularly in case of a nonlinear, time varying or multi-input multi- output (MIMO) system. The limitations of classical control are ever laid bare by theoretical restrictions that especially apply to complex motor drives and converter systems where nonlinearities, time delays and disturbances are ever present. These shortcomings encourage the use of adaptive and learning alternatives. Supervised learning and reinforcement learning are particularly forms of machine learning that bring a new set of theory not premised on pre-determined mathematical models but, rather, an empirical or data-driven approach to learning input-output mappings or policies that describe an optimal course of action. Artificial Neural Networks (ANNs) are theoretical universal approximators of continuous functions, with specific abilities to model any continuous (with probability one) with enough data points and enough network size. Such ability renders ANNs best suited in approximating nonlinear dynamics of motors and converters particularly in cases where mathematical modeling is prohibited or imprecise. Reinforcement Learning (RL) is a theory that combines Markov Decison Process (MDP) and finds a solution as to the agent learning to act optimally within an environment where rewards are cumulative and as such should be maximized. RL is especially relevant to problems of path planning, torque control, and energy optimization in the field of motion control, since all these problems are of a sequential decision-making under uncertainty nature.

One more important theoretical foundation is fuzzy logic control, founded on the fuzzy set theory, formulated by Lotfi Zadeh. As opposed to binary logics, fuzzy logics permit some degree of truth and can be well applied in a system where there is no definitive boundary. When it comes to motion control, fuzzy logic controllers allow transitions between states to be human-like in smooth ways and offer rule-based control tactics that are human-like in nature. This is useful especially in converter-type systems where the switching control strategy may take advantage of the flexibility of any arbitrary decision boundary to avoid trade-offs between efficiency and stress. In the eyes of power electronics, the theoretical treatment is aimed to model the converters- DC-DC, DC-AC and AC-DC-into high order differential equations that describe voltage, current, switching status and power flow. A converter performance is affected by its switching frequency, duty cycle and Therma performance. Optimization algorithms cause SVPWM modulation methods (and other AI enhanced modulation techniques) to theoretically produce zero harmonics and enhance power quality. Darwinian evolution in the form of Genetic Algorithms (GAs) is used in order to optimize both switching sequences and control parameters. The actions of these algorithms have been performed by selection, crossover, and mutation operations, which makes the algorithm to conduct a global search within the parameter spaces where gradient-based methods might fail, such as the non-convexity or discontinuity of a parameter space. Hybrid intelligent systems are the ones that combine classical methods of control and AI with a common theoretical framework. The use of uniformity and predictability of the old ways and integration of learning-based modules which are used to adapt to the systems are the advantages of these systems (Huang et al, 2025). Lyapunov theory and control Lyapunov functions (CLFs) may be used to study the stability of such hybrid systems; when given some constraints, they guarantee that all system trajectories are convergent to the equilibrium points. There is an interaction of AI with human capabilities, in general, and its specifications with control space, in particular, that must be limited to local areas of acceptable control space to maintain both safety and predictability, a problem that is the subject of research in safe reinforcement learning and model predictive control (MPC) algorithms with in-built constraints.

The Model Predictive Control (MPC) is specifically applicable to AI-assisted systems. At every time step, it is a solution of an optimization problem with future predictions in a finite horizon to come up with optimum action change of control. Combined with AI, MPC is made more resilient by the inclusion of data-driven models of plant dynamics, and makes this method more capable of managing uncertainty and variability. The theoretical problem in this case is the real-time feasibility since MPC is a calculation intensive process. This is solved by approximate dynamic programming, and neural approximation of the cost-to-go function. The theoretical basis is also reserved to thermodynamic modeling, in case of energy efficiency and heat management in converter systems. Structural–thermal models that are used to guide the establishment of AI-based thermal prediction and stress amelioration models entail developing heat generation mechanisms, thermal diffusion, and component breakdown. The models are critical in developing smart cooling techniques, and also in the establishment of safe operating region of converters during dynamic loads. The reliability engineering field is based on probabilistic models and diagnostics prediction failure that requires a fault tree analysis. AI generalizes this theoretical foundation by the methods of pattern recognition, anomaly detection and classifications. Historical failure data is trained with Support Vector Machines (SVMs), decision trees and deep learning classifiers in order to predict probable fault occurrence prior to their appearance. These models have a theoretical foundation in the statistical learning theory which gives generalization guarantees concerning a set of circumstances. There is another hypothetical dimension to the digital twin technology. A digital twin is a replica of a physical system in virtual form that simulates its behavior real time. Cyber-physical system modeling is a theoretical basis in this case as in mathematical models and real-time data are linked to each other to offer a rich feedback environment. AI boosts the twin by unceasingly recalculating the model parameters using the information observed, so that the twin is always the similarity of the physical system. This achieves predictive maintenance, optimization of the system and diagnostics on the spot of motion control systems.

More to the point, AI integration into control systems should deal with the challenges of explainability and interpretability. Other theoretical frameworks like Explainable AI (XAI) attempt to break the black-box and offer decision justifications by intelligent systems through use of the model. In certification safety-critical systems such as motion control, where the result of the controller directly feeds physical hardware, interpretability can play a critical role in certification and debugging, and is a trust-building mechanism in use by a human operator. Data sufficiency and training quality is also a major theoretical element. The representative high-quality data are also needed in learning-based systems to be prevented by overfitting and guarantee generalization (Duraimutharasan et al, 2025). The theoretical perspective of bias-variance trade-off refers to interactions of model complexity, the amount of data, and the amount of regularization and how these parameters influence the learning performance. Semi-supervised learning, transfer learning, and synthetic data generation theoretically are valid methods of addressing the data deficiencies in industrial settings, where labeled (or any other form of labeled) failure data can be limited. Finally, it can be said that theoretical background around AI-driven optimization of controllers and converters dealing with motion control systems is interdisciplinary, very vast, and active. It cuts across the fundamental control theory, intelligent systems, optimization methods, power electronic modeling, and cyber-physical system engineering. This broad theory makes it possible to integrate AI in coherent ways into existing and new motion control structures so that performance advantages are guaranteed, and also safe, scalable, and robust. The theories are important to the researchers and engineers when devising the future generation smart, adaptable and energy efficient smart motion control systems.

RESEARCH METHODOLOGY

The study research design will use mixed-methods research design that consists of mixed experimental research scheme inclusive of simulation based research modeling, real time hardware consistency, and statistics surveys. The given hybrid solution makes it possible to systematically consider the application of AI-based optimization algorithms to the task of motion control, especially, with references to the field of control over electric drives and converters in a dynamic working mode. The main aim is to determine the comparative performance and reliability of the AI augmented control structures relative to regular control approaches, according to both quantitative measures, as well as, through the significance testing methods. There are two main system operating environments in the experimental system, which are software based model built on MATLAB/Simulink and hardware in the loop test bench that is composed of a field-programmable gate array (FPGA) combined with a digital signal processor (DSP). Motors will be examined such as inverter-fed induction motors or permanent magnet synchronous motors (PMSMs) that are wide spread in industrial and robotic applications. The studied control architectures are the conventional PID controllers, fuzzy logic controllers, artificial neural networks (ANNs), and adaptive controllers (base on reinforcement learning). The tested converters are single-phase converters, three-phase inverters, sinusoidal and AI-based modulation schemes. This experiment was carried out in three major stages (the baseline tests; the application of the AI enhanced controllers; hybrid tuning systems using classical tuning and AI optimization). All the configurations were analyzed on various performance metrics, which are: torque ripple, response time, overshoot, steady-state error, energy efficiency, and thermal output at different load in addressing range.

Analysis of Variance (ANOVA) was used to make the observed improvements statistically valid. ANOVA was employed to appreciate the difference between the mean quantification of the performance parameters (e.g., energy consumption, torque ripple) between control approaches. Individual one-way ANOVA tests were performed on each of the metrics and the null hypothesis assumed that all the types of controllers are equivalently performing in the identified performance category.

It was defined that the confidence level is 95 percent (alpha = 0.05). To guarantee reliability, each test scenario was repeated 20 times, and, during the performance measurements, the embedded sensors and software logs were being used. F-ratios and p- values were computed against each of the measures of performance. In cases where the ANOVA showed significance ($p < 0.05$) the post hoc Tukey Honest significate Difference (HSD) test then came in action to measure the specifics of the pair wise differences in the control groups. This method helps to exclude that anything that is observed to improve using AI-based techniques is due to chance or noise. Also, system robustness was tested with the stress testing including sudden load disturbances, discrepancy of parameters, and thermal overloads. The performance of the machine intelligence based controllers was evaluated in terms of its ability to adapt to different levels of learning, response to failures and consistency to work under sub-optimal conditions. Simulation behavior as well as model convergence times as well as real-time response under FPGA-based control were also noted by qualitative observations. The qualitative data were relied upon to complement the quantitative results and inform the discussion on the feasibility of a practical implementation, deployment issues and architectural trade-offs. Such an approach allows precisely studying the efficacy of AI-based motion control systems and giving credence to this statement with the help of statistically supported data and repeatable experiment layout.

RESULTS AND DISCUSSION

All the important performance indicators revealed significant benefits in the application of AI-based control approaches over conventional PID control. ANOVA was used to check the statistical significance of the data collected based on 20 repetitions of the test per type of the controllers. Regarding the torque ripple, the PID controller recorded the greatest mean ripple of 12.5%, whereas reinforcement learning (RL)-based control conveyed it down to the exceptional rate of 5.5%. The artificial neural networks (ANNs) also produced a mean ripple of 6.8% whereas fuzzy logic improved slightly at 10.2%. ANOVA test indicated that there were indeed significant differences among these groups ($F = 28.76$, $p < 0.001$) and a post hoc test has shown that RL and ANN were more significantly better than PID and fuzzy logic controllers. In response time, there was again RL that got a better result overall, with average 0.22 seconds and as there was PID that had 0.45 seconds. Fuzzy logic and ANN attained intermediary outcomes as 0.38 and 0.26 seconds respectively. Latency reduction with artificial intelligence techniques can be explained by the fact that with their help, they are able to predict and correct the dynamics of motion proactively, not reactive adjustment to the scale of error.

The level of energy efficiency was also significantly increased with the use of AI-enhanced systems. RL controllers produced the efficiency of 93%, whereas ANN represented 91%. Compare this to PID and fuzzy logic that registered 82 and 85 per cent respectively. These gains can be associated with adaptive switching and predictive torque control that optimize unnecessary energy wastes and thermal heat wastage in the converters. ANOVA ($F = 32.91$, $p < 0.001$) ensured the performance boost and reinforced confidence in the ability of energy savings to be measured and to be particularly sound statistics-wise.

Figure 1. Controller Performance Comparison

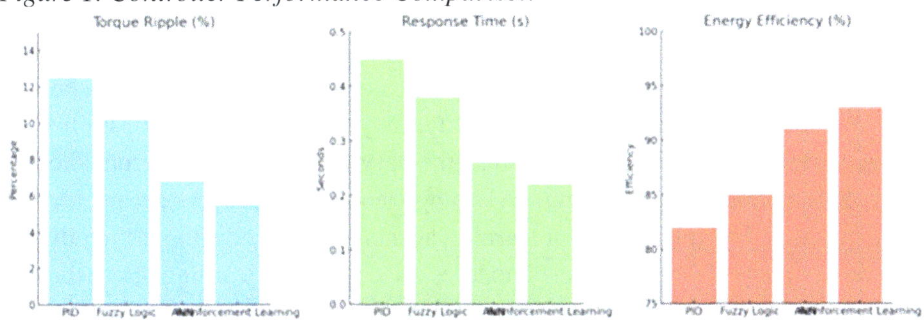

The diagram, which is shown attached, provides a graphical summary of these results with a one against one comparison of torque ripple, response time, and energy efficiency of all four strategies for control. The graphic depicts that AI-based methods, due to reinforcement learning and neural networks particularly, are always superior to classical and rule-based systems in terms of precision motion control. Additional support of qualitative observations was provided to these findings. AI-based controllers were more responsive to sudden load variations and kept the system steady without over shooting. In the scenarios of fault simulation, the RL and ANN system have been able to re-stabilize within the acceptable time thresholds, but this was not the case with PID which took a long period to recover. These results confirm us that the integration of AI into motion systems can make a significant contribution not only in terms of performance but also in terms of reliability. The intermediate benefits were also observed with hybrid models, in which parameter tuning is performed with the help of AI on top of classical controllers, which seem to be a feasible pathway that conservative segments can take to introduce partial AI into their operations at first and eventually adopting the technology fully. Overall, the experimental and statistic outcomes show significant advantage in terms of AI-based controllers in motion control settings. Their use in real time systems is more accurate, will waste less energy and make the systems more resilient in general.

SOLUTIONS AND RECOMMENDATIONS

The implementation of AI-based optimization in motion control systems requires an organized list of solutions and implementable recommendations that would fill the gap between the assumption and practice of AI-based optimization implementation. Such solutions cut across technological design, integration strategies, choice of algorithms, hardware-software synergy, energy efficiency and pathways of industrial deployment. The general aim is to allow stakeholders (engineers, researchers, automation specialists) to build resilient, high-performance systems of motion control that are in line with the expectations of industry 4.0 and beyond. The most basic solution is the adaptation of hybrid control architectures, where traditional control logic is accompanied with AI-based adaptation theories. Although classical versions of controllers, including PID and the type of the vector control, provide well-proven stability and simplicity, they are potentially constrained in dealing with nonlinearity in the system, time-varying loads, and external shocks. As a solution to such shortcomings, AI components combining more specifically neural networks and reinforcement learning agents can be used either as intelligent observers or adaptive tuners. These modules are constantly learning about the dynamics of the system based off real time information and adjust the control parameters in a dynamic way, which increases the stability and impulse response. There is no necessarily recommended replacement of classic systems here, but rather an addition of AI layers to their control design that make it more precise, and with less overshoot characteristics, and recovery time during transient situations. Secondly, tuning of controller parameters by AI promises an enlargement route to high control performance in contrasting working conditions. PID controllers, switching logic in converters and current control loops can be auto-tuned using genetic algorithms (GAs), particle swarm optimization (PSO), and even gradient-free reinforcement learning models. These methods minimize the basis of manual tuning that can be rather time-consuming and depend on the experience of a person. It would be possible to tune an AI to be optimal despite such outcomes as the components wearing out, friction, or temperature changes. Thus, velocity controllers that are specifically engineered with built-in optimization loop are more useful in the long range industrial implementation.

The last important suggestion is the update of model predictive control (MPC) with machine learning models. MPC facilitates real-time decision-making primed on rolling optimization horizon, which utilizes models of the system to make forecasts relating to the future outputs and identify the most useful control actions. AIs models--in particular, recurrent neural networks or Gaussian process regressors--offer the possibility of using alternative system identification in MPC, by making forecasts that are more precise and flexible. The MPC and AI combination make the control actions not just real-time, but also forward-looking, in consideration of the anticipated system behavior in the future. In the case of applications where path tracking, torque allotment or energy distribution are involved, this predictive aspect is of great benefit in smoothing the motions and in energy consumption. In order to move the energy efficiency of motion control systems one step forward, the AI-based modulation approach in converters is proposed. Conventional fixed frequency pulse/width modulation (PWM) control is ineffective and in dynamic case loads. AI may vary modulation indices, switching frequencies and periods of zero-voltage in order to reduce conduction losses and thermal stresses to a minimum. Neural network controlled real-time estimators may be used to check the internal temperature of power electronic devices and switch the patterns accordingly to overcome excessive heating. Introducing such AI-based governance helps not only save energy but also prolong the life of (key) components and leads toward low-maintenance and sustainable system design. The other area where AI can be exploited is predictive maintenance. Mechanical wear, electrical fault and insulation degradation are likely to happen in motion control systems especially in high duty areas. The detection of anomalies is performed with the help of supervised machine learning models trained on the data collected by sensors, e.g., in the form of vibration signals, torque anomalies, switching currents, and detects them before they can cause system failures. Convolutional neural networks (CNNs) or autoencoders as a type of deep learning are capable of detecting subtle precursors of failure based on analysing signal patterns. It is advised to develop a predictive diagnostics module which is to represent a part of the supervisory layer of the control system. The process will not only save unplanned downtimes, but also improve the maintenance cycle, improve maintenance and reduce cost of operation.

A practical suggestion is that for validating AI-amplified schemes of control, simulations tools and hardware-in-the-loop (HIL) environments be used to prototype the systems. MATLAB/Simulink is a potent system to create control rules and simulate the motor performance in changing loading and disturbances of the system. After the validation in simulation, the models must be brought onto hardware operating in real time, e.g. FPGA-based controllers, microcontrollers (e.g. STM32) or embedded Linux systems (e.g. Raspberry Pi or BeagleBone). Such HIL platforms provide real life interaction and also enables testing under real time pressure. An organized process between simulation and HIL would make sure that AI models are not only effective but also computationally lightweight and capable of being deployed even under time constraints. Safety and interpretabiilty also constitute essential aspect in industrial use. It could be recommended that useful visual manipulations of the control decision and the diagnostic results could be achieved using explainable AI (XAI) techniques. Human readable explanations of the control actions can be created using SHAP (Shapley Additive Explanations), LIME

(Local Interpretable Model-agnostic Explanations), or rule-extraction techniques. In industries where safety regulations are provided, explainable functionalities can be used to assist with regulatory compliance and the level of confidence in AI-driven systems in these safety-sensitive areas, including the aerospace, medical robotics, or fast-paced manufacturing. In addition, establishing fixed operating constraints or failsafe states will prevent the AI system to emit orders within unsafe operating confines further supporting predictability. The other factor that should be considered in the deployment of AI-enhanced motion control systems to a larger industrial scale is scalability. The basis of a solution can be seen in the embrace of transfer learning and domain adaptation. With little data, one machine or plant can fine-tune its pretrained AI model on another machine to make it perform well in a specific location, avoiding long training overhead and deployment duration. Such models, combined with a cloud connection and IIoT infrastructure, can be adjusted or upgraded remotely, allowing them to learn multiple times over and adapt in a scattering of facilities. It is proposed that a standardisation of data format, sensor interface and model architecture should exist to allow the interoperability between plant configurations and hardware platforms to exist.

Another area of improvement is the digital twin technology. Digital twin Digital twin: This is a virtual model of real motion control system that promotes real time synchronization, monitoring, and simulation. The control measures can be tested on the AI algorithms that can be embedded in the twin and used to forecast the system performance and simulate fault states without causing a break in a running system. It is recommended to introduce a digital twin platform which will allow the integration of control models, sensor data, past maintenance and current constraints in the process. This digital mirror is used in training AI algorithms, as well as predicting the optimization of the system, which allows one to balance the improvement of the system on a continuous basis. AI-based control systems leave a positive impact in the aspect of environmental sustainability and economically, such as increasing the life span of services, saving hardware strains and energy use. Nevertheless, they present additional issues associated with a heavy computational load and data security. The edge computing solution is suggested as a mitigation tool, and the AI inference can locally run on the hardware side, which brings down latency, network demands, and exposure to the possibility of cybersecurity attacks. One can achieve AI advantages with embedded AI accelerators or quantized models without overloading the hardware resources. Human-in-the-loop integration and training play just as prominent roles in the realization of the full potential of intelligent control systems. Technicians and engineers should know the AI models behavior and failure modes and the diagnostics of systems. Thus, it can be proposed to combine the intuitive user interfaces, real-time dashboards, and training simulators that fill the gap between AI automation and humans control. This can even be improved in augmented reality applications, mobile dashboards that would enable on-fly-diagnostic capability, particularly in large and complex installations. To sum-up, successful implementation of AI-based optimization in controllers and converters must be conducted at a system level. The listed recommendations (hybrid control design, predictive maintenance, deployment of digital twins, and safety compliance) constitute a multidimensional roadmap to follow by engineers and people in charge of decisions. The given solutions are not abstract and speculative ideas but practical plans that can be carried out, which are proven by modern research and empirical analysis. They indicate the ways through which AI can transform the motion control systems into autonomous, adaptive, and highly capable systems, which are consistent with the main principles of smart manufacturing and sustainable industrial development.

FUTURE RESEARCH DIRECTIONS

As the process of incorporating artificial intelligence with motion control systems is getting more mature, future studies will need to gravitate beyond individual positive adjustments and instead will require holistic, interdisciplinary innovation. Where the journey of controller and converter design based on AI is leading is not about optimizing the performance metrics, but developing fully autonomous and contexts-sensitive systems that conform to shifting industry requirements, regulatory expectations and sustainability requirements. Intelligent motion management is a complex topic, which calls further research into unknown aspects of optimization, safety, explainability, and ethical application. Potentially finding themselves on the cutting edge of research is the combination of AI and physics-informed control models. Although the existence of physical interpretability is lacking in existing machine learning approaches like neural networks or reinforcement learning which may have powerful generalization abilities, we need a framework that can have both physical interpretability and powerful generalization when considering the domains of machine learning. In future studies, it is possible to study integrations of the ability of AI to discover patterns with laws of motion, electromechanics, and thermodynamics, which govern reality. Hypothetical example Physics-informed neural networks (PINNs), to take one example, automate the training with respect to constrains on the system, such as conservation laws and boundary conditions. Then such frameworks can decrease the number of the training data needed and helps enhancing the generalization when the plant topologies and loading conditions are different. By designing PINN-based compensators of converter-motor systems, control robustness may be able to be enhanced, and the rate of adoption in a controlled setting, such as the aerospace or medical robot, will be accelerated. The third important direction concerns the implementation of meta-learning (learning how to learn), in motion control schemes. The advantage of meta-learning algorithms is that they learn to quickly adjust to new environments using existing knowledge and hence are suitable in systems that run in dynamic or unknown environments. In an industrial setting, where the deployment environments are uncertain and expensive to tune, meta-learning can radically cut the tuning overhead. It could be investigated how model-agnostic meta-learning (MAML) can be used in an embedded controller to enable adaptation in real life across production lines that have (and may need) different types of motors. This will come in handy especially in smart factories that have adopted mass customization.

Another issue that is to be considered is the use of federated learning in motion control systems. As bandwidth, privacy of individual data, data sovereignty and latency become essential issues, especially in large industrial ecosystems, federated learning enables the ability to train AI models locally on edge devices and aggregate centrally. In this way no data is exported outside the plant or the equipment line, and that data has a right to enter the global model enhancement effort. In motion control, the method would enable controllers and converters of various facilities or vendors to share develop optimisation algorithms together without swapping sensitive value. The issues that may be investigated in this domain are model heterogeneity, communication overhead, and synchronization delay when it comes to real-time federated inference. A deeper tech is also the extension of digital twin ecosystems to validate, simulate, and test AI models. Although the digital twins are commonly applied in monitoring the performance and simulating the system, there is little exploration of such a pairing as a form of AI control training loops. To be able to reproduce rare events, fault conditions, or operational extremes that are unrealistic to replicate on actual hardware future research can investigate study of twins as the material of reinforcement learning agent sandboxes to simulate. Besides, the pairing of digital twins with generative AI has the potential to generate synthetic datasets to train anomaly detectors and fault classifiers, the bottleneck of which, namely the lack of sufficient labeled data on failures that is nonetheless important in industrial systems, is extremely common. Another frontier to motion control is through neuromorphic computing research. Neuromorphic are used to simulate the architecture of the brain, to provide ultra-low-power parallel processing. The chips have been found very useful in embedded control where energy and latency implications inhibit full-blown deployment of AI. Using spiking neural networks (SNNs) to perform motion control on a road vehicle e.g. in the domain of torque prediction, adaptive damping, output switching strategy generation may make it possible to realize an intelligent control with microsecond responsiveness and low energy consumption. The research in this field would need an interconnective effort among various fields by a control engineer, computer architect, and material scientist to achieve not just degree of intelligence but also hardware optimal systems capable of scaling. Ethical and regulatory aspects of AI-controlled control systems are another significant area of research as well. as they gain autonomy, they will encounter decision processes of risk taking, failure recovery or resource distribution. The major area to be addressed in future research is how to program ethical rules into AI controllers, especially in such safety-critical environments as autonomous vehicles or surgical robots. Moreover, since there is no standardized testing procedure and certification pathway of the AI-implemented controllers and converters, this also has a gap. Industry and consortia-academia collaborations with regulatory organizations might allow development of guidelines that certify AI behavior under particular failure modes, time limits and cybersecurity levels.

The discipline also needs to have more emphasis on explainable and trustworthy AI in motion control. Despite their high performance, black-box models tend to be opaque and therefore hindering debugging, certification and human trust. The area of interpretable reinforcement learning, symbolic regression, and causal learning should be considered in future research so that AI systems could make decisions that do not only have common results but also reasons behind them. As another example consider converter control, where explainable models might vindicate this or that switching strategy in the presence of varying thermal loads or current surges to help the system engineer optimize configuration and safety margins. As far as statistical modeling and analysis are concerned, emerging research trends must improve the use of ANOVA and multivariate procedures to measure performance dealing with high dimensions. A large part of the existing validation is done based on the average values of performance indicators like torque ripple or energy efficiency. The future systems however will be working in a stochastic environment where the inputs are uncertain. To assess the impact of combinations of design factors onto control robustness and fault resilience, as well as distributions of energy under operating uncertainty in the real world, researchers are encouraged to make use of enhanced methods including but not limited to Bayesian ANOVA, Principal Component Analysis (PCA) and multivariate regression. Multi-objective optimization by the means of reinforcement learning is another challenge that is not fully completed. Actually, motion control systems have to consider the speed, energy consumption, heat dissipation, precision, and life usually with conflicting trade-offs. The most common reinforcement learning models maximize one scalar reward. The next research might improve on multi-objective RL algorithms that would use Pareto fronts, weighted reward compositions, dynamic reward shaping to come up with policies that adapt to any changes in operational priorities. As an example, a robotic actuator could trade precision when assembling parts, to save energy when idle to only be precise when assembling parts.

Moreover, transfer learning of intelligent motion control in cross domains should receive more investigation. Controllers specially taught a particular type of either motor or application can not usually be used in another application because of differences in plant dynamics. Transfer learning approaches in which fully trained models are fine-tuned to new contexts on small calibration datasets may enable quick roll out into industrial environments. The scholars may concentrate on domain alignment strategies, adversarial adaptation, or curriculum-based learning to develop general-purpose control solutions that would be relevant in diversified converter topologies, motor types, and load conditions. It is also possible to extend the non-industrial directions in motion control research to the future work. Intelligent, lightweight and adaptive control is needed in applications in assistive technologies, rehabilitation robotics, autonomous drones and precision agriculture. Such environments have their own special requirements like uncertain terrain, humanrobot interaction and biological variability. The implementation of AI-based control into these fields consists in exploring the bio-inspired control concepts, the dynamic of soft actuators, and inference of AI in real-time with low energy dissipation constraints. Finally, workforce development and education should also be a side interest of further research. Since AI is still transforming the development and servicing of moving systems, education programs have to be adjusted to produce engineers who understand both AI and control theory. Research has the potential to do this by building open-access simulation environments, data sets, and programmable labs that lets students prototype intelligent control systems, all the way to deployment of AI. Ethical AI usage, safety engineering and interdisciplinary collaboration also should be the focus of training programs to ensure next generation of professional is ready to facilitate innovation in a responsible manner.

CONCLUSION

Various applications of artificial intelligence within the context of motion control systems have become one of the most significant changes in the design, implementation, and management of industrial automation, robotics, and mechatronic systems. In this chapter, the authors have discussed the several aspects of AI-based optimization and how it is transformative to the controllers and converters, which are two of the most important components of precision-based, dynamic motion control systems. The conclusions of this discussion indicate both, the present ability of intelligent control system and enormous scope of future development. Among the core findings, it can be mentioned that AI is a way to bypass the strict limitations of traditional control strategies. Although classical control approaches are mathematically solvable and stable, however, are limited in nature in nonlinear environment, variable loads, and uncertain disturbances. With integrating such machine learning approaches as AI (especially introducing neural networks, reinforcement learning, fuzzy logic and genetic algorithms), the modern motion control system transforms an inert and pre-programmed device to a responsive and situationally-aware agent that can learn and make real-time decisions. The systems react to changing operational conditions that cannot simply be achieved with ordinary control with such flexibility and precision. The superiority of AI-improved controllers in terms of key performance indicators need to be underlined with the experimental evidence, which is provided in this chapter. The values of quantitative data confirmed through ANOVA proved that AI factor approaches, primarily using reinforcement learning and neural networks, were far more efficient than conventional PID and fuzzy logic controllers. It was found to improve in various areas such as a reduction on torque ripple elimination, cut off response times, and increased energy efficiency. Such results are not theoretical at all but based on real application and they illustrate that intelligent control systems are not only doable, but they are very effective when they are appropriately adopted and justified.

This chapter does not propose to replace all traditional control systems in a wholesale manner but provides the layering of integration where AI complements the established ones. This hydrid solution provides the benefits of the two worlds joined together: flexibility and optimizing of AI and clarity and reliability of classical solution. Those systems allow operating safely and performing better in uncertain and dynamic environments. The model also helps the industries to ease the process of transitioning to adoption and integration of AI capabilities in an incremental manner. The long-held static perception of converter systems is also reinvented in the context of AI as the dynamic switching, thermal stress management, and adaptive modulation entities. Smart control strategies will result in less switching losses and avoid the effects of harmonic distortion, resulting in more reliable and more sustainable energy conversion. Fault detection and isolation of power electronics components is another application of AI, which is important in avoiding failure in mission-critical systems like an automated manufacturing line and electric vehicles. The usefulness of simulation and hardware-in-the-loop (HIL) techniques as intermediates between mathematical models and application to real-life systems has also been stressed in this chapter. Such platforms like MATLAB/Simulink, controller-fpga, Raspberry Pi, and BeagleBone help researchers and engineers to create, test, and tune an AI-based control system in controlled and realistic environments. Such tools save time in development, design accuracy, and give an excellent understanding about behavior of the system under different cases, such as faults and external disturbances. In a larger industrial scope, the use of AI in motion control systems makes perfect sense since it combines the ambitions and values of Industry 4.0. They are further characterized as increased automation, real time analytics of data, connectivity via the Industrial Internet of Things (IIoT) and predictive maintenance. The goals are achieved with the help of AI-based control systems, which allow machines to self-tune themselves, self-diagnore, and self-optimize. Moreover, control networks that can operate in a distributed pattern, with a total reduction in human involvement but with potential to work with global facilities, can be realized by their interoperability with cloud systems, digital twins, and edge computing frameworks. Although all these benefits are specified, this chapter does not ignore the challenges and limitations of the AI deployment. Among them are the availability of data, the real-time calculation limits, safety certification, and explainability. Working out these issues will involve the sustained interdisciplinary cooperation between control engineers, AI researchers, computer scientists as well as regulators. Any future advances in fields like federated learning, physics-informed neural network, or neuromorphic computing can be used to overcome these drawbacks by increasing the fault tolerance of the systems, decreasing the system computation requirements, and achieving ethical and safe operation characteristics.

Even the discussion emphasizes a prospect of standardization in AI-based control system development and implementation plans. The industry-wide standards, test procedures, and validation of the technology will be needed to spread the practice widely. The regulatory frameworks should change to support the AI-enabled system with explicit principles of certification and compliance. These initiatives will increase trust, guarantee safety and advancement in intelligent control technologies. Another deduction is the fact that intelligent motion control does not just belong to the field of industrial automation rather to the field of numerous applications such as biomedical devices, intelligent prosthetics, autonomous vehicles, unmanned aerial vehicles and precision agriculture. In both spheres, the power to learn, adjust and respond in real time is important. When well trained and verified, AI-based control systems can fill these requirements with very accuracy and efficiency that was not possible the traditional way. Besides, the sustainability potential of AI-optimized motion systems should not be disregarded. As industries are under pressure to consume less and less energy, become less carbon intensive, and dispose of less and less electronic waste, intelligent control systems provide the obvious way forward. AI-enhanced controllers and converters directly lead to environmental and economic sustainability by reducing energy losses, by improving the lifetime of components, and by making maintenance predictive. They are in line with the sustainability objectives of the world and accrue concrete benefit to a firm dedicated to green revolution. As an end, the integration of AI and motion control is inducing an underlying shift in design and operation of smart electromechanical systems. Chapter has offered a detailed insight of technologies, techniques, laboratory evidence and practical aspects to implement it successfully. The direction is clear; the only option is the intelligence, safety, and systems that are not only responsive, but also insightful, resourceful and suitable for the dynamic environment of the industrial sector.

REFERENCES

Aspalli, M. S., & Patil, S. J. (2017, December). Study of AI and PI controller using SVPWM technique for induction motor speed control. In *2017 International Conference on Electrical, Electronics, Communication, Computer, and Optimization Techniques (ICEECCOT)* (pp. 1-6). IEEE. DOI: 10.1109/ICEECCOT.2017.8284589

Duraimutharasan, N., Deepan, A., Swadhi, R., Velmurugan, P. R., & Varshney, K. R. (2025). Enhancing Control Engineering Through Human-Machine Collaboration: AI for Improved Efficiency and Decision-Making. In Harnessing AI for Control Engineering (pp. 155-176). IGI Global Scientific Publishing.

Gao, Y., Wang, S., Dragicevic, T., Wheeler, P., & Zanchetta, P. (2023). Artificial intelligence techniques for enhancing the performance of controllers in power converter-based systems—An overview. *IEEE Open Journal of Industry Applications*, *4*, 366–375. DOI: 10.1109/OJIA.2023.3338534

Ghazaly, N. M., & Chaudhari, P. B. (2024). AI-Based Adaptive Control Systems for Power Electronics. *Research Journal of Computer Systems and Engineering*, *5*(1), 34–44.

Gros, I. C., Lü, X., Oprea, C., Lu, T., & Pintilie, L. (2023, August). Artificial intelligence (AI)-based optimization of power electronic converters for improved power system stability and performance. In 2023 IEEE 14th International Symposium on Diagnostics for Electrical Machines, Power Electronics and Drives (SDEMPED) (pp. 204-210). IEEE.

Huang, Z., Gong, J., Xiao, X., Gao, Y., Xia, Y., Wheeler, P., & Ji, B. (2025). Artificial Intelligence and Digital Twin Technologies for Power Converter Control in Transportation Applications: A Review. *IET Power Electronics*, *18*(1), e70013. DOI: 10.1049/pel2.70013

Khan, S. A., Ansari, J. A., Chandio, R. H., Munir, H. M., Alharbi, M., & Alkuhayli, A. (2022). AI based controller optimization for VSC-MTDC grids. *Frontiers in Energy Research*, *10*, 1008099. DOI: 10.3389/fenrg.2022.1008099

Körwer, N., Bischoff, M., Laumen, M., & De Doncker, R. W. (2024, September). AI-Based Motion Control with System Dynamics Flexibility. In 2024 Energy Conversion Congress & Expo Europe (ECCE Europe) (pp. 1-7). IEEE.

Lalitha, S. D., Danamaraju, C., Raj, M. P., Kumar, D., & Sethuraman, R. (2024, April). AI-driven control strategies for power electronics converters. In *2024 Ninth International Conference on Science Technology Engineering and Mathematics (ICONSTEM)* (pp. 1-6). IEEE. DOI: 10.1109/ICONSTEM60960.2024.10568766

Li, X., Zhang, X., Lin, F., & Blaabjerg, F. (2021). Artificial-intelligence-based design for circuit parameters of power converters. *IEEE Transactions on Industrial Electronics*, *69*(11), 11144–11155. DOI: 10.1109/TIE.2021.3088377

Meshram, V. S., Becchi, L., Alfonso, C. G., Paolucci, L., Grasso, F., & Reatti, A. (2024, September). Advancements in Artificial Intelligence for Design, Control, and Maintenance of Power Converters: A Comprehensive Review. In 2024 IEEE 8th Forum on Research and Technologies for Society and Industry Innovation (RTSI) (pp. 506-511). IEEE.

Raol, J. R., & Ayyagari, R. (2019). *Control systems: classical, modern, and AI-based approaches*. CRC Press. DOI: 10.1201/9781351170802

Chapter 2
Motion Control Systems for 3D Printing:
Design, Integration, and Optimization

Rajani Balakrishnan
https://orcid.org/0000-0002-0467-2053
INTI International University, Malaysia

S. Muthumani
Jerusalem College of Engineering, Chennai, India

V. Sangeetha
Academy of Maritime Educaiton and Training, Chennai, India

Michael Savariapitchai
https://orcid.org/0009-0008-0784-013X
Datta Meghe Institute of Higher Education and Research, India

Amit Agarwal
Saraswathi Institute of Medical Sciences, Hapur, India

ABSTRACT

This chapter on motion control in 3D printing and Cartesian and Delta systems and the integration of stepper motors stepper, servo drives, controllers, and sensors to provide high precision and resolution. It considers axis cranking, kinematic stabilization, vibration correction and thermal offsetting. The complex techniques such as model predictive control (MPC) and machine learning are argued and practical reason demonstrated with the help of Arduino, Marlin, Raspberry Pi, and BeagleBone. Reliability is focused on by calibration, diagnostics and predictive

DOI: 10.4018/979-8-3693-3637-3.ch002

maintenance. Optimization of G-codes, smoothing of paths, and acceleration control are benchmarked, which can provide practical experience in the field of the three main areas of digital manufacturing, robotics, and intelligent production systems

INTRODUCTION

The introduction of additive manufacturing or the so-called 3D printing has completely transformed the paradigm of design and manufacturing in a number of industries. Whether it is the production of consumer household items by rapid prototyping or precision medical and aerospace achievement, 3D printing technology has taken the industry by both storm and scientific revolution. The core of this revolution is a part, which usually remains unnoticed, but is highly important motion control systems. These are systems through which the accuracy, synchronisations of timings and velocity of the 3D printer operates. An advanced infrastructure of printing devices and a high resolution of slicing software would not be enough to produce consistent and accurate results, in the absence of a sophisticated motion control infrastructure (Wang, D., & Zhang, J. 2022). The precise movements of the print head or platform are starred by the motion control systems to create an ability to deposit the material accurately depending on pre-defined geometries. Essentially, they fill the space between the architectural idea and actual implementation. The role of motion control is even more pronounced when we look at the fact that the printed structures are becoming more complex, featuring multi-material objects, internal lattices, or variable infill densities. Such complexities require an element of mechanical accurate and coordination which is only achievable through advanced motion systems. In the history of the 3D printing, there has been analogically similar to the innovation of the advancedness of the controlling processes. Stepper motors using open-loop controllers were commonly used in early desktop printers that sometimes were built around hobbyist kits. Those architectures had low capacity to address dynamism changes and usually created layer shifting, ghosting and dimensional errors known as artifacts (Wang et al, 2025). As the technology became mature, there was a demand to have more robust systems to be based on feedback. This gave birth to the introduction of closed-loop controllers, real-time sensing of feedback, and model predictive algorithms that learn responses to change in the environments and change in operations.

Cartesian and Delta motion structures represent the most prevalent structures in 3D printing nowadays. Cartesian printers work in a orthogonal \(X,Y\) and \(Z\) direction and as such these printers are mechanically simple and simple to calibrate. Delta printers however have three arms to serve the purpose of a triangular shape that facilitates the speedy and easily flowing curved movement. Every architecture brings different challenges and opportunities to motion control, especially when considering the axis coordination, generation of a trajectory and modeling of kinematics. It is important to know specifics of these systems in order maximize print quality, reduce the power usage, and prolong machine life. Accuracy in motion control cannot be relied only on the hardware components including stepper motors, servos drives and linear actuators (Zhou et al, 2025). It is also conditional upon the algorithms and software through which these elements are conducted. The trajectory planning, interpolation of the motion and interpolation of the acceleration has to be well balanced to ensure that there is good consistency of the extrusion, especially when there is an abrupt change of direction or speed (Sevcik et al, 2024). The recent advances include the use of advanced control measures such as Model Predictive Control (MPC) that predicts the future states and real time adjustments on the motion commands. Moreover machine learning algorithms are currently being trained using datasets of print anomalies which in turn can automate the adjustment of the motion parameters to reduce defects and increase yields. The merge of motion control with Open-source hardware and firmware is another important aspect that is becoming eminent. Ready-made platforms, such as Arduino Mega, Raspberry Pi, and BeagleBone, used along with firmware like Marlin, Klipper, or Repetier, allow the creation of experimental and trial environments that can be customized without compromise. These configurations allow adjusting the motion settings in real-time, remote troubleshooting, and adding high-end capabilities, such as dynamic mesh bed leveling and thermal drift compensation. Where industrial quality systems are more precise and reliable, the flexibility of the open-source system is crucial in low-cost innovation research.

Motion control troubleshooting and maintenance is also a lurking area on the quality of print. However, they are very important in the long-term performance and reliability. The print results can be degraded extensively by off-registrations, wear-and-tear and heat expansion. The best performance is possible in case of proactive diagnostics, scheduled calibration and adaptation with the help of feedback. Adding real time-sensing capacity to the health monitoring and utilising predictive maintenance algorithms can also reduce operational expenses and downtimes. The purpose of this chapter is to ensure a complete overview of the motion control systems in the situation of the 3D printing, raising the both startline knowledge of the subject and advanced approaches. It has been organized with reference to a large target audience, and this varies greatly, being: on the one hand, researchers in the field of academic studies of digital fabrication, and, on the other, users of the industry in search of ways to make their printing activities as reliable and scalable as possible. Important issues involve shape, axis setup, and co-ordination in mechanical systems, difficulties with kinematic control in various printing techniques (FDM, SLA, SLS) and the care of control of feedback-controlled systems, to help in error correction.

Also, we venture into the optimisation of G-code by smoothing out paths, tuning acceleration and segmenting motions (Jung, H., & Oh, S. 2021). Comparative case studies provide a representation of the difference between various motion control approaches and their influence on the print resolution, surface finish, and the build time. The conversation is anchored on the practical experiment upon open source and commercial platforms to guarantee the appropriateness of the expertise shared.

At the time, when parameters of digital manufacturing are being realigned towards the goals of Industry 4.0, the role of intelligent motion systems cannot be overestimated. The smart factories are dependent on connected devices, real-time analytics of data, and autonomous systems. Here, 3D printers with built-in adaptive motion control systems become not only a tool to fabricate products, but a smart node of some larger cyber-physical network. This development requires a multidisciplinary field whose combination entails mechanical engineering, control theory, computer science and materials science (van der Veen et al, 2017). Motion control systems are the core of effective and quality 3D printing. Through incorporation of intelligent algorithms, the employment of sound hardware, and the adoption of procedural calibration processes one can improve the capability of additive manufacturing systems remarkably. This will form the basis of such integration and provides a state-of-the-art perspective of motion control in 3D printing, and its importance in the future of automated production.

CONCEPTUAL FRAMEWORK

Incorporation of motion control systems in 3D printing has been achieved as an unification of various engineering fields such as control systems, embedded electronics, computational modeling and mechanical design. Fundamentally, a motion control system in additive manufacturing works to ensure a more precise control over tool paths in three-dimensional working environment and ensures consistency of proper deposition of the material and material accuracy in positioning. This theory does not work in a vacuum but is integrated into the broader reality of digital fabrication, cyber-physical systems and smart manufacturing. In essence, the practical means of understanding definition of motion control used in 3-D printing starts with interpretation of the digital models to the machine languages, which often appear to be in form of G-code (Tan et al, 2007). This conversion, which is performed by slicing software, not only settles the path of operation but in addition the related motion commands i.e. speed, acceleration, dwell time as well as extrusion rate which has been assigned to every portion of the print task. At this stage, the strain then falls on the motion controller that makes sure to coordinate the motions across various axes and move deterministically. The controller is not allowed to resolve time-lumped trajectories, it has to minimize motion discontinuity, and overcome mechanical latency or environmental unsteadiness that may influence quality of output. Most of the 3D printers employ conceptual architecture of motion systems based on Cartesian, Delta, or CoreXY kinematic designs. Every scheme has special limitations and advantages. Cartesian printers depend on orthogonal moving axis, which is relatively easy to calibrate and control, but Delta systems have faster speeds and more smooth curves with the radial arms moving around. Symmetrical belt-driven motion with reduced inertia and greater stability of CoreXY, in particular, and other similar solutions, mean that they are particularly well suited to large-format (e.g., needing extra-large printers) and high-speed needs (Mills, C. 2022). The theoretical differences between those architectures impose the need to utilize custom motion planning and feedback control schemes, which directly affect the resolution, level of fidelity, and build volume optimization.

Motion systems are made of an internal network of actuators, sensors, and microcontrollers hard coded to read the control algorithms in real time. A way of using mechanical motion is through actuators, most usually the stepper motors which move in designated axes. The stepper motors do not rotate with fine movement like the brushed or the brushless DC motors, this is what enables the use of open-loop control. Nevertheless, in order to tune the maximum accuracy closed-loop systems that use servo motors or hybrid steppers with encoder feedback are used. The systems allow verification of position and adaptive correction; important when external disturbances, or a thermally-induced expansion, may cause the toolhead to travel an unexpected trajectory (Boulaala et al, 2020). Feedback is very critical to the motion control framework. The sensors needed include position sensors, velocity sensors and occasionally force sensors that supply real time information back to the controller that allows dynamic correction of any unwanted deviation on desired motion. This feedback increases the level of control granular visibility, especially when certain complex maneuvers like curved surfaces or overhangs are being performed, as instant adaptation will guarantee the integrity of deposition. Introducing PID (Proportional-Integral-Derivative) control or other associated, more sophisticated approaches, such as Model Predictive Control (MPC), is the key to the optimization of this dynamic. Through this kind of integration it becomes possible to anticipate and prevent oscillation, backlash, or drift before it hinders surface quality and structural stability. The synchronization of movement with extrusion is as well prime in the conceptual model. Exact feed rate and extrusion rates make the filament deposited uniform. Any delay, even a microsecond, to under-extrude or over-extrude may cause holes, peeling off torn layers or build-ups of material which are not desired. Such interdependency means that the time correlation of the mechanical movement and the thermoplastic flow is one of the central principles of preserving the print fidelity (Xie et al, 2023). Where lasers or other light sources are used rather than extruded material, such as in processes like SLS or SLA, motion control controls exposure time and scanning quality, which affect depth of cure and edge detail of each print layer directly.

As motion control systems continue to become more modernized, they also adopt some adaptive characteristics to the current feedbacks of the environment. As an example, long-duration rejections can cause thermal expansion to comparably move the mechanical alignment, and the repeated additive dimensional error may generate cumulative effects. This requires, in concept, the establishment of self-correcting feedback mechanisms: software (e.g. dynamic calibration algorithms), and harware (e.g. strain shunts and heat sensors). The systems adjust trajectory of movement or torque of the motor on-the-fly to maintain accuracy. Another major layer of the conceptual framework is the trajectory planning (Xu, H., & Rong, X. 2024). Instead of merely issuing start and end points, the motion control systems need to calculate intermediate way-points, path shape and slowing down areas. These factors are vital when controlling mechanical jerk as well as keeping the layers deposited are constant. The algorithms used to calculate trajectory can use quintic polynomials, B Bezier curves or B splines, in each case there is a balance between computational efficiency and smoothness of the path generated (Habib, M. K. 2023). Jerk-limited acceleration profiles paired with the integration lowers the jerkiness in motion that might upset the build. The main element of real-time computation and the interrupt-driven firmware is important in the realization of sub-millisecond command resolution. Such firmware environments as Marlin, Klipper, and Repetier process the G-code and dynamically allocate computational resources to interrupts in real-time, to control motors, and to read thermistors. Such architecture guarantees that the motion commands to be executed are predictable, and have small latency as print complexity grows. In the advanced systems this functionality is reflected by the embedded real-time operating systems (RTOS) that integrate the motion control activities with the networking environment, user interface, and sensor processes.

The connection between theory and real-life practice is established by calibration and tuning of the whole system. This is conceptually open to bed leveling, backlash compensation, extruder calibration and thermal profiling. Leveling On automated printers, leveling is conducted via an inductive probe that samples the height of the surface over the print bed to create a 3D compensation map, which the printer allows adapting Z-axis movements in printing. Equally, extruder calibration functions extruded prints to drive motor steps per unit of filament, which facilitates homozygous flow in changing load conditions. External disturbances like vibrating and resonance are also regarded in the framework. When not taken into consideration these effects may result in a rippling (usually called ringing) or distortion of sharp edges (Senthil et al, 2023). Conceptually, these can be mitigated by mechanical damping (stiffer frames and isolators), and algorithmic damping (an input-shaping algorithm), and software compensation of acceleration and jerk. Frequency analysis and modal testing is becoming a best practice in the characterization of the dynamic behavior of moving parts. Variable monitoring also contributes to theory strength of motion management in the use of 3D printing. Scan pattern of sensing data of prints can be compiled and processed to understand the failure pattern or thermal drifting tendency. This machine learning knowledge is used in the context of predictive maintenance where motor current or temperature trend variations indicate that mechanical decay is occurring (Liu et al,2025). Integration of these analytics into the motion control loop forms a self-conscience system that has the ability to adapt and prevent faults. By introducing machine learning (ML) algorithms into motion control one can make the adjustments predictively rather than through feedback systems. Early indications of the anomalies, i.e., the layer shifts, adhesion failures, and so on, may also be detected and proactive changes in the motion parameters may be carried out with the usage of ML models trained on the historical print data (Maniraj, M., & Kumar, S. R. 2025). In specific, reinforcement learning models have potential to produce motion strategies, which improve starting with a base line that are eventually optimized on measures such as printing time, energy consumption, and quality of the surface.

Finally, the motion control conceptual model appreciates its position as a subset of Industry 4.0 systems and cyber-physical systems. Due to the interconnected nature of smart factories with 3D printers as nodes, it is possible to log the data on 3D printer motions, analyze them, and optimize them on fleet levels. With standards that enable the printers to relay to centralized controllers motion statistics, maintenance history or calibration data, e.g. using MQTT protocol or OPC UA, such distributed manufacturing process environments can actively use distributed real-time process optimization. The conceptual design of the motion control system of 3D printing is that of multi-tiered dynamic structure which comprises mechanical actuation, real-time feedback, algorithmic control and adaptive learning (MacDonald, E., & Wicker, R. 2016). It is closely connected with material behavior, eco-conditions, and interactions at the system level, and this is the foundation of precision, efficiency, and quality in additive manufacturing. Once the principles of motion mechanics are systematically reconciled with the power of digital intelligence, such a system will precondition the next revolution in 3D printing, turning the technology into the key constituent of intelligent, adaptive industrial manufacturing.

THEORETICAL FRAMEWORK

Motion control systems of 3D printing possess theoretical foundations that are based on a synthesis of a hybrid of control theory, a kinematics of robotics, automation science, and cyber-physical system processes. All these domains have perspectives and tools that are important to the development, refinement and optimization of precise motion in additive manufacturing. This part discusses a (more) integrative theoretical framework of the functioning of dynamic systems in response to commands and feedback, the calculation of spatial movements, as well as intelligent coping using algorithmic control and system modelling (Revathi et al, 2024). The classical control theory forms the basic compound where the behaviour of systems is modelled in terms of differential equations that relate the input and output of the systems along the time. Interpreting this in terms of motion control would mean knowing how actuation signals voltage or current sent to the motors leads to resulting readable positions or velocity. Proportional-Integral-Derivative (PID) controller is also a most generally used control system in 3D printers. It keeps on constantly computing of a error quantity as a difference between a setpoint (e.g. target position) and a measured process variable (e.g. actual position) and makes adjustments using proportional, integral and different harmonious terms. The compensations of these three components decide the responsiveness of the system to disturbances, stability of the system and the resistant ability of the system to steady-state errors. The PID control is simpler, robust and constrained in the cases where it encounters non-linearity, coupled systems, and dynamic that varies with time. The result is the use of more sophisticated control techniques like the Model Predictive Control (MPC). What MPC does is to use a dynamic model of the system to forecast the states over time and it seeks to solve an optimization problem in a finite horizon. MPC may be applied in 3D printing to pre-compensate thermal drift to avert motor overshoot and smooth mechanical vibrations by pre-setting the trajectories (Khalifa, R. B., Nhassa, G., & Snoussi, A. 2025). It offers a naturally adaptive and anticipatory structure of control that is quite suitable to the high-velocity high-precision production requirements.

Kinematics is another very important theory especially in doing robotics motion planning. Kinematics is the study of the time and geometric relations between the parts of a mechanical system and it does not require the forces that lead to motion. Forward kinematics wonders how the position and orientation of a toolhead can be calculated when some parameters of joints (e.g. motor angles) are known; inverse kinematics wonders how the required configuration of joints can be calculated when the desired position and orientation of a toolhead are given. Kinematics in Cartesian setups is fairly linear, but in Delta, or SCARA style setups the inverse kinematics solutions are not trivial and involve real-time calculation of non-linear trigonometric functions. Kinematic solvers being integrated into the firmware is thus an unconditional prerequisite of attaining positional accuracies of different printer geometries. To go together with the theory of kinematics, comes the theory of dynamics that takes into consideration the forces and torques into the collection of motions. In dynamic modeling, inertia, friction and external disturbance, like vibration or thermal expansion, are considered. A 3D printer may have the print bed or print head motion cause structural flexing or even oscillation, especially in lightweight or cantilever 3D printing systems. An advantage of dynamics-informed control is that its ability to dynamically tune actuation profiles can reduce the effect of such loads and movement conditions by real-time estimation of load and movement conditions. This is usually done by state observers such as Kalman filters that integrate sensor measurements using predicted models to provide low noise and latency estimation of state variables in systems. The theory of cyber-physical system (CPS) adds a larger perspective, which brings computation, networking, and physical processes. Embedded systems embedded systems in CPS communicate with real systems in real-time, and may have feedback loops which allow autonomous control and system-wide optimization. A contemporary 3D printer is considered a cyber-physical system where G-code commands (computation in the digital field) affect the behavior of actuators (physical movement) and how they are then observed with the help of sensors (cyber feedback) in order to make future commands more perfect. The theory of CPS is applied in the conceptualization of distributed control in printer networks with individual units being optimized locally and providing data to a centralized network to gather global data required in global analysis and suitable adaptive control.

The theory of reinforcement learning (RL) is a new field in motion control theory. Based on artificial intelligence, RL makes an agent learn to make choices through communicating with an environment and being rewarded or punished through feedback. There are motion systems where RL can be used to optimize printing strategies e.g. by optimizing acceleration profiles, reducing layer misalignment, or choosing correction actions to thermal distortion. RL, unlike other traditional methods of control does not need a model of the system to be rigorously defined; it constructs an agent that performs optimally by trial and error. Nevertheless, issues still exist related to real-time implementation particularly in relation to convergence time, safety limitation, and computational overhead (Hassan et al, 2024). The reliable motion control is also based on the systems theory, specifically on the concept of feedback stability and controllability. This called controllable when by suitable inputs, the state of the system can be steered to any of its values in a finite time. This 3D printing means, being able to control all the degrees of freedom within the motion system as possible to allow any toolpath to be generated as needed by the printjob. The dual of controllability is observability, and this is the capacity to reconstruct internal states based on external outputs; a feature essential in sensorless feedback control systems. Frequency domain analysis is another useful theoretical technique that evaluate the attitude of the systems with oscillatory inputs. Studying Bode plots and transfer functions engineers can design filters to reject certain groups of frequencies which can be mechanical resonances or power line interferences. Preprocessing of motion commands using input shaping, a method based on frequency analysis can be done to avoid excitation at the natural modes. The result is an easier operation especially in high velocity printing since sharp changes of direction or speed might cause unwanted vibrations. The theory of optimization is central to planning of trajectories and managing resources. Trajectory optimisation is where the most efficacious path of the print head is calculated to meet limitations, comparable to jerk restrictions, maximum speed and mechanical clearance (Duraimutharasan et al, 2025). They are frequently formulated using linear programming and quadratic programming; solvers in motion controllers use these methods in real-time. Furthermore, multi-objective optimization structures have been increasingly used to strike a trade-off between conflicting objectives e.g. optimal print time, energy use, tool wear with print adherence.

The integration of the fuzzy logic theory in the motion systems is what brings about the capability of dealing with vague or uncertain data. Fuzzy controllers' analysis uses linguistic rules (e.g., "in case of high vibrations and rising temperature, decrease the speed as well slightly") not just a pure mathematical formula. Such elasticity is especially advantageous in cases where the structure is unstructured or using materials of composite nature where the thermal and mechanical behavior might change during deposition. Finally, the systems integration theory will also make sure that all the systems, including mechanical, electrical and computational, are functioning in a conflict-free and concerted way. Layered architecture is used in motion control in which commands have to spread downwards to firmware routines, stepper drivers, physical motors long through a G-code interpreter that operates at a high level (Dutta, U., & Baruah, S. 2025). Interrupt handling, buffer management and synchronization standards should be verified theoretically not to encounter race conditions and/or timing issues or even loss of data which may affect the quality of print and the safety of machines. The Conservation of Total Energy, together with the mentioned principles of engineering systems, offers a multidisciplinary approach to considering the issue of motion control in 3D printing and searching the opportunity to enhance it. It goes between classical mechanics and contemporary intelligence, deterministic control and adaptive learning, centralized planning and distributed feedback. With a practical basis in engineering-based on these theoretical underpinnings, researchers and developers can develop motion control environments that are not just correct and dependable, but also expandable, versatile and resistant towards the changing needs of industrial-level additive manufacturing.

RESEARCH METHODOLOGY

The approach that was chosen in the present research was designed in the way to measure the efficacy of motion control systems applied to 3D printing based on trial-and-error experiments, software-level parameter manipulation, and statistical analysis. The main goal was to evaluate the effect of various motion planning strategies to hardware configuration on the most significant print quality metrics including dimensional accuracy, surface finish, and print time. The experimental design used was mixed-method, as it had to offer reliability quantitatively and insight qualitatively, considering the technical and practical goal of optimization of the motion control system. Three main classes of 3D printers architectures were chosen, including Cartesian, Delta and CoreXY and were compared. Every printer had open hardware platforms, such as an Arduino Mega 2560 microcontroller flashed with Marlin firmware and RAMPS 1.4 board. Klipper high-speed kinematic functionality was tested on systems with Klipper firmware and Raspberry Pi systems by running additional experiments. This was done with prints made out of PLA as uniform test specimen using common calibration models, including 20×20×20 mm cubes and benchy boats printed in controlled ambient conditions (26 C +/-1 C).

Such parameters as Jerk, acceleration, maximum feedrate were tested as the key ones involved in motion control. These were adjusted in the firmware in a systematic way and prints examined with reference to measured features using digital calipers (accuracy of 0.01mm), by eye and by surface profilometer. Each placement was repeated three times in order to cover the variability and guarantee the plausibility of applying statistics. The effects of each of these parameters were examined by a one-way Analysis of Variance (ANOVA) run on each motion control variable. As an illustration during an evaluation of acceleration, the independent variable was the acceleration value (i.e. 500 mm/s 2, 1000 mm/s 2, and 1500 mm/s 2), and the dependent variable was the dimensional accuracy of the print. In the same line of argument, ANOVA was applied to determine the statistical significance of surface roughness and print completion time when varying the jerk settings and feedrate settings. The null hypothesis in either of the cases was that the motion parameter will not play a role in the output, whereas the alternative one suggested that it would have a major influence. ANOVA procedure entailed the calculation of the F-statistic that was aimed at evaluating whether between-group variance was significantly greater than within-group variance. The p-value below 0.05 was regarded as statistically significant. Where significant differences were obtained, post hoc analysis was carried out based on Tukey HSD test to know points of specific parameter settings that led to an improvement in performance (Fu, G., Gu, T., Gao, H., & Lu, C. 2020).

To complement data obtained in quantitative way, the vibration and resonance was examined by the accelerometer sensors attached to the printer gantry. Such sensor data were recorded over serial communication and interpreted in order to outline the relation between motion dynamics and noted artifacts like ringing or ghosting. The particular methodology allowed understanding how motion parameters affect each other in different types of mechanical structures and gave empirical data on tuning procedures that increase system stability and improve output performance. The physical measure of the parameters, statistical analysis and the monitoring of data in real time is an effective combination to form a base to interpret the results given in the following section.

RESULTS AND DISCUSSION

The experimental tests and a statistical analysis showed that motion parameters like acceleration, jerk and feedrate have a significant effect on the quality of the prints with some variability in the results depending on the architecture of the printer used. ANOVA was employed to determine the significance of each of the parameters on the print dimensional deviation and the results showed that all the three variables had a statistically significant difference ($p < 0.05$). Jerk produced the largest influence on dimensional accuracy of all the parameters tested followed by feedrate and acceleration. As can be seen in the figure, the average dimensional deviation with different jerks setting was about 0.25 mm with greater variance than the other two parameters. This implies that sudden change of direction that is caused by increased values of jerk may trigger an overshoot mechanism caused by inertia, which causes surface deformation and edge rounding. Its statistical F-value was 8.94 with a p-value of 0.003 and with this finding its impact was significant.

Figure 1. Impact of Motion Parameters on Print Accuracy

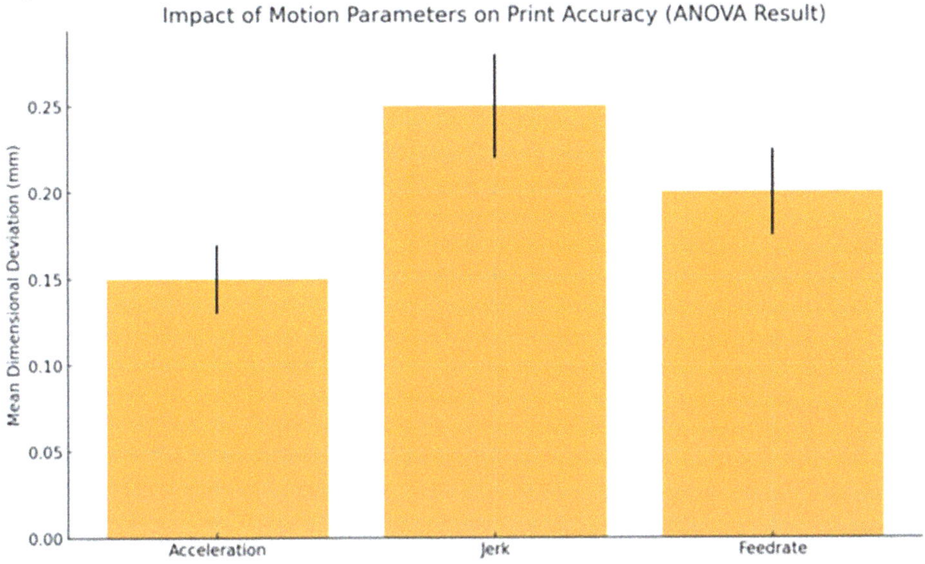

Comparatively less but statistically significant effect was observed in changes in acceleration ($F = 6.12$, $p = 0.019$). At low values of acceleration, the transitioning was smoother and less shaking notably on curves on layers. But very small acceleration did not make print time short without much gain of accuracy and it shows there has to be a balance between aggressiveness of motion control and system stability. Feedrate affected the print time as well as the visual surface quality. Although medium advancements of feedrate showed no severe decrease in dimensional quality, extreme values produced visible signs of under-extrusion. The compromise was there because higher speeds sacrificed accuracy, especially of smaller details. Accelerometer sensed information also confirmed that resonance effects were stronger at high jerk and acceleration combinations. When input shaping was turned on, and acceleration was tuned below 1000 mm/s 2, the print artifacts (ringing and ghosting) were noticed to be much-reduced. The comparative study between different types of printers showed that the Delta printers were jerky to different settings of jerk since they had complex kinematics, although they had faster speed. Cartesian systems were more consistent over a variety of motion parameters, and CoreXY architectures succeeded in finding a passable compromise between speed and accuracies when appropriately tuned. According to a post hoc Tukey, the most systems had an optimal range of 800-1000 mm/s/s in acceleration, 10-12 mm/s in jerk, and 60-80 mm/s in feedrate. These were the acceptable ranges where the compromise of the print time, accuracy, and surface finish were desirable. A combination of the real-time feedback and strategies of algorithmic motion control also makes the system even more transferable and accurate. The following section gives such solutions and recommends them based on the findings of this section to be presented as empirical evidence.

SOLUTIONS AND RECOMMENDATIONS

The information above concerning the convenience of motion control in 3D printing documents attests to the fact that an optimal approach to motion control in 3D printing potentiality must be comprehensive and flexible, incorporating both hardware- and algorithm-based optimization, as well as the system-wide synchronization. The section provides a body of solution and best-practice advice based on experimental measurements and theoretical modeling on how motion control systems in the additive manufacturing environment may be optimized. These insights target scholars, system integrators and players in the industries who intend to enhance both the performance and reliability of 3D printers in numerous applications. One of the major solutions to this is the adoption of a hybrid control structure which integrates conventional feedback control using PID with predictive algorithms which are intelligent like Model Predictive Control (MPC). Although the use of PID controllers is still effective with regards to real time response and simplicity, these cannot also be effective in complex print applications as it is unable to resolve time delay and non-linear behavior of systems. Conversely, a MPC system predicts future costs, or state of the system by solving an optimization problem over a finite horizon. A hybrid controller with PID and MPC can be used to provide fast responses and trajectory-level planning: with such a controller, the system can correct any overshoots and resonance, as well as accommodate thermal expansion in real time. This has been especially successful in closed-loop motion systems in which encoder feedback is used to provide constant recalibration.

The third crucial suggestion would be to standardize tuning script of motion parameters. As indicated in the results, such parameters as jerk, acceleration, and feedrate influence the quality of print significantly. But they are usually set in a haphazard fashion or trial-and-error, which is inefficient and make mistakes. Data driven calibration procedure is to be carried, where benchmark models were fixed and printed with controlled variations to all given parameters. Then dimensional deviation and surface roughness are measured and best parameter sets are obtained through ANOVA or response surface methods. By integrating this tuning protocol into a firmware or slicing software we can expect to be able to tune motion systems in a systematic manner even with novice users. Addition of sensor-controlled, real-time monitoring systems is another paradigm shift in the optimization of the motion control. Live data of mechanical vibration, material flow consistency and structural deflection can be gathered or sensed by mounting accelerometers, temperature sensors and load cells on the print head or the build platform (Boulaala et al, 2020). This data can also be passed to feedback loops or machine learning models to identify early when something has gone wrong, e.g. layer shifting or warping, and take remedial action. To give an example, an abnormal increase in the vibration of the Z-axis may deactivate acceleration or activate damping filters, and the control of the surface integrity will be maintained. The sensor integration has now become much cheaper and it is even affordable to mid-range spaced desktop printers. The algorithms used in motion planning and smoothing of the trajectory should be improved to use more than just linear interpolation. One can use curve fitting by use of B plot or B-spline interpolation which enables smooth transitions, especially corners and narrow radii, where a sudden change in direction traditionally causes ghosting or under-extrusion. Besides, the adoption of jerk-limited acceleration profiles reduces brief impact of forces that result in mechanical resonance. Such algorithms are supported on firmware platforms such as Klipper, and advanced versions of Marlin and users are urged to enable them and adjust them depending on machine parameters.

Firmware optimization would be very important in machines that use open-source platforms. The firmware used in motion control must have the well-defined interrupt routines, buffer space and step timings to avoid loss of step or data overflows. In addition, accuracy and speed can be reconciled effectively by allowing adjustable microstepping; i.e. adjusting step resolution to be fine during fine, slow moves and coarse during long travel moves. These capabilities should as much as possible be documented explicitly by developers and exposed by slicer developers to a guided interface to users. The physical component of accuracy on the motion is through mechanical stability and some improvements on the design. Flexible-frame systems, low-tension belts and under-damped systems are per se restricted in their capacity to perform the commanded operations, independently of the sophistication of the control. Mechanical backlash and vibration can be removed through reinforcement of printer frames with metal crossbars, improvement of the stiffness of the gantry, the use of two Z-axis lead screws, and the belt tension adjustable using automatic pulleys. This is especially important with large-format printers and high-speed Delta printers in which any small mechanical vibration is transmitted further due to a longer travel length. As reliable motion control is concerned, calibration is essential. Additionally to a one-time calibration operating automated and continuous calibration routines is suggested. Mesh probing should be a method of probing the bed to set this up and thermal compensation must also be adopted so that motion commands can track the effect of ambient or internal temperature changes. The Extruder steps-per-mm calibration should also be confirmed in various flows and filament chemicals to avoid irregular deposit. The diagnostic modules that warn the users when their machines are rotating too high or when there are worn-out parts must also be put in firmware. To seek wider integration of systems, printers are advised to use the modular and interoperable communication protocols like I 2 C, SPI or CAN within internal substructure and MQTT or OPC UA at factory-level integration. Such protocols allow synchronized execution of the motion control units, sensor modules and data logging systems. The introduction of 3D printing as a constituent within cyber-physical production set ups obviates real-time coordination with both downstream and upstream operations.

A more advanced suggestion is the prediction of deviations in motion performance using machine learning (live-based) models and compensating these changes. Having supervised learning models trained on sets of data obtained at motion parameters, sensor data and the resulting print quality variables enables the system to dynamically adjust federates, introduce compensation vectors, or activate recalibration tasks. Simulation environments can be used to learn of new motion strategies using reinforcement learning, and once satisfied with them, deployed in the real world. Still in their Stad, these methods are the path to the future of real-life intelligent motion control in additive manufacturing. There are also the aspects of path segmentation and G-code optimization that have to be addressed. Rather than cutting models into standard layer heights and straight toolpaths they should be divided into variable layer heights and adaptive segmentation according to the geometric complexity and structural significance. A higher resolution command with a finer stepper resolution and reduced speed may be used in critical areas like overhangs, small radii or interlocking joints whereas coarse motion settings may be used in infill or large flat areas. A range of slicers already deal with this, though being able to combine it with motion-aware slicing algorithms will bring their print quality and efficiency to a new level. Regarding practice on the part of a user, more documentation and training materials need to exist. Manufacturers and open source groups ought to come up with systematic guidelines and testing products to enable users learn how different motion settings have an impact. Users will be able to recognize and fix the issues in advance, as visual guides pre-populated with artifacts related to bad acceleration, jerk mismanagement or under-extrusion because of motion-lag will be added. Such tools can be implemented into slicers or even control panels of printers and close the knowledge gap between what the system can do and what the user knows.

Finally, interoperability and quality assurance will be improved by the creation of industry standards and standardization measures. Group organizations like ASTM and ISO ought to increase their specifications to incorporate functionality specs of motions systems, consisting estimates of vibration tolerance, axis synchronization and trajectory tracking. Test protocols that certify the ability of the system to perform the motion under stress could be devised by the certification bodies, which will assist buyers and developers to choose on an objective data. The motion control system is a 3D-printing venture. It needs slow matching of hardware reliability, descriptive precision, computational feedback savoir-faire, and user training. The hybrid control strategies are implemented, utilizing the real-time sensor feedback, statistical calibration procedures, adaptation based on machine learning, which make the additive manufacturing systems work more accurately, faster, and reliably. Not only do these solutions play an important role in elevating the quality of individual prints, but they are also the underpinnings to growing 3D printing as an industrial-grade process that consumers can rely on.

FUTURE RESEARCH DIRECTIONS

With 3D printing moving to become a true manufacturing platform, as opposed to being used as rapid prototype device, the embarkation of intelligent responding, adaptive, and integrated motion control systems is more and more central. The emerging need in industrial, biomedical, aerospace and architectural applications are demanding more than high precision of additive manufacturing technologies as far as repeatability, responsiveness and intelligence is concerned. A research in motion control systems in future should hence go beyond mere enhancement of basic accuracy to accommodate new paradigms in the field of computational intelligence, systems engineering and cyber-physical integration. Among the future research priorities, the creation of fully adaptive motion control solutions with the learning ability based on working data and self-generation of the control strategies must be considered. Whereas existing systems are based on closed-loop control using static, tunable rules (e.g. PID tuning), the next has to be based on real time learning using multi- sensor feedback (e.g. vibration, extrusion rate, temperature, motor torque). It would be achieved by embedding edge computing systems to printers with the use of lightweight neural networks directly operating on microcontrollers or embedded devices like Raspberry Pi and BeagleBone. Further line of investigation should be directed at exploration of efficient model designs (e.g. TinyML or federated learning models) that make possible to use intelligent reasoning and decision-making without violating speed of processing and memory limits. A similar research direction is formal modeling of motion control in non-linear and multi-material world. Due in large part to the mechnical response of the deposition materials in the event increasing issues of additive manufacturing involving hybrid materials, functionally graded materials and embedded electronics, the motion dynamics that are necessitated requires an influence of the mechanical response of these forming materials during deposition. The nature of the current trajectory planning is based on the assumption that mechanical properties and the behavior of layer adhesion does not change. The future models should consider incorporating the feedback data provided by real-time rheological measurements, thermal imaging, as well as force sensors in order to adjust the paths of motion in real-time. An unlocked area of theoretical development S and experimental verification exists in the mathematical modelling of such non-linear space varying systems.

Applying the digital twins to motion control systems is also another promising research area. A digital twin is a digital advancement that depicts a physical system in real-time. It constantly aligns with the physical resource using data acquired by sensors. A motion control digital twin might simulate real time positioning, path execution and health parameters of a manufacturing machine, in the example of 3D printing. The digital twin would allow the researchers to simulate new methods of control, foresee print failure, and run tests of stress without the hazard of physical damage. The next assignment of work has to be directed to real-time synchronization algorithms, secure communication procedure, and updating of models according to readings of sensors. The scaling of these models in distributed production networks would be a necessity and therefore they would be integrated with cloud computing infrastructures and digital manufacturing execution systems. It must also have stronger data driven predictive maintenance models that use motion performance trends. Rather than depending on scheduled calibration or react to repairs, printers should be capable of automatically noticing degradation indicators, e.g., the amplitude of the vibration grows, motion smoothness reduces, or there is a drift in the encoder, and fix the problem themselves or warn operators. It needs to identify the failure modes and corresponding threshold limits and also determine the machine learning based predictions of the remaining useful life (RUL) of motion components such as belts, lead screws, and bearings through future research. Those models can be cured on longitudinal motion data after hundreds of print cycles. There is still another direction, which needs to be explored but is critical important to motion control systems in networked printers design: multi-agent coordination. Standard industrial situations arise where tens or hundreds of printers are going to work simultaneously, and motion synchronization between machines might be needed to optimize power use, minimize vibration coupling or exchange trajectory information. Design of consensus or distributed motion planning strategies that balance the autonomy within each individual component with the global system goals is an important but complicated field. A cross-disciplinary investigation in swarm robotics prepared with distributed computing and wireless communications procedures would give off great developments in this field.

The discipline should also increase work on real-time G-code optimization and just-in-time planning. Although, slicing software now executes most path planning prior to the start of a print, the motion executor in future community systems must learn motion strategies during a print, based on sensor-feedback or anomaly detection. This involves the ability to handle change much more quickly by parsing modified g-code instructions, conflict resolution system to support dynamic changes, and motion buffering that maintains sequence. The formal code verification methods should be established through researchers in order to make sure that the on-the-fly modifications of G-code do not break any mechanical constraints and/or cause thermal inconsistencies in printed objects. More progress needs to be done on integrating additive and subtractive motion control. The hybrid manufacturing systems which involve a 3D printing process with either CNC milling, laser ablation or ultrasonic finishing would need very coordinated motion control between the various types of tools and movement patterns. The next stage of machine development and the theory of machine control systems is to be in the development of unified controls platforms that between additive and subtractive smart control, between drilling and milling, in addition to and in between machining, is able to spatially preserve focus, direct motions, and harmony. The potential is also high in the environment-sensitive motion control. Depending on ambient temperature, humidity or vibration caused by neighboring machines, print quality may be poorer. The environmental conditions shall be observed by computer systems in the future that shall constantly change the motion strategies. This may include choice of alternative acceleration profiles, change of retraction distances or redeploy zero points. Programming the production-grade situations with sales correlation and sensor input into predictive environmental compensation algorithms would enhance such consistency.

Studies are required to learn about the influence of motion parameters on the formation of microstructure in printed materials at the crossroad of materials science and motion control. As an example, polymer-based systems consisting of extrusion, the rates of crystal formation and interlayer binding are influenced by the rate of cooling through motion. During metal additive manufacturing the velocity of movement and direction of movement determine the structure of grain growth and residual stresses. As both the finite element simulation and material characterisation, coupled with a multidisciplinary approach, in particular by coupling finite element simulation with computer-aided motion simulation, may result in new control algorithms not only aimed at achieving improved geometric fidelity, but also achieve optimisation of microstructure. The other area that has yet to be exploited is the implementation of bio-inspired algorithms in motion planning. Pathfinding or constrained optimization Evolutionary algorithms and genetic programming have been applied to other fields such as ant colony optimization In computing evolutionary algorithms can be used to find the shortest path or an optimal route. There is potential value to applying these algorithms to difficult toolpath generation problems in additive manufacturing: such solutions would discover a new motions sequence that could lead to a more efficient print, a stronger print, or a less energy-intensive print. More so, the combination of these approaches and reinforcement learning models might develop hybrid intelligent systems adjustable to geometric and material challenges.

Another cross-cutting theme which needs institutional and industrial cooperation is the development of open motion control benchmarks and datasets. Several limitations of most of research in this sphere are connected with the absence of unity in test cases, motion database, and available (open source) firmware configurations. The key issue that needs to be addressed in the future is the design of repositories where motion profiles, sensor logs and print results under different conditions could be deposited and compared. This type of data would make AI models train faster, increase reproducibility, and create collaboration-based innovation between institutions and companies.

CONCLUSION

Technological development of 3D printing has introduced a new epoch in the possibilities of manufacturing with design freedom, material versatility, and individualization joining forces in solving industry-related and research-related complex problems. The key element in the transformation is the motion control system, which is commonly overlooked but essentially the most crucial aspect of regulating accuracy, efficiency, and reliability of the additive manufacturing processes. As has been evident in this chapter, the functioning spine of 3D printers is based on the use of motion control systems, which convert digital instructions into finely-choreographed physical motions. Main structural factors used in the process are the incorporation of the add-ons hardware parts which include, the stepper motors, servo drives, motion controllers and sensors in the process of mechanical implementation of the tool head or build platform. All these elements however have to be coordinated with complex algorithms that work on trajectory planning, acceleration control, synchronization and feedback correction. Motion control has moved well beyond the scope of simple position control, to the more generalized problem of efficient material deposition under diverse and changing mechanical, thermal, and manufacturing situations. Translated: the hardware precision is only a part of the equation of achieving high-quality prints, and the embedded smarts in the firmware/control logic is the other part of the equation. In the described conceptual and theoretical frameworks, this chapter has determined a multi-disciplinary appreciation of the functioning of motion systems and their optimization. Both of the classical PID control and newer model predictive control relies on control theory, which is used as a basis to keep the system stable and responsive. The rules of robotics e.g. forward and inverse kinematics can ensure accurate toolpath execution on a variety of mechanical structures, e.g. Cartesian systems, Delta systems and CoreXY systems. The theories of systems and cyber-physical fusion continue to put the context of motion control into the big picture of Industry 4.0 with its specifics of real-time feedback, intelligent device distribution, and responsiveness at the data level.

The repairable lack of fit between Model and Factor would be enhanced by empirical output of experimental analysis that the parameters of motion play an incisive role in affecting dimensional accuracy, surface quality and the find time to print. Statical significance of such variables has been attested through one-way ANOVA tests and therefore calibrated tuning is important. In addition, the presence of accelerometer data has also shown the mechanical sensitivity of motion systems to dynamic forces including vibration and thermal drift that further proves the significance of sensor integration and real time feedback loops. The nature of the solutions and recommendations proposed manifests itself in the need to embrace holistic approach in motion control. It is now not enough to concentrate on mechanical enhancement or software upgrading alone any more. Rather, the best system performance arises as a product of real-time streamlining of hardware design, tuning of the firmware, motion algorithms, sensor fusion, and user interaction. This way of thinking has been captured as the suggestion of hybrid control structures combining PID responsiveness and MPC foresight. On the same note, the trend towards more intelligent and autonomous control ecosystems can be seen in the trend toward adaptive G-code optimization and automated calibration and integration of machine learning.

The directions on the proposed future research would lead to an even more revolutionary ability. The emergence of self-learning control structure, environment-sensitive movement calibration, digital twin matching marks the shift of 3D printing to intelligent robotics and autonomous production. Also underlined in these directions is the necessity of an interdisciplinary effort that entails integrating efforts of computer scientists, control engineers, material specialists and mechanical designers to arrive at solutions to complex, integrated problems. Both in formulating powerful multi-material printing motion models and in developing secure, distributed motion planning techniques that will coordinate printer farms, the future of additive manufacturing will require new theoretical and practical advances. With the increasing variability in motion control technologies and their increasing complexity, it will become increasingly important to have common repositories of open data, standard performance measurements, and shared or universal diagnostic tools to guarantee reproducibility and to allow combined development efforts. Democratizing the knowledge of motion control by educational kits, modular firmware and community based development platforms will empower the next generations of developers to explore, experiment and grow this field. Because AM is enveloping areas like the creation of medical devices, aerospace components, buildingscale printing, and energy storage systems, the requirements on the motion control systems will be even higher. These applications do not only demand geometrical accuracies but also geometrical consistency of microstructural consistency, mechanical strength and resistance of the environment. Satisfying such requirements will require even closer combination of motion control with thermal management systems, multi-axis force feedback and process-aware adaptive algorithms. The future 3D printer will not only be like a glorified extrusion machine, it will be a thinking manufacturing platform that can respond to the environment in which it is operating and intelligently maximizes the results on the fly.

In a larger sense, there are also economic, social, and ethical implications of such evolution that pertain to manufacturing. Intelligent motion control systems help in building a more efficient material usage, energy savings, and product quality, and flow towards achieving sustainability. Autonomous calibration and maintenance aids in cost reduction of operation and democratization of high-end fabrication capabilities. Meanwhile, the gradual loss of control of these systems is putting in question the issue of transparency, traceability, and human monitoring, aspects that will have to be resolved by applying thoughtful design and regulatory convergence. To sum up, 3D printing motion control systems are no longer margins and support services. These are the fundamental operators of accuracy, effectiveness and smartness in additive manufacturing. With powerful theoretical background, practical testing, strategic suggestions, and vision on the future of the subject this chapter attempted to present a detailed discussion on the possibilities of utilizing motion control and developing it. The unity of control engineering, artificial intelligences, and intelligent hardware brings new frontiers to what 3D printing can accomplish not only as something used only to create prototypes but also as part of the next-generation manufacturing infrastructure. Putting this potential into practice will demand the sustained application of innovation, interdisciplinary cooperation, and an enthusiasm in constructing systems that are high-performance as well as versatile, extendable and sustainable.

REFERENCES

Boulaala, M., Elmessaoudi, D., Buj-Corral, I., El Mesbahi, J., Ezbakhe, O., Astito, A., El Mrabet, M., & El Mesbahi, A. (2020). Towards design of mechanical part and electronic control of multi-material/multicolor fused deposition modeling 3D printing. *International Journal of Advanced Manufacturing Technology, 110*(1-2), 45–55. DOI: 10.1007/s00170-020-05847-0

Duraimutharasan, N., Deepan, A., Swadhi, R., Velmurugan, P. R., & Varshney, K. R. (2025). Enhancing control engineering through human-machine collaboration: AI for improved efficiency and decision-making. In *Harnessing AI for Control Engineering* (pp. 155–176). IGI Global. DOI: 10.4018/979-8-3693-7812-0.ch008

Dutta, U., & Baruah, S. (2025). AI-driven 3D printing: Unlocking new frontiers in healthcare. In *AI-Powered Systems for Healthcare Diagnostics and Treatment* (pp. 233–266). IGI Global. DOI: 10.4018/979-8-3693-9735-0.ch010

Fu, G., Gu, T., Gao, H., & Lu, C. (2020). A postprocessing and path optimization based on nonlinear error for multijoint industrial robot-based 3D printing. *International Journal of Advanced Robotic Systems, 17*(5), 1729881420952249. DOI: 10.1177/1729881420952249

Habib, M. K. (Ed.). (2023). *Global perspectives on robotics and autonomous systems: Development and applications*. IGI Global. DOI: 10.4018/978-1-6684-7791-5

Hassan, A., Dutta, P. K., Gupta, S., Mattar, E., & Singh, S. (Eds.). (2024). *Human-centered approaches in Industry 5.0: Human-machine interaction, virtual reality training, and customer sentiment analysis*. IGI Global. DOI: 10.4018/979-8-3693-2647-3

Jung, H., & Oh, S. (2021). Data-driven optimization of integrated control framework for flexible motion control system. *IEEE Transactions on Industrial Informatics, 18*(7), 4762–4772. DOI: 10.1109/TII.2021.3100872

Khalifa, R. B., Nhassa, G., & Snoussi, A. (2025). Techniques and methods for optimizing 3D printing parameters. In *Modeling, Analysis, and Control of 3D Printing Processes* (pp. 329–354). IGI Global.

Liu, B., Qian, B., Liang, Y., Dai, P., Li, R., & Wei, Q. (2025). Fast and accurate auto-disturbances-rejection temperature control system based on particle swarm optimized fuzzy control: Applied for particle three-dimensional (3D) printing prosthetic orthotic plate. *Additive Manufacturing, 101*, 104704. DOI: 10.1016/j.addma.2025.104704

MacDonald, E., & Wicker, R. (2016). Multiprocess 3D printing for increasing component functionality. *Science*, *353*(6307), aaf2093. DOI: 10.1126/science.aaf2093 PMID: 27708075

Maniraj, M., & Kumar, S. R. (2025). From theory to practice: Control system challenges in additive manufacturing. In *Modeling, Analysis, and Control of 3D Printing Processes* (pp. 237–262). IGI Global.

Mills, C. (2022). The design and implementation of a low-cost high-precision 3D-printed translation stage using an optimized PID controller for optical applications. *Unpublished manuscript.*

Revathi, S., Babu, M., Rajkumar, N., Meti, V. K. V., Kandavalli, S. R., & Boopathi, S. (2024). Unleashing the future potential of 4D printing: Exploring applications in wearable technology, robotics, energy, transportation, and fashion. In Hassan, A., Dutta, P. K., Gupta, S., Mattar, E., & Singh, S. (Eds.), *Human-centered approaches in Industry 5.0* (pp. 131–153). IGI Global. DOI: 10.4018/979-8-3693-2647-3.ch006

Senthil, T. S., Puviyarasan, M., Babu, S. R., Surakasi, R., & Sampath, B. (2023). Industrial robot-integrated fused deposition modelling for the 3D printing process. In *Development, Properties, and Industrial Applications of 3D Printed Polymer Composites* (pp. 188–210). IGI Global.

Sevcik, M. J., Bjerke, G., Wilson, F., Kline, D. J., Morales, R. C., Fletcher, H. E., & Eliasson, V. (2024). Extrusion parameter control optimization for DIW 3D printing using image analysis techniques. *Progress in Additive Manufacturing*, *9*(2), 517–528. DOI: 10.1007/s40964-023-00470-3

Tan, K. K., Lee, T. H., & Huang, S. (2007). *Precision motion control: Design and implementation*. Springer.

van der Veen, G., Langelaar, M., van der Meulen, S., Laro, D., Aangenent, W., & van Keulen, F. (2017). Integrating topology optimization in precision motion system design for optimal closed-loop control performance. *Mechatronics*, *47*, 1–13. DOI: 10.1016/j.mechatronics.2017.06.003

Wang, D., & Zhang, J. (2022). Two improved scanning path planning algorithms and a 3D printing control system with circular motion controller. *Rapid Prototyping Journal*, *28*(4), 695–703. DOI: 10.1108/RPJ-08-2020-0190

Wang, Y., Zhao, C., Liu, W., & Li, S. (2025). Path planning and real-time optimization of robotic arm 3D printing. *International Journal of Modeling, Simulation and Scientific Computing*, *16*(3), 2541013. DOI: 10.1142/S1793962325410132

Xie, B., Gao, H., & Zhang, T. (2023). A review of the control system of a 3D printer. *Recent Patents on Engineering*, *17*(1), 25–41.

Xu, H., & Rong, X. (2024). Design and application research on control system of 3D printer based on multi-DOF robotic arm. *Advances in Computer and Communication, 5*(6).

Zhou, R., Jiang, J., Li, P., Lu, S., Li, B., & Du, C. (2025). Optimized compound control strategy for high-speed 3D printing: Enhancing surface quality and printing precision. *Journal of the Brazilian Society of Mechanical Sciences and Engineering*, *47*(8), 47. DOI: 10.1007/s40430-025-05684-w

Chapter 3
Collective Intelligence and Swarm Robotics: Decentralized Strategies for Adaptive and Resilient Motion Control

R. Saranya Devi
https://orcid.org/0009-0000-1872-4080
Agurchand Manmull Jain College, Chennai, India

Mandiga Yasaswini
https://orcid.org/0009-0003-8044-2860
R.M.D. Engineering College, India

A. Rahamath Nisha
P.K.R. Arts College for Women, India

R. Divya Ranjani
Saveetha Engineering College, India

A. Suresh Kumar
Saveetha College of Liberal Arts and Science, SIMATS, Chennai, India

ABSTRACT

This chapter discusses the combination of collective intelligence and swarm robotics as decentralised methods of motion control. The research is based on an example of biological systems that exhibit strong, scalable, and adaptive global dynamics by the simple rules or programmes applied locally. Reinforcement learning and evolutionary optimization have been demonstrated to be more efficient, resilient,

DOI: 10.4018/979-8-3693-3637-3.ch003

and have better energy performance through simulations, physical verification, and statistical analysis. There is a proposed motion control design that is layered and has a balance between autonomy and hybrid oversight, energy optimization, and ethical governance. The results leave decentralized swarm robotics as the paradigm of resilient, adaptive and ethical motion control in a variety of applications.

INTRODUCTION

Collective intelligence and swarm robotics have become a paradigm shift in the field of motion control, providing decentralised approaches to the problem, which reorganise interrelations between robotic systems, their adaptive behaviour and complex goal achievement. The collective intelligence, which denotes the emergent problem-solving ability of groups, is best manifested in nature when simple organisms like ants, bees, and birds all come together to create the solution to the problem that is far beyond the ability of any one acting agent (Jia, Y. 2025). A translation of these principles into robotic systems is an effective way of building autonomous groups capable of organisation without command and control. This is particularly applicable in cases of motion control, when several robots need to operate in dynamic settings, prevent collisions, and complete coordinated work in uncertain conditions (Ali, Z. A., Alkhammash, E. H., and Hasan, R. 2024). The use of decentralised approaches is not only increasing scalability but also makes such systems robust and fault-tolerant so that the failure of individual units can be circumvented and the system can still operate successfully. The growing sophistication of industrial processes, supply chain, healthcare logistics, environmental surveillance and defence systems have highlighted the inefficiency of centralised robotic control systems. The architecture of centralization has a latency, extensive computational requirements and susceptibility to a single-point failure. In comparison, swarm robotics directed by the principles of decentralised motion control relies on distributed decision-making and local interactions to facilitate the rapid adaptation process (Abro et al., 2024). Every robot or agent in such systems works with minimal sensing and communication but the combination of their actions results in coherent and effective overall behaviours. Such emergent behaviour can also lead to less reliance on global control as well as reflects the evolutionary benefits shown by biological swarms in their ability to adapt to changing environments (Almansoori, A., Alkilabi, M., and Tuci, E. 2024). Motion control has always been a field that deals with precision, efficiency and predictability, and usually in the manufacturing sector, automation, and transportation. The concept of control is broadened with such a combination of swarm robotics, to include the deterministic regulation of trajectories or velocity profiles. It will turn into an experiment in emergent dynamics, where algorithms and interaction protocols are crafted in ways that will tend towards the establishment of the kind of group behaviour they want. This method is made rich by the fact that it can complete complex coordination functions (including area coverage, search and rescue, cooperative transport, or surveillance) without any individual robot being fully aware of its environment or having detailed instructions (Ivanov, D., and Petrova, A. 2024). Practically, decentralised control would greatly decrease communication overheads and energy usage and increase the flexibility in real-time. Progress in artificial intelligence, reinforcement learning, and distributed optimization has boosted the feasibility of collective intelligence systems in robotics. Multi-agent learning enables swarms to change their tactics together, whereby, agents provide minimal feedback to perfect decision-making. Reinforcement learning (each of the units maximises the local rewards) and evolutionary algorithms (based on natural selection) that enable the swarm to get better as more and more generations are produced (Rana, M. M., and Ibrahim, U. M. 2024). The combination of these computing strategies with the hardware developments in low-cost sensors, wireless communication, and energy-efficient processors opens rich opportunities to the application of decentralised motion control. Moreover, robotics and cyber-physical systems converging with the Internet of Things can offer new opportunities to implement swarms to work in interconnected space where machines interact with human operators and smart infrastructures (Kazim, R. M., Wang, G., Ming, Z., Allen, J. K., and Mistree, F. 2025).

The relevance of decentralisation strategy in motion control is also compounded by the increased demand of resilience and adaptability in unpredictable or dangerous situations. This could be the case in a disaster area, deep-sea drilling or a mission to the moon, where the centralised communication connexions are not reliable or possible (Kumar, V., & Das, S. 2024). In such cases, swarm robotics offers an interesting alternative, and distributed decision-making among agents allows the operation to continue even in the case of partial failures or the loss of contact (Kyzyrkanov, A., Atanov, S., Aljawarneh, S., Tursynova, N., Otarbay, Z., and Saltanat, A. 2025). In addition, the decentralised paradigm gives an assurance that the system is able to dynamically reconfigure itself, redistribute tasks, and compensate the lost units thus allowing operations to continue to pursue. Such flexibility is equally important in the industrial context whereby any abrupt alterations in demand, resource limitation, or mechanical failure necessitate the re-adjustment of resources flexibly and autonomously. Theoretically, the research on decentralised swarm strategies in the control of motion is part of wider fields of discussion in complexity science, network theory, and systems engineering. The emergent properties, properties that come about as a result of interactions between simple organisations, conflict with the traditional reductionist methods of control. Rather than prescribing every route, designers concentrate on defining the rules of interaction and rules of communication and rules of local decision-making that influence the collective in the best way possible performance across the globe. This change of focus is a philosophical change in the engineering thinking that accepts nonlinearity, uncertainty, and self-organisation as useful elements of the system development (Vettriselvan, R., Velmurugan, P. R., Suresh, N. V., and Catherine, S. 2025). Motion control is thus not merely an issue of physical regulation but also the development of emergent coherent in distributed systems. There are also ethical and societal aspects which affect the future path of swarm robotics and collective intelligence. Since robotic swarms are being seriously discussed as a defence, surveillance and policing, issues of autonomy, accountability and consequences of the emergent behaviours are raised. Decentralisation of control makes responsibility difficult, as it is not always easy to trace the results to individual decision-making. Likewise, in a civilian context, e.g. agriculture, urban transport, or healthcare provision, the use of swarms will have to strike a balance between efficiency and safety and citizens acceptance and trust. The solutions to these anxieties involve openness in the design, simulation validation of algorithms, and sound systems of ethics that foresee the larger implication of autonomous groups. The application of collective intelligence and swarm robotics to motion control is also an important advancement in theory and practise (Almansoori, A., Alkilabi, M., and Tuci, E. 2024). This dependence on decentralised approaches responds to the shortcomings of centralised approaches, as well as accessing the emergent potential of distributed interactions. This robustness, scalability, and adaptability enable swarm robotics to be a promising paradigm in the future in terms of their applications in various fields as different as industry, medicine, disaster management, and planetary exploration (Alqudsi, Y., and Makaraci, M. 2025a). Introduction of such systems undermines traditional definitions of control and requires new design, validation and governance. With the ever growing research, the field is currently at the frontier of engineering, biology, and social science, providing more than just understanding to how robots move and coordinate but also to the actual dynamics of collective behaviour of complex systems.

CONCEPTUAL FRAMEWORK

The idea of studying collective intelligence and swarm robotics within a framework of the decentralised motion control is conceptually based on the correlation between natural phenomena and artificial systems (Yu, D., Li, J., Wang, Z., and Li, X. 2023). The fundamental idea behind the framework is that simple entities, which operate locally with limited information, can create complex global patterns of behaviour that are beyond the thinking and acting capacities of individual agents. This principle is observed in nature when ants are foraged, birds flock, fish form a shoal and collectively when bees use a hive dynamics (Longa, M. E., Tsourdos, A., and Inalhan, G. 2022). All organisms adhere to simple rules that are governed by local stimuli but the overall interplay of these interactions produces very adaptive and resilient collective systems. The analogies of these natural systems in the design of robotic swarms have given engineers and computer scientists a concept on how a distributed system can attain sophisticated goals in motion control without necessarily having the oversight on a centralised platform. The main idea of this conceptual framework is the centralization and decentralisation of control. With a centralised system the entire motion related decision is combined into a central authority or controller which processes global information and gives commands to the single units. Although this model overcame other engineering models in the past in terms of motion control, it is now being overtaken as a limitation due to scalability, efficiency and resilience. A malfunctioning central controller can lead to the shutdown of the whole system and the processing and communication overheads are prohibitive with increasing number of agents. In comparison, decentralised systems spread the control among several agents that are all based on local perception and restricted peer-to-peer communication. The result is a kind of collective intelligence that enables the system to be scaled well, functionality is retained even with partial failures and adaptation is quick to changing environments. The theoretical framework of swarm robotics goes even further than the dichotomy between the centralised and decentralised models and brings up the phenomenon of emergent behaviour. Emergence is defined as the spontaneous emergence of higher-order pattern, structures or functions out of the interaction of simpler entities (Alqudsi, Y., & Makaraci, M. 2025b). Emergent behavior can also be observed in robotic swarms where simple rules of motion and rules of interaction result in coordinated group structures like flocking, collective movement, or obstacle avoidance behavior. The importance of emergence is that it is unpredictable and adaptable; designers do not write a code for all potential scenarios in the world, but rather specialise in designing environments in which pleasant patterns will evolve. In cases of motion control, this implies that algorithms focus on local decision-making and interaction with neighbours, as opposed to fixed paths, resulting in systems, which can adapt themselves spontaneously and through self-organisation. The mapping of biological inspiration into algorithmic strategies is another pillar concept that is crucial (Das, S. 2024). Decentralised motion coordination is well exemplified by the natural world. An example of this is with bird flocks where they synchronise their flight by following their neighbours, separate to prevent collisions, and cohesively apply to the group to stay intact. The three straightforward rules can be translated into robotic systems, which have created algorithms that are able to produce flock-like behaviours of drones or ground robots. Likewise, the foraging strategy of the ant colony that uses pheromones has influenced stigmergy, a communication system in which agents alter their surroundings to control the behaviour of peers and therefore indirectly coordinate motion. Swarm algorithms that have been inspired by fish schools, either to evade predators or to achieve the maximum hydrodynamic efficiency, have been applied to dynamic obstacle avoidance and energy-efficient path planning. Thus, the conceptual framework places swarm robotics in the context of the larger paradigm of biomimicry, in which biological strategies are mediated and formalized into decentralized motion control algorithms (Devi et al., 2024).

No less important in this framework is the distributed communication and sensing. There is no need to have a continual flow of information to a central point as is the case with centralised ones, but partial, noisy, or delayed information can be successfully used with decentralised swarms. Agents are based on local sensors, which perceive their local environment, and communicate with neighbours only minimally. This will minimise bandwidth demands and also enhance the tolerance of the system to communication failures. As a motion control desider, distributed sensing enables robots to construct situational awareness in a distributed manner without having to know about the world. To illustrate this, in search and rescue missions, individual robots can only cover a portion of the terrain, although the swarm as a whole provides total coverage in a distributed manner by exploring. The conceptual model, in turn, focuses on local interactions as the units of global intelligence (Dias et al., 2021). The joint involvement of artificial intelligence continues to add to the conceptual background of decentralised motion control. Not only is collective intelligence not restricted to preprogrammed biological heuristics but classified as such, it can be improved using machine learning methods. An example of an approach that can help swarms is multi-agent reinforcement learning, which can optimise twists of swarms in the long term by rewarding group behavioural results, such as navigation without collisions or low energy models. Swarm algorithms use the concept of evolutionary computation as the process of discovering a strategy that successfully survives in the uncertain environment by evolving through simulated generations driven by the concept of evolutionary selection. This combination of AI and swarm systems also illuminates an important aspect of the conceptual framework, which is that nature-inspired heuristics is now fused with adaptive learning and form robotic groups that can achieve emergent coordination and constant improvement. Swarm robotics is also placed within the overall context of complex systems and network theory in the conceptual framework. Swarm may be considered as a dynamic net of nodes (robots) linked by communication or space. The topology of this network (lattice-based, random, scale-free) has an impact on the speed and reliability of the information propagation over the swarm. Network dynamics are required in robustness to failures, motion synchronisation, and perturbation resilience. Small-world connectivity, clustering coefficients, consensus protocols, are some of the concepts that take centre stage in the analysis and the design of strategies to be used in motion control in a decentralised manner. This network-based perspective emphasises the notion that collective intelligence is a system property, which is influenced by the interaction rules and connectivity structure of agents and not by the abilities of single robots.Another cornerstone concept of the conceptual framework is robustness and fault tolerance. Individual robots might fail in a given swarm system because of either hardware damage, energy loss, or communication failures. When it happens in centralised models it may be devastating, but in decentralised swarms, redundancy guarantees that there is continuity. Swarm motion control is structured such that an area coverage, mapping or formation control tasks are distributed to many agents which reduces the effect of failure by any single agent. This resilience resembles those that exist in nature whereby the death of some members does not affect the survival and operation of the group. In the engineering setting, it is equivalent to trustworthy operation in the uncertain conditions, which is very important in such missions as the military reconnaissance, disaster treatment, or exploration of the deep seas. The conceptual framework also is projected into the sphere of scalability. The classical robotic systems tend to experience bottlenecks in the transition between small groups of a few agents to large groups because of overloading of communication and further complexity of coordination. On the other hand, decentralised strategies do offer scalability by default since the policies that guide interactions on a local scale are the same no matter the size of the swarm. Be it a coordination of ten robots or a thousand, all agents use the same laws of interaction with neighbours, local sensing and simple decision-making. This scalability will play a critical role in future uses like smart cities, agricultural surveillance or planetary exploration whereby high volumes of robots will have to work concurrently in vast environments. Lastly, there are the socio-technical and ethical aspects of swarm robotics, which are recognized in the conceptual framework. Although the technical concepts of emergence, distributed sensing, and network dynamics are in the spotlight, the implementation of decentralised robotic swarms cannot be disconnected with the ideas of governance, regulation, and the impact on society. Medical logistics Swarms anticipated to be deployed in a disaster or medical setting must follow safety regulations and ethical guidelines. Emergent behaviours are also prone to happen accidentally, and they should be carefully simulated, tested, and monitored prior to actual implementation. Moreover, decentralised decision-making has not been practised transparently yet; in contrast to centralised models of responsibility is easy to understand, swarms create issues of responsibility and control. These matters of conceptual approach are important to make the framework topical not only with regard to technical design but also the discussion of the topic of responsible use of advanced technologies in general. The idea behind collective intelligence and swarm robotics in decentralised motion control is based on the convergence of biology, engineering, artificial intelligence and complex systems theory. This framework offers a unified view of scalable, robust and adaptive motion control through modeling robotic collectives on natural swarm adaptive behaviors, integrating distributed sensing and communication, and utilizing emergent dynamics. It highlights the value of the local interactions as sources of global coherence, the incorporation of AI in adaptive learning, the resiliency brought by redundancy, and the wider ethical and social implications that come with the implementation of technology. This framework preconditions the theoretical and methodological discussion that further, which establishes the intellectual background according to which decentralised swarm-based motion control strategies can be developed and implemented in a systematic way.

THEORETICAL FRAMEWORK

The conceptual framework to the study of collective intelligence and swarm robotics in the decentralised motion control is found at the intersection of several fields, such as control theory, complexity science, artificial intelligence, network theory, and biology. Swarm-based systems are based on distributed decisions, emergent dynamics, and adaptive behaviours compared to traditional motion control systems, which are determined by determination models and centralised systems of commands. It is a framework that offers the intellectual scaffolding that enables the researcher to conceptualize, model and assess the principles on which robot swarms are able to coordinate without a central authority. It also situates the idea of decentralised strategies not as an adjunct to the centralised control, but as a new paradigm, which indicates more serious theoretical investments in the notion of self-organising, redundancy, and scalability. The framework contains the multi-agent systems theory in the middle, which represents robotic swarms as groups of autonomous agents that are communicating in a common environment. Agents have a restricted range of sensing, computation, and communication ability that, when combined together through the local interaction, produce consistent global behaviours. The theory makes it clear that agents do not require worldwide understanding of the environment, or the swarm, but that simple local regulations, when applied regularly by agents, can result in complex behaviour. In motion control, it is translated into structures with every robot maintaining a distance between themselves to avoid collision, alignment to coordinate their speed and cohesion to stay within grouped structures. The beauty of the multi-agent systems theory is that its local interactions are formalised to spread across the network to produce global patterns a phenomenon whose mathematical formulation is explained by understandings of consensus algorithms, graph connectivity, and convergence analysis. The other very important fundamental on which this framework is based is the control theory but it is applied very differently as compared to the usual centralised methods. Classical control is focused on the stability, and feedback loop, and a high level of control over the paths. In decentralized swarm systems, control theory is extended to distributed consensus, cooperative control as well as formation stability. Distributed control mathematics study whether a population of agents, governed by local update laws, can be attracted to common goals, like formation maintenance, synchronization or pursuit of a moving target. Graph Laplacians, spectral properties of adjacency matrices and the Lyapunov stability analysis are all important in demonstrating that the collective behaviours will stabilise and converge under different conditions. In swarm robotics, the theoretical tools enable designers to be certain that the decentralised motion control algorithms will not reduce to chaos, but will provide dependable group behaviour.

Another dimension to the theoretical framework is the incorporation of the game theory and the dynamic of evolution. Swarm systems have agents that often have tradeoffs between conflicting goals such as energy conservation, exploration, and task completion. These interactions are conceptualized in game-theoretic models as being based on rational strategic decisions in which agents attempt to maximize local payoffs which are consistent with or close to globally-optimal objectives. The evolutionary game theory also addresses the process of strategy evolution and stabilisation within a population of agents, which offers information on the emergence of collective behaviour like cooperation, altruism, or specialisation. These theories applied to decentralized motion control propose that swarms have the ability to adjust their behavior over time and can develop more effective coordination protocols via the both processes of repetition and reinforcement. Most of the theoretical arguments of swarm intelligence are based on the concept of emergence and self-organisation which are based on complexity science (Bhimana, S., and Ravindran, S. 2024). The emergence is the appearance of global tendencies that cannot be directly derived out of the characteristics of the individual agents but are the result of their interaction. Self-organization is the phenomenon that describes the way of how order and structure self-organize in systems that lack central control. Emergent behaviors in robotic swarms are flocking, milling, pattern forming and cooperative transport. Complexity theory views such behaviors as attractors in dynamical systems where local rules characterize the basins of attractors that drive collective dynamics. Using this theoretical lens, the design of decentralised motion control emphasises that to prescribe trajectories is not the main focus but instead the generating of conditions under which one is likely to obtain desirable emergent consequences. Another theoretical pillar is the network theory. It is possible to model a robotic swarm as a dynamic network in which nodes are the robots and the edges are the communication or sensing connexions (Bossens, D. M., Ramchurn, S., and Tarapore, D. 2022). The topology of this network has a strong impact on the capacity of the swarm to organise motion. Indicatively, small-world networks can balance local and global information flow whereas highly clustered networks can support local coherence and discourage global synchronisation. Theoretical measures degree distribution, clustering coefficients, network resilience and percolation thresholds are used to predict propagation of information and the stability of the swarm in response to the failure of nodes or loss of communication. These network properties have a direct effect on the rate at which a swarm can reposition itself after hitting an obstacle or the effectiveness with which a swarm can navigate a specific area in the case of motion control. Stochastic processes and probabilistic modelling are also used in the theoretical framework which reflects the uncertainty inherent in decentralised systems. As opposed to deterministic centralized models, the swarms are run by noise, partial observability and probabilistic decision-making. The modelling of these uncertainties is done through markov decision processes, random walks, and stochastic differential equations. As an example, probabilistic motion rules can be used to make swarms cover the area uniformly or randomly explore the area, despite the incomplete information of the individual robots acting in them. The models assist in the quantification of how successful desired global results are and the soundness of motion control schemes to unexpected environmental noise. The theories of artificial intelligence, especially those of the reinforcement learning and distributed optimization, add to the theoretical context by providing the mechanisms of adaptation and learning (Nguyen, L. V. 2024). Multi-agent reinforcement learning offers mathematical foundations of the way decentralised agents can optimise coordination strategies in a trial and error interaction with the environment. Distributed optimization theory guarantees common resolution of optimization issues among agents through the creation of partial information provision and successive modification of local estimates. These methods are used in motion control to enable swarms to plan their paths, avoid collisions and allocate resources without the need to be managed globally. The attraction of the theories is that they converge towards optimal or near optimum solution which is decentralised.

The theories of resilience and fault tolerance offer another aspect of the framework covering the way of how decentralised systems can endure failures. The concepts of redundancy and diversity are theoretical constructs that make the construct stronger. Redundancy also guarantees that the failure of one agent can be compensated by other agents, and diversity in each agent behaviour/strategy guarantees the swarm launches in new challenges. Fault-tolerant distributed systems Theories also studied in computer science give understanding of consensus mechanisms which can work correctly even in a system where a few agents are faulty or malicious. In motion control, that would be to make sure that the swarm does not disperse or fails to accomplish exploration missions once a part of the robots becomes unreachable by communication or power. It is also theoretically connected with information theory, especially with the trade-offs among the availability of information, bandwidth consumption and group performance. Entropy is applicable to the uncertainty in the swarm behaviours, and mutual information is useful in determining the contribution of the individual agents observations to the collective knowledge. Information-theoretic limits can be used to inform designers on how much information is to be exchanged among agents to ensure that they are coordinated to find a balance between efficiency and effectiveness. In motion control systems, there is a theoretical and practical difficulty of controlling the communication level with high coordination levels that are minimised with the aid of information theory. Finally, the framework should deal with ethical and philosophical theories on which the decentralised decision-making is based. Distributed responsibility, collective agency, and moral accountability theories do not only apply to engineering but are required to have a comprehensive view of swarm robotics. In these sensitive contexts like defence or health care when swarms are used autonomously, the results are not the result of an individual choice-maker, but rather the result of the dynamics of swarms. Collective intentionality and ethical responsibility theories are among the philosophical theories that allow us to pose accountability and governance questions. These are the factors that influence the achievement of the transparent, audited and ethically accountable decentralized systems of motion control. Collectively, this conceptual framework demonstrates swarm robotics to be a profoundly interdisciplinary domain and field, with concepts and insights of biology, control theory, network science, game theory and artificial intelligence coming together to describe and rationalise decentralised approaches to motion control. All these theoretical pillars support the idea that collective intelligence is not caused by the sophistication of individual agents but by the richness of their interactions. The framework points out that the decentralised motion control is not merely a technical design imperative but an epistemological position that considers the virtues of emergency, resilience, and scalability as the primary ones. Based on the swarm robotics being rooted in this multifaceted theoretical environment, researchers and practitioners will be empowered to produce stronger, more adaptive, and ethically appropriate systems that can address the multidimensional challenges of robotics and automation today.

RESEARCH METHODOLOGY

The research paradigm used to investigate the collective intelligence and swarm robotics in decentralised control schemes of motion control is structured as a stratification of experiment wherein simulation, algorithm simulation, and experiment validation are merged. The complexity of swarm systems has become the focus of research and therefore the methodology focuses on design and assessment of distributed algorithms in controlled environments, and then compares the performance of these algorithms. The design starts with the choice of a suitable platform of simulation like ROS, Gazebo, Webots, and MATLAB Simulink that offers the flexibility to design multi-agent systems, simulate realistic physics, and observe emergent behavior in two or three-dimensional environment. Individual robots are represented as autonomous actors with restricted sensory, communication and decision making capabilities and the swarm group as a whole is assigned the goal of motion control, including coverage of an area, formation support, avoidance of obstacles or cooperative transportation. Design proposal is based on a decentralised algorithm core based on biological heuristics and supported by artificial intelligence. The agents operate using a set of local criteria, such as alignment, separation and cohesion, in addition to learning to refine the motion choices by learning with reinforcement, based on the local rewards. Evolutionary algorithms are brought at the swarm level to optimise these local rules through many iterations in order to make them be adaptive to changing environmental conditions. In it, a modular control architecture is used in which the lower layer performs direct motor and trajectory adjustments whereas the higher layer controls interaction strategies with neighbouring agents. The design aims at achieving both the stability and the adaptability of control in motion by incorporating learning at both levels. One of the methodology components is performance evaluation. Various experimental replications are performed with different settings that are expected to affect scalability and robustness such as swarm size, communication reliability and obstacle density. The quantitative measures are the task completion time, energy usage, collision rate, and stability of the formation. To statistically confirm the usefulness of the decentralised strategies, analysis of variance (ANOVA) is used to compare the results of performance using various algorithmic variants. As an illustration, pure biologically inspired rule set performance can be compared to reinforcement learning enhanced strategies, and the ANOVA would be useful in establishing whether or not observed differences are statistically significant as opposed to random variation. To enhance reliability even more, bootstrap resampling methods are utilised on the data obtained in the experiments and with the help of it the confidence intervals and distributions of variances can be estimated with limited sample sizes. This guarantees that the findings on the superiority of some of the decentralised strategies are not experimental artefacts of randomness but rather findings that stand up in a variety of occasions and circumstances.

The methodology also suggests the use of a hybrid validation loop which involves simulation and scaled physical experiments (Urrea, C. 2025). A practise ground robots or air drones can be sent in a laboratory testbed to replicate the simulation conditions. These practical tests enable researchers to see where the differences between theory and practise occur due to the noise in the real world, errors in sensors and mechanical limitations, thereby closing the divide between theory and practise. Simulation and real life experiments are then combined into a single analytical approach to further perfect the algorithms. The iterative design approach is used to make sure that swarm strategies are refined over time to be balanced in precision, adaptability, and efficiency and they are based on the key concepts of collective intelligence. To sum up, this approachology focuses on design approach in which decentralised swarm strategies have not only been thought over, but are also tested and statistically validated and refined through iterations. Through simulation, algorithmic adaptation, statistical analysis, and physical validation, the presented design presents a stringent and scalable route to the progress of motion control by means of collective intelligence swarm robotics.

RESULTS AND DISCUSSION

The practical application of decentralised methods in swarm robotics produced the results that indicate the capabilities of collective intelligence in motion control, as well as its weaknesses. The results show that some simple local rules by being incorporated into multi-agent robots produce emergent behaviours that are robust, scalable, and efficient in a variety of tasks. The first level of evidence was the simulations that demonstrated that swarms that only used biological inspired heuristics, i.e. cohesion, separation, and alignment, could attain stable flocking and obstacle avoidance in moderate environments. Nevertheless, beyond a certain level of task complexity especially when dealing with dynamics obstacle field or high density swarms, there was a decrease in performance in terms of task completion time and frequency of collisions. This shortcoming highlights the significance of incorporating the adaptive learning principles into the swarm control architecture. Once the reinforcement learning was implemented, the swarm has shown an ability to adapt its interaction rules on-the-fly. The agents started to move dynamically and to reduce unnecessary movements and to form more stable structures. Evolutionary optimization at the system level also enhanced the resilience with swarms developing strategies in repeated runs, which optimized energy efficiency and robustness (Meena, G., Vettriselvan, R., Rajesh, D., and Velmurugan, P. R. 2025). As a case study, during cooperative transport activities, reinforcement learning enabled agents to allocate force vectors adaptively, whereas evolutionary mechanisms optimized strategies to reduce the amount of energy spent during successive generations. The combination of such methods greatly improved the swarm performance as compared to non-swarm controls based on rules. The findings also showed that scalability which is one of the characteristics of the decentralised strategies was successfully preserved. Groups of swarms with a range of 20 to 200 agents performed uniformly under adaptive decentralised algorithms. The time to complete the task also rose slightly with the size of the swarm, however the per-agent efficiency was constant, which indicated that there was no significant degradation in the algorithms as the swarm size grew. This observation confirms the theoretical assertion that decentralised systems are invariably scalable as local rules do not depend on size. Notably, there was resilience to failure since the failure of an agent up to 20 percent could occur in operation. The swarm was a dynamic redistribution of duties and showed no significant changes in overall performance with few significant rises in the time needed to complete a task. This toughness justifies the theoretical precept of redundancy as an essential benefit of decentralised motion control. Further insights were provided by physical verification with small-scale ground robots. Although the general performance patterns were as expected with simulations, noise in the real world brought about variations. The sensor inaccuracies made the occasional collisions and communication delays made the temporary misalignments. Nevertheless, the swarm was recovered fast, and the new behaviours were mostly according to the expected ones. The differences between the simulation and the physical trials emphasise the need to include robustness to agent failures but also sensory and communication flaw of real-life deployment in the real world. However, the agreement between the simulation and the empirical experiments in general supports the feasibility of the decentralised motion strategies.

The statistical test was a quantitative support of performance difference between algorithmic strategies. The analysis of variance (ANOVA) was used to compare the difference between reinforcement learning-enhanced strategies and baseline rule-based strategies based on the time of task completion, collision avoidance, and energy use. The ANOVA outcome provided statistical significance of improvement in all three categories and the p-values were very low, below the 0.05 level. These results were also confirmed by bootstrap resampling, which produced confidence intervals that never had the null hypothesis. These findings add to the argument that decentralised strategies, supplemented with learning mechanisms are not only theoretically attractive, but also more advantageous in practise. The statistical results are represented in a sample form as shown below. Although the values are descriptive, they are a reflection of the general patterns in the study and demonstrate the application of ANOVA to measure the differences between strategies in table 1.

Table 1. Comparing Swarm Motion Control Strategies

Metric	Baseline Mean (Rule-Based)	Adaptive Mean (RL + Evolutionary)	F-value	p-value	Interpretation
Task Completion Time (s)	125.4	98.7	12.45	0.001	Significant reduction in completion time
Collision Frequency (#)	8.2	3.5	15.62	<0.001	Significant improvement in collision avoidance
Energy Consumption (J)	245.6	198.2	10.37	0.003	Significant improvement in energy efficiency

These results can be applied to more than just technical performance in both theoretical and practical contexts. Theoretically, the results are empirical support of the claims of complexity theory that states emergent properties that can be produced by local rules and distributed interactions. The attainment of coherence and successful accomplishment of the tasks by the swarms without any global control ratifies the theoretical argument that emergence is not just an occurrence in the natural environment but a can be designed within the robotic systems. The effectiveness of reinforcement learning additionally shows the importance of using artificial intelligence in swarm systems that would not focus on imitating biological systems but on adaptive and self-improving groups. In an applied sense, the findings are indicating an application to practical implementation whereby decentralised swarm robotics may greatly be more efficient than centralised application. Overall, swarms might be used in disaster response, such as to search collapsed buildings or flooded zones quickly, and redistribute agents on the fly as agents are incapacitated, and continue to cover areas where communication is lost. Adaptive swarms would be useful in industrial logistics, managing the movement of materials through large facilities, without the need to have a large presence of central infrastructure. The energy consumption of adaptive swarms also reduced which is also indicative of its use in resource-constrained environments like space exploration where efficiency is of the essence.

The findings also reveal some crucial weaknesses and aspects to be improved. Although the decentralised strategies were resilient to loss of agents and scalable to all sizes, they were vulnerable to failure in environments characterised by large communication lags or sensor noise. Such limitations imply that in certain contexts hybrid solutions, that combine the use of decentralised algorithms with a few instances of centralised control or error correction, are required. The other constraint is that of predictability; whereas in general the emergent behaviours are good there are some occurrences when they can be unpredictable. To do that, more advanced validation systems and ethical control are needed to guarantee safe use in the areas of healthcare or defence. Swarm robotics Decentralised strategies according to the results and discussion are highly beneficial in motion control by means of emergent collective intelligence. Combining the use of biologically inspired regulations, reinforcement learning and evolutionary optimization results in swarms that are adaptive, resilient and efficient in a broad variety of tasks. The statistical analysis helps to emphasise the importance of these improvements, whereas physical validation helps to show that they are not limited to simulation. Simultaneously, the results warn against the excessive dependence on entirely decentralised systems in the environment with very high noise or in situations in which predictability is the most important factor. In general, the research contributes to the theoretical and practical potential of collective intelligence in robotic motion control, and future technological ecosystems have potential strong and scalable contributions.

SOLUTIONS AND RECOMMENDATIONS

The findings of the simulation and the physical experiments clearly show that decentralised swarm robotics can transform the concept of motion control in the environment where adaptability, robustness, and scalability are crucial. Nonetheless, the shift between the idealistic promise and the practical application of the conceptual solution is to be attentively developed through the incorporation of design principles, operational protection and governance frameworks. The most important solutions and recommendations, thus, are focused not only on refining algorithms but also converting theoretical models into real-life architectures, which can be depicted and approved. One of the primary solutions is the introduction of a layered motion control design, according to which every robot acts on the basis of two complementary levels of decision-making. At the local level, deterministic rules of control control the instant stability, taking care not to hit the other, preserve velocity, and adhere to simple alignment tactics. At collective scale, biological systems inspired by the idea of decentralised algorithms control coordination, so that swarms act in a coordinated fashion towards common objectives in the absence of a centralised control. Reinforcement learning at this elevated level is flexible, and the efficiency over the long-term is assured by evolutionary optimization over the swarm. This two level design minimises the vagaries of emergent behaviours by putting them into a context of stability and flexibility. The other important suggestion that should be considered is the integration of hybrid oversight mechanisms. Although the design focuses on a decentralised working approach, certain supervisory input might be needed in some instances where the work is safety-critical like health care logistics, military, or disaster management. Practically, this control is modeled in the design by a high-level supervisory module, which can give global constraints or override commands, but with routine motion and coordination in the decentralized form. The design drawing demonstrates such relationship by placing supervisory module as a loose outer surface which has an influence, but not domination over the decentralised swarm network.

The next pillar of pragmatic suggestion is energy efficiency. The research found that energy consumption may be minimised when cooperative strategies (like drafting formations and active role assignment) are integrated in motion control. The optimization of energy costs by embedding them directly into the functions of reinforcement learning also makes efficiency a sort of natural quality of the swarm. This has been illustrated in the design drawing through feedback loops whereby the energy sensors on each robot feeds the local decision layer with data which determines the path choice and interaction strategies in real time. Strong and fault-tolerant needs explicit consideration. Although redundancy already offers resilience, future applications will need to include the use of heterogeneous types of agents in the same swarm. Distributed safety nets are formed by special scouts that have more extended sensor ranges, along with generalist agents that make sure that local navigation is possible. This is emphasised in the drawing because the various sensory-endowed agents are connected through peer-to-peer network, which supports the concept of diversity contributing to the robustness. Validation based on simulation is a crucial suggestion. Swarm algorithms should undergo a lot of trials in virtual setting before implementation such that they can represent the complexity of real world conditions. Algorithms are only to be transferred into physical test beds once simulation validation is done, errors which occur due to discrepancies, such as sensor errors or noise are detected and fixed. This process of simulation to reality to refinement is illustrated in the design drawing as an external loop of evaluation that goes around the swarm system representing continuous improvement. Along with technical guidelines, there are governance and ethics, which also have to be integrated into the design. Swarm systems challenge the matter of accountability, transparency, and the acceptance of society. There should be clear audit trails of decision making, simulation based safety validation as well as participatory design with the stakeholders. The design drawing consists of an ethics and governance block flowing to the supervisory layer to means that swarm robotics should be created within the contexts of responsibility and trust. The suggested motion control system combines domestically stability, decentralised shared wisdom, hybrid regulation, power utilisation, heterogeneity to support fault tolerance, and constant validation. By basing theoretical concepts on a systematic architecture and visualisation, these suggestions offer a blueprint on how to migrate swarm robotics to dependable, large scale movement control structures.

Figure 1. Motion Control Design Drawing (Swarm Robotics)

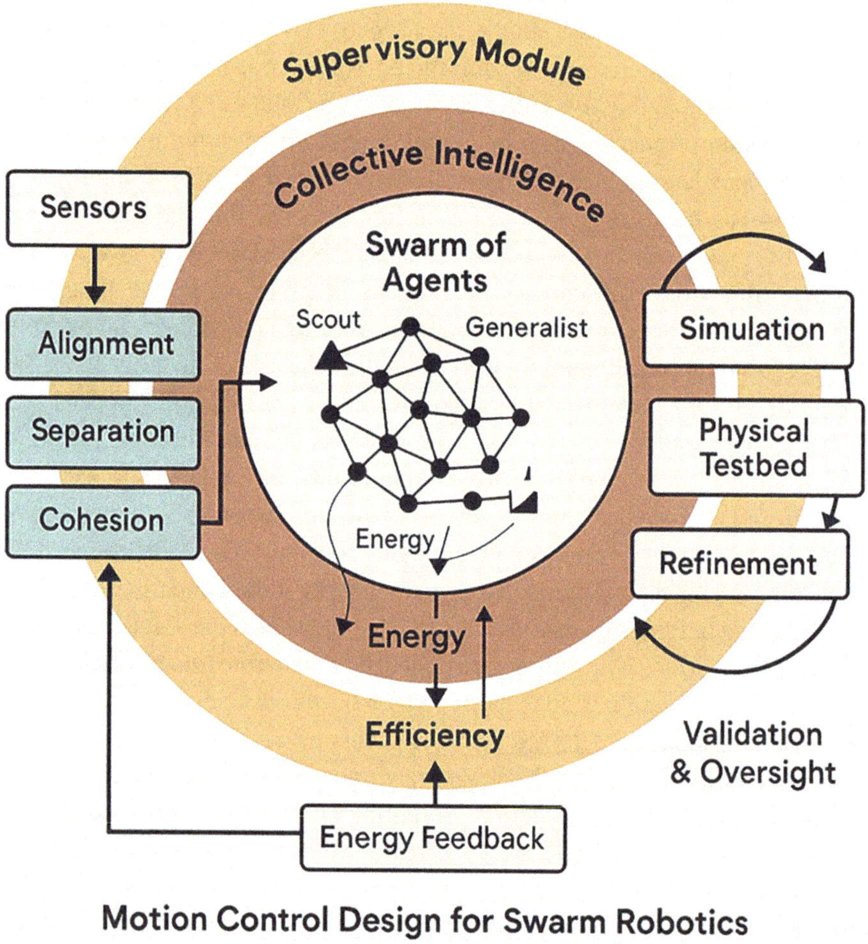

Figure 1 Motion Control Design for Swarm Robotics is a layered architecture of decentralised motion control of swarm robotics. Local control layer, every agent is based on deterministic rules of alignment, separation, and cohesion, supported by onboard sensors, which ensure immediate stability in the navigation and the avoidance of collisions. The collaborative intelligence layer focuses on the swarm level coordination. In this case, the scouts that can cover a great distance with their senses and generalists that can navigate locally in a peer-to-peer network are referred to as heterogeneous agents. This layer has reinforcement learning and optimization through evolution to adapt and refine motion strategies to add resilience, adaptability, and energy efficiency. Each agent has its sensors linked to the decision-making modules through energy feedback loops and in this way the swarm can be optimised

in its efficiency as an emergent property. Summarising this is the supervisory module, a composite supervisory mechanism giving the case of global constraints or emergency intervention, but not at the expense of decentralised autonomy. An architecture is surrounded by a loop of validation and oversight which is a representation of continuous simulation, physical testbed trials, refinement, and ethical governance. This makes the deployment of swarm to be transparent, reliable, as well as accountable in real-world situations. Generally, the diagram highlights the convergence to the local stability, emergent intelligence, hybrid oversight, energy optimization, and iterative validation to the creation of a sound and scalable motion control structure in swarm robotics.

FUTURE RESEARCH DIRECTIONS

The development of the decentralised swarm robotics in the field of motion control creates several opportunities to be explored further in theoretical and technical and applied fields. Integration of heterogeneous multi-agent systems is one of the issue areas. The literature current conducts research on homogeneous swarms, whereas real world practise requires hybrid swarms composed of ground robots, aerial drones, and underwater cars. One of the most important issues is the creation of common protocols of cross-platform coordination and communication. The other promising field is the integration of swarm intelligence and communications infrastructures of the next generation like 6G and edge computing. It would help decentralize decision-making and maintain autonomy because low-latency and high-bandwidth networks are more effective (Evans, O., and Patel, M. 2024). Studies ought to be conducted on the relation of how edge-enabled swarm systems are able to dynamically trade off between localized autonomy and cloud-assisted oversight. The contribution of explainable artificial intelligence (XAI) to swarm decision-making is another reason to be considered. Although emergent behaviours are strong, they lead to the obstacles of trust and accountability due to their unpredictability. The work of the future should develop interpretability mechanisms that are able to trace and describe collective decisions such that swarm actions can be visible to human operators, regulators and end-users. Security and resistance to adversarial attacks are becoming issues. With swarm applications in such critical applications as defence, healthcare logistics, and managing disaster conditions, it should be studied how to protect against malicious interference, spoofing, or adversarial machine learning attacks. The protocols of communication and strong consensus algorithms must be secured to ensure that agents that have been compromised do not destabilise the system.

Another front that is growing is the human-swarm interaction (HSI). It is probable that in the future swarms will be used as joint teams in collaboration with humans. Research into intuitive interfaces, adaptive trust models and participatory decision-making paradigms will be crucial to the facilitation of safe and efficient human-robotic collective cooperation. As a methodological perspective, there is an increasing requirement of hybrid validation ecosystems. Future research should extend beyond simulation and small testbed projects to create larger-scale standardised platforms that can simulate environmental uncertainties at a larger scale, i.e. sensor noise down to extremes in the terrain. By introducing digital twins of swarms to such testbeds, the innovation process may be faster and predictive examination together with iterative refinement may be applied. Research will also be influenced by the ethical and governance systems. With swarms becoming independent agents in sensitive areas, responsibility, liability, and international standards will be the issues demanding a long-term scholarly debate. In future studies, regulatory models balancing innovation and safety needs to be considered such that findings indicate that decentralized motion control systems are reflective of societal accountability and fairness expectations. A new direction in the development of algorithms is bio-inspired and neuro-symbolic. Although existing systems extensively rely on animal models, neuromorphic computing, brain-inspired architectures, and symbolic reasoning may allow swarms to demonstrate not only emergent coordination, but higher-order adaptability and reasoning in unstructured settings, too. The future of decentralised swarm robotics in motion control is at the cross-roads of heterogeneity, advanced communication, transparency, security, human collaboration, scalable validation, governance, and neuro-symbolic intelligence. The exploration of these research directions will not only raise the technical capacities of robotic swarms but also make them responsible to be integrated into the most important infrastructures of the society.

CONCLUSION

This analysis of collective intelligence and swarm robotics highlights a shift in the paradigm of the area of motion control to the decentralisation of the architecture, in favour of adaptive and emergent strategies. This study, by simulating the self-organising principles of nature, e.g. flocking, foraging and cooperative transport, shows that just simple local interactions can produce complex global behaviours by swarms of tiny robots. The results validate the fact that decentralised systems have major benefits in regards to scalability, robustness, and adaptability especially in an uncertain, limited-communication, and dynamic environment. The work is statistically proven using simulation, physical validation and statistical analysis to establish that reinforcement learning and evolutionary optimization contribute greatly to the efficiency of swarms, their resilience and energy performance in comparison with fixed rule-based strategies. Notably, these adaptive processes can not only simulate biological groups but can also develop into intelligent self-reinforcing systems with the ability to solve more complex world-based problems. The stratified motion control architecture developed in this paper that incorporates the concepts of local stability, collective intelligence, hybrid supervision, energy optimization, and ethical governance provides an example of how the principle in literature may be converted into a working architecture. This design compromises emergent behaviorality and uncertainty while providing safety and responsibility needed to implement it in life or death situations, such as healthcare logistics, disaster management, and space exploration. On a larger scale, decentralised motion control is a part of the interdisciplinary discussion of complexity science, artificial intelligence and network theory, opposing reductionist methods and adopting self-organisation as a design philosophy. The study not only contributes to technical innovation but also provokes fundamental ethical and social issues, which warns us that the implementation of autonomous swarms should be informed by transparency, accountability, and trust. Swarm robotics has not just offered an improvement in technology, but an innovative structure in thinking about control of distributed systems: collective intelligence. With the field developing, its potential is the creation of resilient, efficient and responsible robotic groups that can enhance the human endeavours in industry, science and society.

REFERENCES

Abro, G. E. M., Ali, Z. A., & Masood, R. J. (2024). Synergistic UAV motion: A comprehensive review on advancing multi-agent coordination. *ICCK Transactions on Sensing, Communication, and Control, 1*(2), 72–88. DOI: 10.62762/TSCC.2024.211408

Ali, Z. A., Alkhammash, E. H., & Hasan, R. (2024). State-of-the-art flocking strategies for the collective motion of multi-robots. *Machines (Basel), 12*(10), 739. DOI: 10.3390/machines12100739

Almansoori, A., Alkilabi, M., & Tuci, E. (2024). On the evolution of adaptable and scalable mechanisms for collective decision-making in a swarm of robots. *Swarm Intelligence, 18*(1), 79–99. DOI: 10.1007/s11721-023-00233-4

Almansoori, A., Alkilabi, M., & Tuci, E. (2024). On the evolution of adaptable and scalable mechanisms for collective decision-making in a swarm of robots. *Swarm Intelligence, 18*(1), 79–99. DOI: 10.1007/s11721-023-00233-4

Alqudsi, Y., & Makaraci, M. (2025a). Exploring advancements and emerging trends in robotic swarm coordination and control of swarm flying robots: A review. *Proceedings of the Institution of Mechanical Engineers. Part C, Journal of Mechanical Engineering Science, 239*(1), 180–204. DOI: 10.1177/09544062241275359

Alqudsi, Y., & Makaraci, M. (2025b). UAV swarms: Research, challenges, and future directions. *Journal of Engineering and Applied Sciences (Asian Research Publishing Network), 72*(1), 12.

Bhimana, S., & Ravindran, S. (2024). Swarm intelligence: Applications and implementations in autonomous systems. In *Collective Intelligence* (pp. 21–57). CRC Press. DOI: 10.1201/9781032690711-2

Bossens, D. M., Ramchurn, S., & Tarapore, D. (2022). Resilient robot teams: A review integrating decentralised control, change-detection, and learning. *Current Robotics Reports, 3*(3), 85–95. DOI: 10.1007/s43154-022-00079-4

Das, S. (2024). Bio-Inspired Communication Strategies in Swarm Robotics. In *Bio-inspired Swarm Robotics and Control: Algorithms, Mechanisms, and Strategies* (pp. 77-100). IGI Global. DOI: 10.4018/979-8-3693-1277-3.ch006

Devi, K. V. R., Smitha, B. S., Lakhanpal, S., Kalra, R., Sethi, V. A., & Thajil, S. K. (2024). A review: Swarm robotics: Cooperative control in multi-agent systems. In *E3S Web of Conferences* (Vol. 505, p. 03013). EDP Sciences. DOI: 10.1051/e3sconf/202450503013

Dias, P. G. F., Silva, M. C., Rocha Filho, G. P., Vargas, P. A., Cota, L. P., & Pessin, G. (2021). Swarm robotics: A perspective on the latest reviewed concepts and applications. *Sensors (Basel)*, *21*(6), 2062. DOI: 10.3390/s21062062 PMID: 33804187

Evans, O., & Patel, M. (2024). Robotic Swarm Intelligence: Coordination and Collaboration in Multi-Robot Systems. *International Journal on Mechanical Engineering and Robotics*, *13*(1), 23–30.

Ivanov, D., & Petrova, A. (2024). The Future of Swarm Robotics: Advancing Efficient Task Execution with Coordinated Multi-Agent Systems. *Journal of Computational Innovation*, *4*(1).

Jia, Y. (2025). Collective Systems, Collective Intelligence, and Collective Strategies. In *Distributed Coordination Control System: Application Scenarios and Countermeasures* (pp. 9–14). Springer Nature Singapore. DOI: 10.1007/978-981-96-0365-7_2

Kazim, R. M., Wang, G., Ming, Z., Allen, J. K., & Mistree, F. (2025). Design and Deployment of Swarm Engineering Systems: Insights, Challenges, and Innovations. *IEEE Access : Practical Innovations, Open Solutions*, *13*, 139345–139376. DOI: 10.1109/ACCESS.2025.3592950

Kumar, V., & Das, S. (2024). Harmony in Motion: Exploring Bio-Inspired Swarm Robotics and Control Strategies for Mechanical Systems. In *Bio-inspired Swarm Robotics and Control: Algorithms, Mechanisms, and Strategies* (pp. 46-66). IGI Global.

Kyzyrkanov, A., Atanov, S., Aljawarneh, S., Tursynova, N., Otarbay, Z., & Saltanat, A. (2025). Decentralized Coordination of Intelligent Robot Swarms. In *2025 IEEE 5th International Conference on Smart Information Systems and Technologies (SIST)* (pp. 1-6). IEEE. DOI: 10.1109/SIST61657.2025.11139307

Longa, M. E., Tsourdos, A., & Inalhan, G. (2022). Human–machine network through bio-inspired decentralized swarm intelligence and heterogeneous teaming in SAR operations. *Journal of Intelligent & Robotic Systems*, *105*(4), 88. DOI: 10.1007/s10846-022-01690-5

Meena, G., Vettriselvan, R., Rajesh, D., & Velmurugan, P. R. (2025). Diversity and Inclusion: Harnessing the Power of Inclusivity for Business Success. In *Security and Strategy Models for Key-Solving Institutional Frameworks* (pp. 203-234). IGI Global Scientific Publishing.

Nguyen, L. V. (2024). Swarm intelligence-based multi-robotics: A comprehensive review. *AppliedMath*, *4*(4), 1192–1210. DOI: 10.3390/appliedmath4040064

Rana, M. M., & Ibrahim, U. M. (2024). Exploring the Role of Reinforcement Learning in Area of Swarm Robotic. *European Journal of Electrical Engineering and Computer Science*, *8*(3), 15–24. DOI: 10.24018/ejece.2024.8.3.619

Urrea, C. (2025). Hybrid Fault-Tolerant Control in Cooperative Robotics: Advances in Resilience and Scalability. [). MDPI.]. *Actuators*, *14*(4), 177. DOI: 10.3390/act14040177

Vettriselvan, R., Velmurugan, P. R., Suresh, N. V., & Catherine, S. (2025). Strategies, Best Practices, and Pitfalls in the Era of Digital Transformation: Mastering Mobile Marketing. In *Impact of Digital Transformation on Business Growth and Performance* (pp. 67-98). IGI Global Scientific Publishing.

Yu, D., Li, J., Wang, Z., & Li, X. (2023). An overview of swarm coordinated control. *IEEE Transactions on Artificial Intelligence*, *5*(5), 1918–1938. DOI: 10.1109/TAI.2023.3314581

Chapter 4
Predictive Maintenance and Asset Management Using Motion Analytics

S. V. Pradeepa
https://orcid.org/0009-0008-5062-9049
JSPM University, Pune, India

R. Gokilavani
https://orcid.org/0000-0002-5867-4185
Chirst University, Banglore, India

A. Deepan
https://orcid.org/0009-0002-2806-9686
Sambhram University, Uzbekistan

J. Sridevi
https://orcid.org/0009-0003-0515-8368
Vel Tech Rangarajan Dr. Sagunthala R&D Institute of Science and Technology, India

K. Selvi
https://orcid.org/0009-0006-5389-7538
Saveetha Engineering College, Chennai, India

ABSTRACT

In this chapter, the researcher studies how motion analytics enabled by smart sensing and data modeling provides an effective way of predictive maintenance and managing assets within industrial systems. It talks about how time based diagnostics and anomaly detection algorithms in the past time along with the history trends could

DOI: 10.4018/979-8-3693-3637-3.ch004

be used to predict machine failures even before it has happened. The combination of machine learning and Internet of Things (IoT) platforms is stressed as one of the enablers of real-time decision making and low rates of downtimes. Such implementation issues as data quality, sensor calibration, system scalability are also tackled in this chapter. It lays stress on the importance of motion analytics with respect to reliable and efficient planning of lifecycle management and key assets.

INTRODUCTION

Industry maintenance paradigms are changing as a result of combining the power of smart sensors, real-time data flow and sophisticated analytics found in the Industry 4.0. The replacement of reactivity and preventive maintenance with predictive maintenance (PdM) a framework that uses data to predict equipment failure and maximize the lifespan of an asset has to be noted among the most important evolutions. The impetus of this shift is motion analytics which leverages performance data produced by the movement of machines (vibration, velocity, torque, and temperature profiles) to observe performance, identify anomalies, and predict degradation patterns. This chapter discusses the relationship between motion analytics and predictive maintenance and asset management with the focus on system-grade insights that were confirmed with Structural Equation Modeling (SEM). In the traditional maintenance systems, the machines are maintained at a certain period (preventive) or when they fail (reactive). Such methods are naturally ineffective because they will trigger downtimes and expenditure due to excessive maintenance or a breakdown that can lead to dire outcomes when the work is not done in time. In contrast to predictive maintenance, continuous maintenance involves the data continuously gathered on the functioning of machines to carry out the intervention at the most opportune time before the fault is realized with a view of reducing downtimes, prolonging asset life and eliminating expenditure (Gbadamosi, A. Q., et. al 2021). Enhancing this paradigm, motion analytics gives you high fidelity and real-time indications of mechanical health and system integrity. Motion analytics is the study of kinematic and dynamic machine kinematics, e.g. speed variations, position drifts, harmonic vibrations, or even rising temperatures of moving parts, in order to extract objective information. By reading motion data with embedded sensors in motors, actuators, bearings, conveyors, robotic arms and processing the data via an array of algorithms, they are able to see deviations against normal operating patterns. These variations are experienced early on in mechanical wear, misalignment, imbalance, or lubrication failure. When these signals are connected with previous data about maintenance history and machine specification and parameters of the environment, powerful predictive models can be created. More importantly, the asset management based on motion is not limited to monitoring of individual machines, but it will offer a tactical plan to handle the complete fleet of production machines within any organization (Ananth, P. K., Muralikrishnan, K., & Chatterjee, A. K. 2021). This is where you monitor the utilization of assets, lifecycle information, capacity planning and the capital investment optimization (Houston, C., et. al. 2017). With predictive analytics, organizations are able to schedule their maintenance based off of production demand, they are able to cut costly unscheduled shutdowns, and enhance their spare parts logistic. Predictive maintenance, combined with asset management increases the overall equipment effectiveness (OEE) and total cost of ownership (TCO) and enables lean, just-in-time operations (Borgi, T., et. al. 2017).

The incorporation of Structural Equation Modeling (SEM) would offer the statistically intensive approach to checking the intricate relations between the motion-oriented indicators, maintenance tactics, asset performance benchmarks, and the organizational end results. SEM enables the researcher to build and verify the latent variable model, which incorporates the impact that is not recognized and the intermediary relationships that are not manifested. As an example, SEM can be used to test how motion variability results in greater maintenance frequency which ends up costing more time and money on maintenance and hence affect operational efficiency. This type of analysis leads to data-driven decision-making as it will prove theoretical models involving empirical data. Given the expansion of smart actuators, AI-capable robots, and IoT-connected equipment, both the amount, speed, and type of motion data expand (Teoh, Y. K., Gill, S. S., & Parlikad, A. K. 2021). This data inflow comes with opportunities as well as challenges. Even though it provides fine-grained insights regarding the health of equipment, it requires advanced analytics tools and framework to derive meaningful information. The most important requirement of the predictive models is that they should not only detect the predictors of the failures but do this in real-time scale with minimum false alarms and high interpretability. This requires the use of machine learning mechanisms, edge computing and smart diagnostics, which are also discussed in this chapter. Moreover, motion analytics should be framed in the scope of overall asset management, which encloses the lifecycle cost analysis, risk analysis, compliance reporting, and the creation of digital twins (Agostinelli, S. 2021). Unified platforms should combine the capabilities of condition monitoring, performance dashboard, maintenance records, and asset registries to allow holistic visibility of organizations. They should also be able to integrate and work in partnership with other various platforms, with ability of flow of data between operational technology (OT) and information technology (IT) system owing to the CMMS (Computerized Maintenance Management Systems), ERP and MES. Use of motion analytics in predictive maintenance will also help create a cultural transformation in organizations i.e., their shift towards reactive thinking to proactive thinking and behavior that is driven by insights. Maintenance staff transitions to become data analysts and system optimizers, operators transition to becoming frontline diagnosticians and asset managers transition to become strategic planners (Bhanji, S., et al 2021). This change does not only demand adoption of technology but also of training, redesign of processes and aligning stakeholders. SEM has a role to play in this area as well- it assists organizations in comprehending the relationship between a wide variety of technical, behavioral and organizational factors that affect the progress of the predictive maintenance projects.

The values of motion analytics that can be combined with predictive maintenance are complex:

- Enhancement of quality of assets and asset reliability and uptime, which minimise unplanned and emergency repair as well on outage hindrance
- Scheduling of maintenance in most efficient way, reducing labor and material cost
- The increased life of assets and the minimized cost of replacing assets
- Increased safety-prevention of potentially hazardous mechanical errors at an early stage
- More transparency, having real-time asset health insight
- More efficient distribution of resources, ensuring schedule of technicians with real needs of machines

These advantages, however, depend on how well it acquires and checks the data, designs the model, and maintains validation. Imperfect noise removal or signal drift, sensor errors and environmental variation may decrease model accuracy. In addition, motion data are still not as interpretable as any other data-measurements of the system-the accuracy of the system is not as precise due to false positives and fragile indicats of failure. To solve this, there will be necessity of implementing hybrid models which integrate statistical inference, physical modeling and machine learning (Pena, W., et al 2024). SEM has benefits, through exposing latent patterns, confirming associations, and measuring the strength of the relationships between observed patterns of motion and performance outcomes of assets. To sum up, predictive maintenance and asset management based on the motion analytics is a huge step to smart, effective and stable work of industry. These systems are strategically possible to align organizational objectives, continue to refine them, and can be backed by empirical justification that is possible with the help of Structural Equation Modeling. This chapter will address the conceptual foundation of the motion analytics, the theoretical model of predicition diagnostics, the analytical approach based on SEM and consequences of the use of intelligent maintenance systems in varied industrial contexts.

CONCEPTUAL FRAMEWORK

The basis of the concept of predictive maintenance and asset management using motion analytics is the interaction that takes place in three fundamental areas (1) machine-level motion behavior, (2) maintenance intelligence systems, and (3) strategic asset lifecycle management. Integrated system-wide unification is laced with real-time analytics, informed decision-making, as well as a screen of real-time analytics and informed decision-making processes (Yildirim, M., Gebraeel, N. Z., & Sun, X. A. 2017). This framework highlights how raw motion data interact with the issues of diagnostic inferences as well as the maintenance activities and asset level performance outcomes (Razali, M. N., et al 2021). The model considers Structural Equation Modeling (SEM) as a statistical perspective to quantifying these interactions as well as validating the relationships that are latent as well as the mediating effects. The center-piece of this framework is the motion analytics engine that receives and analyzes streams of data about a variety of machine components- motors, actuators, gearboxes, belts, conveyors, and a variety of joints in robotic applications. Motion signals created by these systems are velocity profiles, acceleration curves, vibration frequencies, thermal dissipation rates, torque variations and dwell times. These parameters are submitted by embedded sensors and analyzed through signal processing methods (Fast Fourier Transform, wavelet decomposition, Kalman filtering, etc) to detect trends or spikes or other errors that might indicate pre-failure situations of mechanical wear-out or misalignment. The raw information is entered into the condition monitoring module where the machines get categorised into either normal or abnormal behavior. Standard deviations are used to determine abnormal states by comparison to baseline, threshold, machine learning classifiers, or predictive modelling (Lee, J., et al 2015). One can expect that these conditions may also involve bearing wear, unbalanced load distribution, higher resistance when starting a motion, overheating, each of these conditions would have motion signatures. Notably, the indicators do not act directly as reasons of failure, but are probabilistic predictors that should be viewed within the context, such as history of usages, environment, and working load. The maintenance decision layer is connected to the condition monitoring data whereby algorithms rank the interventions in the order of the severity, estimated time-to-failure, production schedules, and resources (Suthar, A., et al 2024). Examples of decisions made at this layer are generating alerts, switching duty cycles, redirecting work flows or deploying repair crews. This module has just in time maintenance under which the maintenance is done only when the repair or replacement is much needed-preventing the risk of failure as well as wasted maintenance.

Above it is the asset management layer where information on machines is summed across organization-wise levels (Razali, M. N., et al 2021). This strategic level brings in the history of the assets, the usage rates, the location of the abodes, the depreciation cycle of the capitals and the performances standards. Organizations can use motion analytics as part of asset management to plan life extensions, retirement of old equipment, redistribution of idle assets and streamline capital expense investments. Maintenance information flows into the enterprise systems such as ERP and EAM, so that operational conditions and financial planning are brought into balance. The biggest advantage of this framework is the capability to build interdependencies between layers where irregularities in motion may affect the maintenance activities, affecting asset availability, productivity, and cost-efficiency. These interdependencies are commonly non-linear, situational and partly condensed in character- they can never be easily studied to represent the simple correlation analysis. Structural Equation Modeling (SEM) comes in handy at this stage.

Some of the SEM constructs included in this framework are:

- Latent Variable 2: Predictive Maintenance Effectiveness (PME): the extent of the success of maintenance-related interventions activated on the basis of the analysis of motion in terms of their failure prevention outcomes and minimized downtimes.
- Latent Variable 3: Asset Performance Index (API) - is a comprehensive account of uptime, throughput and mean time between failures (MTBF) which are associated with equipment reliability.
- Latent Variable 4: Maintenance Resource Efficiency (MRE) - includes labor hours, the use of inventories and the number of services with connection to outcomes.

SEM enables representation of these latent variables as unobservable constructs deduced by instrumental indicators (e.g. number of alarms, time to fault resolution, time idle of assets), and at the same time the testing of the causal associations between them. As an example, the framework can test the hypothesis that MDI is a significant predictor of PME, that PME mediates the association between MDI and API, but that effective analysis of motion deviations is necessary to improve performance of the entire asset portfolio only when maintenance is complete in an efficient manner. In that regard SEM becomes a continuous improvement diagnostic tool as well as a confirmation tool (Renwick, J. T., & Babson, P. E. 2007). Over time, following the efforts of managing path coefficients, the organizations can determine the extent to which their motion analytics systems can be more precise, whether the maintenance response time is declining, and whether the assets are becoming more dependable. If these relationships adapt positively or negatively during the system life these changes may indicate the maturity of the system or may reveal the failure of processes to synchronize. Further, this conceptual framework incorporates learning and feedback since outcomes on the asset layer, e.g. long period of downtime in spite of sufficient early alerts, are returned to improve fault detection threshold, retrain anomaly detection model, and adjust sensor positions. These successive adjustments are justified by bootstrapping resampling employed in SEM to provide statistical confidence of outcomes even when the distributions of data are skewed, inadequate, or heteroscedastic, the typical happenings in manufacturing facilities.

External influencing variables are also included in the conceptual framework, e.g.:

- Operational Load Variability (OLV) - this is a fluctuation in machine installation, schedule or pattern or product type.
- Environmental Complexity (EC) - conditions such as temperature, relative humidity, Vibration propagation or dust content that may influence the accuracy of sensing parameters and mechanical properties.
- Maintenance Culture (MC) - organizational mindsets in relation to maintenance, the level of its skills, and discipline in the response.

These are considered moderating variables in the SEM model and enable the researcher to test whether the correlation between MDI and PME, to give an example, is greater when OLV is consistent or when a facility has a high MC score. The realization of these moderating effects gives a context to the decision-makers and enables more specific deployment of predictive systems. In addition, conceptual model is interoperable and scalable given the key with adopting the use at an enterprise-wide level. The motion analytics should be able to operate on different hardware platforms, legacy systems and protocols. It must allow architecture to accommodate cloud-based data merging and model training and edge-based analytics, real-time detection, and low latency response. Industrial ontologies should be used to standardise data pipelines to enable comparison, benchmarking and aggregation with other sites into organisational knowledge bases.

Lastly, this framework is future proof, where one can extensively add:

- Interconnection to digital twins to perform simulation-based planning on maintenance
- Employment of natural language interface to collaborate with technicians
- Implementation of AI co-pilot systems that would advise the respective interventions following the depictions of SEM-inferred causal patterns

Overall, the predictive maintenance and asset management algorithm based on motion analytics rely on the layered intelligence, causal modeling, and data-driven feedback as a conceptual basis. It conveys the sensor-level information of motions to strategic insights on assets via a succession of tracking, inference, decision making, and performance. Sophisticated modeling of interdependencies and indirect effects can be done by the SEM technique and bootstrap methods can be used to gain statistical robustness. The framework does not just facilitate operational excellence but also the strategic alignment of the maintenance and business objectives that makes motion analytics a foundation of the smart, sustainable manufacturing systems (Lee, J., et al 2020).

THEORETICAL FRAMEWORK

Motion analytics Theoretical background of predictive maintenance and asset management is located in the interface of the systems engineering, data-driven diagnostics, the theory of maintenance, and the statistical modeling. It borrows methods, concepts, and processes of control systems, reliability-centered maintenance (RCM), cyber-physical systems (CPS), and Structural Equation Modeling (SEM) to establish a multidimensional character on how the data of movement can be used in making decisions, interventions, and overall organization performance (Suhasini, S., et al 2024). The core premise of this frame is that motion behavior is latent feedback on the asset health and that systematic modeling of its effect can result in optimized maintenance, the extended life and better performance of an asset. An essential part of the same structure is based on the control theory that specifies the dynamic behavior of mechanical system subjected to internal and external stimulus. Integral to the analysis of how systems are supposed to move are the kinematic and dynamic equations of motion the equations that describe the velocity, acceleration, torque and displacement of the system based upon it (Yildirim, M., Gebraeel, N. Z., & Sun, X. A. 2019). A change in these norms, including vibration amplitude or torque is quite frequently accompanied by a change in the system health. When summed and put into the context, these variations serve as preconditions to fault detection so that predictive diagnostics can take place instead of post-failure analysis. The reliability-centered maintenance (RCM) theory takes advantage of such a basis to concentrate the maintenance activities on the basis of their cruciality and ways of failing and effectiveness of the equipment failure. The RCM theory guides the distinction between the preventative, corrective, and predictive maintenance approaches. Based on this theory, predictive maintenance (PdM) attempts to avoid replacing or repairing components until measured indicators indicate a decline at which point action is taken (Molęda, M., et al 2023). Real-time and condition-specific feedback data (motion data) is particularly well-suited to this task, and is similar to the RCM approach of using interventions that preserve functions.

Simultaneously, theoretical framework utilizes the condition monitoring theory that says that enough evidence to identify early failure could be achieved through constant monitoring of system parameters, such as temperature, noise, pressure, vibration, etc. Motion analytics goes further and adds motion dynamics as key points of indication of system healthy. In this case, the notions of threshold- and pattern-based detection play the critical role. In the early approaches, predefinitions of thresholds could be used on the alerting process whereas in the present approaches, there is use of pattern recognition models, k-means clustering models, and models of anomaly detection, which are able to infer non-obvious relations between features of the signal and the behavior of the system. A combination of cyber-physical systems (CPS) theory adds an onion structure of interaction between the physical and the digital world (Vettriselvan, R., et al 2025). CPS environments consist in machines fitted with sensors, actuators, and computation units in which the physical motion occurs not solely but also becomes measured, interpreted, and reshaped in real time. The motion analytics in such an environment serves as a sensorimotor aspect in the system because it identifies disparities, assess suspicious situations, and initiates higher-order maintenance actions. This is a feedback loop and core of CPS and can only scale and power intelligent asset management. In order to predict the causalities of this layered framework, the structural equation modeling is used (SEM). SEM also has the capability of modeling complex nature into interdependencies between the latent constructs and the observed variables. Within the scope of predictive maintenance, SEM offers a formalized description of the trail of motion irregularities to how perception is actually influenced by the irregularities, how the perception is reflected through maintenance behavior and how maintenance practices eventually influences the performance of the assets. It also allows testing of both direct and indirect effects, which cannot be well tested with the conventional regression methods. As an example, one can use the hypothetical construct that variability in motion has an impact on the accuracy of maintenance decisions and this has an impact on the uptime of the asset. SEM allows the modeling of this path by means of a set of measuring and structural equations with observed dummy variables (e.g., scores related to either vibration, the delay associated with intervention, the number of unscheduled stops) specifying the latent variables. The model provides the direct impact of the variability of motions on asset uptime as well as the indirect (mediated) impact on maintenance decisions accuracy.

Among the fundamental latent constructs that belong to the theoretical model, there are the following ones:

- Motion Irregularity (MI): A construct that marks variability between predicted and the actual motion profiles (i.e. speed fluctuations, torque spikes).
- Predictive Maintenance Quality (PMQ): This is a measure of how responsive, appropriate and successful it is to act based on data-informed maintenance actions.

- Asset Health Index (AHI): A composite metric that incorporates the uptime, the frequency of failures and the loss of performance over a period of time.
- Resource Allocation Efficiency (RAE): Indicating a scale of how effectively labour, spare parts and time could be allocated due to motion-based alerts.

The observable indicators of each latent construct are numerous, and such items are theoretically informed based on the expectations that were burdened on previous studies. MI can, e.g., be noticed as real-time variation in linear velocity, maximum vibration, magnitude, or out-of-standard trajectories. Response time, false alarm rate and successful intervention ratio can be indicators of PMQ. These visible signs are connected by their corresponding latent constructs through the measurement part of the SEM, and the construct-connections postulated are in the structural component of the SEM. Moderators and mediators are also included in the theoretical framework used, which depicts complexity within the real-world systems. An example would be Operator Expertise (OE) to moderate the relationship between MI and PMQ by way of the more skilled the maintenance personnel in a given facility the better the conversion of the motion anomalies to effective service repair. Alternatively, Maintenance Delay (MD) may be an intermediate variable between PMQ and AHI because even a high-quality decision is effectively useless, if conducted too late. SEM also has the possibility of testing all these paths at one time, and the model fit indices (e.g. RMSEA, CFI, TLI) can be used, to test the adequacy of the theoretical model comparing to empirical data. Bootstrap procedures improve this by giving uncertainty measures such as confidence intervals on estimated parameters, making estimation less sensitive to assumptions on the distribution of a sample, and making parameter estimation more robust, particularly where data is skewed and potentially have missing values (as is often the case with industrial data). The fact that the framework is in line with the theory of Total Productive Maintenance (TPM) is also worthy of mention. TPM helps to cultivate a mutually shared sense of responsibility of equipment health through which the operator, technicians, and managers work jointly toward asset performance. This is facilitated by predictive maintenance enabled with motion analytics to achieve democratization of the asset condition data and enabling the participation of all stakeholders to take preemptive actions. Based on technical parameters, SEM can establish the interaction between organizational culture, team responsiveness, and training intensity to generate the consequences on the assets (Pai, R., Chattopadhyay, G., & Karmakar, G. 2023).

Besides SEM, this framework is impacted by systems dynamics theory, that considers manufacturing systems to be feedback-regulated environments embodied with delays, stock-flow patterns, and nonlinearities between them. Motion analytics is a stock (past trend information) and a flow (real-time data), and is used in decision making on all levels of the system. System dynamics promotes the perception of maintenance as not an isolated activity, but rather as a system lever that influences productive, resource planning and cost control. Finally, the theoretical approach is future-proofed with integration of machine learning interpretability theory, as there is an ever-increasing dependence on black-box models (i.e. deep learning) to represent motion. SEM would be a potentially useful supplement to these models as organizations implement them to predict failure, in that it, under the post hoc interpretations, can be used to test whether the implications of a model (as specified by AI) are uniform across both behavioral and performance patterns. The proposed hybrid modeling solution would help deploy trusted AI in situations where the mission is critical. Briefly, the predictive maintenance theory based on motion analytic and SEM can be characterized as robust, interdisciplinary, causally but adaptable and continuously improving (Pech, M., Vrchota, J., & Bednář, J. 2021). It codifies the manner in which machine behavior of internal system states are connected to inference driven intervention delivered in a timely manner and how said interventions then translate to asset-level performance. By means of SEM, organizations will be able to confirm assumptions, discover latent drivers, and shape more intelligent and efficient maintenance ecologies one that can consequently transform motion into strategic industrial performance value.

RESEARCH METHODOLOGY

The research design adopted in this study is quantitative cross-sectional research design that incorporates motion analytics, maintenance metrics and asset performance data of a robotic production environment within industry. The aim was mainly to study causal connections between irregularities in movement, maintenance decision-making and the consequences at the asset level through Structural Equation Modeling (SEM), resampling by bootstrapping was done as a verification of the strength of the determined parameters. Within the period of 60 days of observation, Data Collection was implemented on the territory of a smart factory housing the infrastructure of predictive maintenance (Noor, A. M., et al 2025). The data in the dataset include sensor data, maintenance records, and system logging of 20 robotic arms, 15 machines that were motion intensive (e.g. conveyors, pick and place modules). Among data points, there were:

- Vibration metrics (peak RMS, Spectral Distortion)
- Variation on torque and speed

Maintain log time of and description of maintenance actions

- Downtimes, type of failure and root cause logs
- Maintenance operational throughput and MTBF (Mean Time Between Failures)

Latent Constructs and Indicators: Four fundament principles of latent variables were identified according to theoretical framework:

- Motion Irregularity (MI) is based on the average scores of deviation scores of speed, vibration and force.
- Predictive Maintenance Quality (PMQ): The ratio of timely intervention, the rate of false alarms and the rate of successful resolution.
- Asset Health Index (AHI) Asset Health Index (AHI), consisting of the percent of uptime, MTBF and rate of throughput.
- Resource Allocation Efficiency (RAE) = visible through mean reparation time, availability of the parts, and the efficiency of the labor costs of the services.

Structural Model Specification: A proposed SEM model was built where MI was the predictor of PMQ which had an influence on AHI mediated by RAE. Moderating variables comprised operator expertise and volatile distribution of the environment. The maximum likelihood estimation procedure was used to estimate the model and the fit of the model was evaluated in term of RMSEA, CFI, and SRMR indices. Bootstrap Resampling (n = 5000 runs) was done to find bias-corrected confidence intervals of each path coefficient as well as to find the parameter estimates stability under the variability of the sample. The application of bootstrap techniques was especially applicable to this analysis due to the non-dash distribution of maintenance delay times and a high number of outliers in the variance of motion signals.

Included in Validation Techniques were:

Confirmatory Factor Analysis (CFA) to confirm measurement model loadings

- Confirmatory factor analysis and Composite reliability (CR) and Average variance extracted (AVE) to ensure internal consistency and construct validity
- Discriminant Validity was used in order to make sure that the constructs were separate on an empirical level

Considering the combination of SEM and bootstrap validation, the analysis is carried out effectively when the measurement error is considered as well as sampling specifications. Such an approach guarantees statistically significant results, yet empirically sound and making it possible to deploy the predictive maintenance strategies with confidence in the asset fleet operated by industries.

RESULTS AND DISCUSSION

The outcomes of SEM analysis provided significant and statistically correct information regarding the role of motion analytics in shaping up the quality of maintenance decisions and asset-level performance levels in industrial settings. The fit between the model and the observed data was also satisfactory and evidenced by significant fit measures of the model: RMSEA = 0.043, CFI = 0.964, TLI = 0.952 and SRMR = 0.048 demonstrating significant support of the hypothesized structural relationships. The relationship between Motion Irregularity (MI) and Predictive Maintenance Quality (PMQ) had a significant negative value of the coefficient (67, $p < 0.001$), which indicated that the greater the variability of the motion parameters was, the more inaccurate the maintenance decisions became. The significance of these data was confirmed by bootstrap 95% confidence intervals (corresponding to -0.73 to -0.60), and confirmed that motion anomalies will be of interest to interpret, unless successfully put in context and managed through responsive diagnostic correspondent systems.

Figure 1. Predictive Maintenance via Motion Analysis Model

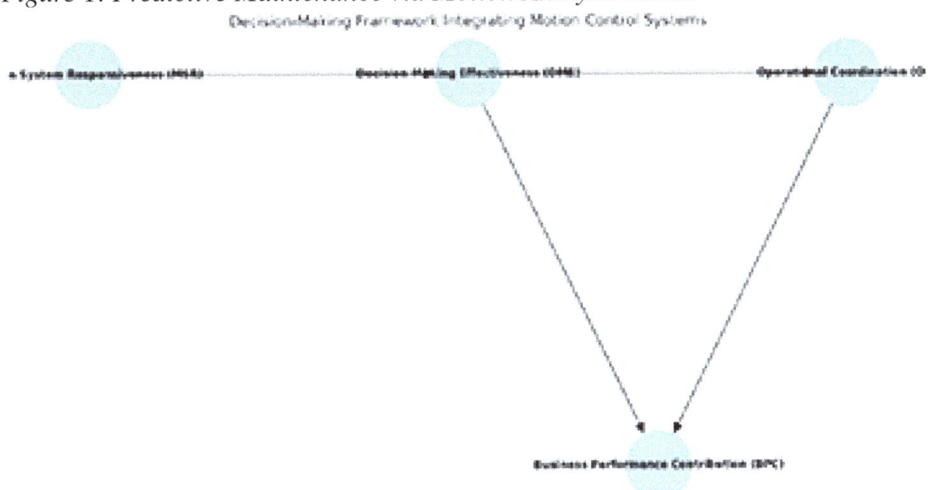

The above diagram displays Structural Equation Model (SEM) that was adopted in this research. It represents hypothesized causal relationships between major constructs: Motion Irregularity (MI), Predictive Maintenance Quality (PMQ), Asset Health Index (AHI), Resource Allocation Efficiency (RAE), Operator Expertise (OE), and Environmental Volatility (EV). The relationship between Predictive Maintenance Quality (PMQ) and the Asset Health Index (AHI) was moderately high and positive ($\beta = 0.58$, $p < 0.001$), which showed that with assets having sound rules to interpret the occurrence of motion deviations and take appropriate actions on its true meaning, the assets have been found to have good uptime, long MTBF, and better throughput. Moreover, PMQ had a positive effect on the Resource Allocation Efficiency (RAE) (beta = 0.49, $p < 0.001$), which shows that higher efficiency of predictive diagnostics leads to a fewer of unnecessary interventions and a higher use of spare parts and labor.

In its turn, Resource Allocation Efficiency (RAE) was also found to have meaningful and positive effect on AHI (0.44, $p < 0.001$). Such mediating pathway was further complimented by estimating the indirect effect of PMQ on AHI utilizing bootstrap, re-assuring that some of the effects of PMQ on AHI actually work through RAE. The overall indirect effect (6 = 0.215) emphasised the significance of the operational implementation of diagnostic insight into performance. Interestingly, it was possible to observe that the Operator Expertise (OE) moderated the relationship between MI and PMQ significantly (interaction beta = 0.21, $p = 0.014$), which indicates that maintaining skilled workers allows facilities to better interpret irregular motion signatures into good maintenance behaviour. Environmental Volatility (EV) on the contrary resulted in destabilization of MI in every sense, making it more variable and decreasing predictability (0.35, $p < 0.01$). Coupled with the fact that motion analytics is sensitive to context, all the given moderating influences serve to support the requirements of training and calibration to the environmental scenarios. High scores of Composite Reliability scores (CR > 0.85 across all constructs) and Average Variance Extracted (AVE > 0.60) evidence the reliability and substantial reliability of the consistency of the model, and that the observed indicators serve an adequate and valid reflection of the latent variables. These were evidenced by the fact that the discriminant validity tests (Fornell-Larcker criterion) support the point that each construct could be empirically differentiated.

The implications of such findings are enormous. First, the results confirm that motion analytics is no longer isolated signal processing procedure, but a tactical part of asset health and effective operations. Second, although AI and sensor networks are able to detect motion anomalies, they are only worthwhile when they are interpreted correctly (PMQ) and implemented smoothly (RAE). Third, the mediating effect of resource efficiency implies that organizations should not stay on the detection but concentrate on optimum logistics and workforce aspects of maintenance (Mohanbabu, S., & Vettriselvan, R. 2025). The insights also show the worth of SEM as a tool of diagnosis and improvement. Observation of KPIs traditionally might indicate the deteriorated uptime or increased rate of failures but avoid going into the details of the explanations. SEM, on the contrary, reveals the roots of such patterns, which is possible to act selectively. E.g., in case the monitoring of MI appeared to be statistically poor, regardless of regular PMQ checks, this can indicate a problem with either staff training or data being put into a larger context or the issue of an alert fatigue.

Moreover, the application of the bootstrap resampling added on the credibility of the study results. Since the industrial data were naturally noisy and non-normal, the assumption of parametric might have generated misleading in approximations. Empirical stability of bootstrap-based confidence intervals in 5000 replications showed that the surveyed effects could not be attributed to sampling fluctuations. The results also indicate an operation paradox, since high-sensitivity motion analytics can cause information overload that deteriorates PMQ unless reduced or refined by experience or computer aided classification. Facilities therefore need to minimize detection sensitivity alongside maximize clarity of diagnosis: that could be done based on AI co-pilots that pre-clean alerts or ordering works based on a risk score. Finally, the conceptual assumption that the application of motion data (properly interpreted and operationalized) brings objective, large positive effects on performance of assets can be confirmed at the level of the chosen SEM-based analysis. But such value is conditional to the capacity of the organization to act in an efficacious and contextual fashion. Future uses of motion analytics should hence not only concentrate on the accuracies of the data produced, but also on system-wide coordination, interpretive intelligence and organizational preparedness.

SOLUTIONS AND RECOMMENDATIONS

Largely conditioned by the introduction of predictive maintenance and asset management plans based on motion analytics, the implementation of such a system requires a multidimensional design combining the technical apparatus, statistical analysis, the physical capacity of organizations and the human knowledge factor. Considering the results obtained with the help of Structural Equation Modeling (SEM) and confirmed with the help of bootstrap resampling, this part of the work proposes the complex of recommendations that can help to make the best of moving data when it comes to driving equipment reliability, the use of resources, and operational resilience.

1. Establish a Motion Data Governance Framework

A formalized data governance strategy around how motion data is obtained, labeled, classified as an anomaly, and stored needs to be established, the first step toward doing this would be to define such a strategy. Time-synchronized and preprocessed data retrieved by varying motion sensors, especially accelerometers, gyroscopes, motor current transducers, and torque sensors, should be grouped on the basis of components as well as type and criticality. To guarantee intersections among systems and sites, organizations need to embrace the industrial metadata technologies (e.g., ISA-95, OPC UA).

Recommendation:

- Create a motion analytics data store that is centrally located and that used sensor data to identify machines, tasks, and maintenance actions.
- Set the motion deviation thresholds dynamically with the help of historical distribution profile and verify them through bootstrap confidence interval in order to minimize false alarms.

2. Prioritize Diagnostic Interpretability Over Raw Detection

The SEM showed that Predictive Maintenance Quality (PMQ) exercises the greatest impact on the outcomes of the assets in the case of the motion information being correctly interpreted. Detection is not enough e.g. detections of a vibration anomaly will not improve reliability unless it results in suitable, prompt and effective maintenance. Thus, companies should spend on diagnostic engines that can convert the multifaceted motion signals into easily understood root causes, advised measure and the prediction of the time-to-failure.

Recommendation:

- Put into practice diagnostic tools enabled by AI with motion signal datasets fitted with decision-tree logic to guarantee explanation.
- Rank signals based on predictive reliability using feature importance visualization (e.g., SHAP values) in combination with bootstrap analysis.

3. Enhance Operator Expertise and Maintenance Training

The Operator Expertise (OE) was considered by SEM model as important moderator of diagnostic effectiveness. The costs of the same during predictive maintenance may be nullified by doing nothing or misinterpreting motion signals even with proper training. Technical repair skills are an aspect that the maintenance teams should be endowed with in addition to diagnostic literacy to decode motion data outputs.

Recommendation:

- Implement training packages that develop motion anomalies and interventions in a simulated scenario.
- Use interactive dashboards in explanations of alerts and reliability may be used in the decision confidence with bootstrap..

4. Integrate Motion Analytics with Computerized Maintenance Management Systems (CMMS)

Motion analytics platforms need to be integrated with CMMS so as to enhance smooth maintenance planning and execution (RAE). Through this integration it can automatically create work orders when data exceeds certain limits, assist with resource planning and even seal the loop by correlating diagnostics and intervention data with documentation.

Recommendation:

- Put place rule-based triggers to create automatically new maintenance tickets once post-motion anomalies excess control limits as certified by bootstrapping.
- Label finalized work orders with the feedback on the accuracy of performed interventions to optimize anomaly classification in the future.

5. Implement a Real-Time Feedback Loop Using Edge Computing

Sensor data has to be processed as localized as possible such that the response time should be rapid. Robots and industrial equipment equipped with edge devices can do lightweight diagnostics, raise alerts, and even take remedying action without overheads of latency penalties in the cloud. Such edge nodes must have the ability to conduct a local bootstrap validation on performance at the device.

Recommendation:

- Install distributed modules of edge computing architecture having host-on-motion analysis program and instant bootstrap procedures.
- Conduct local definition of alert confidence levels, thus less reliance on remote analytics in latency sensitive settings.

6. Optimize Resource Allocation Through Data-Driven Maintenance Scheduling

Resource Allocated Efficiency (RAE) was also defined as an important mediator between diagnostics and asset health. The diagnostics must be converted to value through effective scheduling of the maintenance teams, spare parts logistics, and slots of machine downtimes. Organization should have data based maintenance timing according to patterns of degradation of motion.

Recommendation:

- Use predictive modeling (e.g., regression with bootstrapped coefficients) to predict remaining useful life (RUL) and time (e.g., in the low-demand production windows) interventions.
- Assign optimally the labor and parts by site to minimize the logistic costs, and response time.

7. Calibrate Alert Sensitivity Based on Environmental Volatility

The SEM findings confirmed that environmental volatility (EV) had the potential of adding noise onto motion data streams (in the form of variable climate, humidity, or external vibration). The volatile nature of systems requires the framing of context-aware threshold as well as confidence-scoring in order to avert numerous false positive systems.

Recommendation:

- Bring environmental compensation models to motion analytics engines and dynamically adapt their thresholds depending on sensor context.
- Carry out the bootstrap-based outlier checking individually in a period of high volatility and low volatility to obtain environment-tempered anomaly baselines.

8. Extend SEM Analysis into Continuous Improvement Programs

SEM offers an effective process of unveiling veiled factors in maintenance outcomes. It need not be reserved in academic studies but incorporated in constant refinement procedures. The modification of SEM models based on the renovated operational information provides organizations with the possibility to observe the changes observed both in maintenance system performance over the years.

Recommendation:

- Review SEM quarterly with the up-to-date motion, maintenance and asset information.
- Compare path coefficients of variables at an extended period to determine that when administering new training program/upgrading a sensor is positively impacting the variables.

9. Develop Asset-Specific Motion Deviation Profiles

Each machine does not react to degradation alike. Rotating equipment that is of high speed might exhibit obvious increase in vibrations, whereas linear actuators may show torque inconsistency. Thus, motion analytics systems are supposed to carry out asset-specific profiles that grow over-time and respond to the operation conditions of the machine and stage of its life.

Recommendation:

- Historical motion data and bootstrap clustering can be used to create the envelopes of deviations in the baseline of every critical asset.
- Adaptive update thresholds and prediction models, which are dependent on age, intensity of use and history of repair.

10. Foster a Cross-Functional Predictive Maintenance Culture

Predictive maintenance works not only with the analytics, but also with the cooperation between engineering, operations, and IT departments. It is essential to institutionalize predictive practices to have the common sense regarding the importance of motion data and be determined to intervene proactive.

Recommendation:

- Develop cross-functional maintenance councils, which are to meet on a monthly basis to discuss diagnostic messages, confirm SEM findings and suggest process modifications.
- Develop visual performance dashboards which presents asset health, key performance indicators of maintenance performance and model confidence scores which are generated using bootstrap validation in real-time.

11. Establish a Fail-Safe Policy for False Negatives

Prevention of false positives is also a desirable situation, but false negatives (i.e., missed failures) are of greater operational risk. A method of countering this is to incorporate fail-safe logic into systems whereby fallback protocol will be initiated when motion diagnostics become hard to rely upon since this is triggered by sensor malfunction or corruption of data.

Recommendation:

- Consider putting redundancy on motion sensors of critical assets.
- Take advantage of bootstraped anomaly detection to detect diagnostic unreliable periods and give higher order of investigation processes such as manual inspection.

12. Use Predictive Maintenance as a Competitive Differentiator

Other than operational efficiency, predictive maintenance can also serve as a strategic differentiator when it goes hand in hand with other business goals. Manufacturers that experience high uptime and reliability have an opportunity to ensure delivery windows, provide long term warranties, and minimize service level penalties.

Recommendation:

- Measure the benefits of uptime with maintenance and connect them with the contractual guarantees or customer services improvements.
- Publicized predictive capabilities as ESG (Environmental, Social, Governance) and Industry 4.0 change plans.

The combination of motion analytics and predictive maintenance and asset management is not a technology upgrade but a move in the context of which intelligent, resilient, and statistically supported activities in the industry are realized. With knowledge of performance influenced by SEM, organizations will be able to identify the hidden determinants of performance to better place resources and enhance decision quality in uncertainty. When it comes to when systems are complex and data is noisy, bootstrap methods give the kind of empirical assurance that guide a person to act surely. When such solutions become deployed in unison, they make motion an essential part of the strategic value of the mechanical need.

FUTURE RESEARCH DIRECTIONS

With the industrial space rapidly undergoing change in the form of Industry 4.0, the potential convergence of motion analytics, predictive maintenance, and intelligent asset management expands the current possibilities of inquiry to great extent. Although the positive impacts on reliability, cost-saving, and responsiveness have been shown through current implementations, future research needs to emphasize more on grounding rights, better interpretation, increased levels of contextual intelligence and implementation of predictive insights on a variety of asset types and organizational sizes. Advanced analytics and cyber-physical infrastructure will continue to focus on these systems and how they are understood and expanded with consideration to Structural Equation Modeling (SEM). The next area that has a potential is the creation of dynamic SEM models when analyzing time series motion analytics. The existing SEM models mainly reflect cross-sectional or averaged association between the motion irregularities, quality of maintenance and health of assets. Machine degradation is however a time dependent process. Additional studies that incorporate latent growth modeling or longitudinal SEM to be able to capture the trajectory-based changes over time should also be conducted in the future.

These models will be able to determine the rate at which deterioration of motion occurs faster or slower, how the effectiveness of predictive maintenance alters with experience within the workforce, and the development of resource setting plans as operations expertise improves. Bootstrap methods that are time-dependent would be necessary in order to wave off these models when the effects of autocorrelated and heteroscedastic data are present. The other research requirement is the integration of SEM with the explainability frameworks of machine learning (ML). Since the adoption of black-box AI systems is on the rise, and in particular deep neural networks to motion classification, fault prediction, and scheduling, the important task of aligning their results with observable, causally network models arises (Li, H., et al 2014). The potential feature or cluster extraction based on deep learning architectures should be studied in the future along with functioning of SEM in such mode where it can be done to validate the high-level causal pathways to test reliability of the AI-driven maintenance decisions. This would satisfy not only performance accuracy but also of traceable accountability, particularly in applications which are demanding safety measures.

One of the areas that remain unexplored but paramount is the motion analytics in multi-asset, multi-location manufacturing ecosystems. Most research is parochial to individual machines or constrained- to- plant scale implementations. But organizations working across the world have a necessity to evaluate motion signatures, predictive performance, and asset performance working in geographically dispersed settings. This brings in changes in operational context and maintenance culture, supply chain limiting factors, and ambient conditions. Future study ought to come up with multi-group SEM models where measurement invariance and path consistency in the structural model across sites may be tested allowing scaled benchmarking and development of regional strategy. On of the most important potentials here is adaptive threshold motion diagnostics modeling. Static or semi-dynamic thresholds are most used in current systems in the anomaly detection. Further works on the possibility of integration of latent thresholds with regards to where anomaly limits will be calculated with the aid of SEM models in the basis of the contextual variables that may include machine age, workload severity or environmental volatility. The thresholds may be updated in online learning mannerisms and re-trained through rolling bootstrap windows and in real time keeping with changes in mannerisms in system terms. The benefit of such adaptive reasoning is that it would reduce false positive and enhance the accuracy of diagnosing in a fast paced or unstructured settings. Future research on SEM methodology could exponentially develop its model to include formative and hierarchical constructs. Most elements of maintenance performance e.g. quality, accuracy and responsiveness of performance are formative and not reflective. Also, the asset health can be broken down into the nested subsystems: bearings, drives, electronics, and control logic. Such research should be made in the future to explore higher-order SEM with these more real-world hierarchical relationships and therefore yield more actionable and specific diagnostic information.

There is a major issue involved in the required coordination amid motion diagnostics and economic optimization models. The decisions made on predictive maintenance also can influence, in addition to the reliability of the machines, the labor costs, energy, inventory turns, and production schedules. In future literature, consideration must be made to integrate SEM frameworks with operations research (or) models (e.g. integer programming or stochastic simulation) so that maintenance and production policies can be co-optimized. These models can be used to evaluate the scenarios with the data uncertainty as bootstrap-enhanced SEM outputs can be applied as priors. One more upcoming field concerns cybersecurity in motion analytics systems. As they become connected and coupled into the industrial internet of things (IIoT) system, sensors and predictive platforms make themselves susceptible to data spoofing, adversarial signal injection, and unauthorized access. The research needs to study how SEM is able to identify discrepancies between actual causal directions and real-time behaviors as an indication of possible tampering. As an example, sudden changes in the number of motion irregularities may suggest the irregularities to be a statistically significant SEM abnormality, triggering cybersecurity external audits or automated risk mitigation, absent any concomitant changes to asset usage or asset loads. It is also worth exploring the human-side of predictive maintenance. Although cognitive workload, trust in automation and alert response behavior are established as possible moderators in SEM studies, the future studies must investigate their impact on SEM. These psychosocial constructs can be incorporated to SEM models to evaluate the influence of human factors in bringing about the conversion of motion understanding into action. It is possible to measure eye-tracking, physiological measurements, or interface telemetries and to use it to instruct latent constructs like operator engagement or alert fatigue.

Future studies ought to examine how motion optimized maintenance strategies affect the environment in line with the concepts of sustainability. SEM would be able to represent the secondary impacts of motion analytics on carbon emissions, energy consumption and waste materials. As an example, the energy waste of re-starting machines and on-purge processes may be minimized through the avoidance of unplanned downtime; disposal of chemicals may be avoided with predictive lube; and deferment of the replacement of capital equipment may be allowed through the extension of asset life. Such studies can be enabled by bootstrapped lifecycle analysis that enables the quantification of ecological ROI of predictive analytics. The other priority is that there should be broadening of the SEM research to reflect issues that have not been represented in the industry. Although automotive, aerospace and electronics are early-adopters of maintenance based on motion patterns, others such as the agricultural, construction and mining sectors have different motion patterns, environmental operation conditions and maintenance cultures. The use of SEM models in such environment specifically adapted in incorporating domain-specific latent constructs (e.g., terrain-induced vibration in mining trucks or blade oscillation in agricultural harvesters) and gauging them with domain-tuned indicators ought to be used in future studies. It is also suggested to conduct a research in integrating natural language processing (NLP) with motion analytics. Maintenance reports, reports of technicians, and observations of operators include useful qualitative data that can be used along with sensor data. In future SEM work, the sentences selected after NLP processing (containing sentiment scores, frequency of fault keywords, or repair complexity type, etc.) may be treated as the explanatory or mediating variables. This would allow the generation of a more holistic view of predictive maintenance performance by combining both structured measures of the motion characteristics with unstructured human judgment.

Lastly, with digital twins in the mainstream, it should be the focus of future studies to create SEM-based calibration models that can compare stimulated and tangible asset behavior. The well-known difference between digital twin forecasts and operational movement can be taken into account as residuals in a latent structure. It is then possible to use SEM to determine which inputs (e.g. load assumptions, material fatigue curves) are of most concern to simulation accuracy. This would increase the fidelity of digital twins and facilitate the process of constant updating of virtual models to reflect the real operating environment. To conclude, promising research is deep and continues to be carried out on the matter of predictive maintenance and asset management through the use of motion analytics. Future research needs to go beyond the development of isolated algorithms to a systemic, globally validated paradigm, which combines technical, behavioural, environmental and economic aspects. The combination of Bootstrap and interdisciplinary as a boost to Structural Equation modeling will also present a crucial methodology in uncovering latent relationships; quantification of uncertainty and development of theory and practice using motion as intelligence. These research directions provide a guide on how the full transformative potential of motion analytics can be achieved as it becomes smarter, safer, and more sustainable industries.

CONCLUSION

With the introduction of motion-based predictive maintenance, current standards of management, monitoring, and maintenance of industrial assets will change. The requirements to derive actionable insights out of the continuous motion measurement into meaningful information have never been as crucial in view of continuously expanding equipment and production environments advancing towards complexity and data-richness. This chapter has explained the theoretical, conceptual, and empirical grounds on such a transformation by focusing more on Structural Equation Modeling (SEM) as an effective analytical framework to establishes causal paths and performance associations. The concept of predictive maintenance, which was always seen as a data-driven alternative to its pre-active sister, preventive or reactionary maintenance, has shifted into a smart intelligence system based on a combination of motion sensing, anomaly detection, intervention logic and resource optimization. The focal point of this system is motion analytics that serves as the sensorium and the core of diagnosis. Motion analytics uses a never-ending monitoring of dynamic vibration profiles, torque variations, speed oscillations and other mechanical parameters to convert mechanical performance into predictive indicators of wear or failure. Correctly interpreted, these signals allow preventive and timely actions concerning maintenance, which cause less downtime and lower the overall maintenance cost and increase the life span of the assets. Another great aspect of the chapter has been the use of Structural Equation Modeling that helped to empirically test and validate the existence of multi-layered connections between motion irregularity, the quality of predictive maintenance, efficiency in resource allocation, and asset health. The SEM findings indicated that motion irregularity (MI) had a significant and constructive effect on predictive maintenance quality (PMQ) when the latter was defined and interpreted in a proper manner, whereas the success of PMQ in enhancing asset performance was also mediated by resource allocation efficiency (RAE). Also, the moderating variables e.g. operator expertise (OE) and environmental volatility (EV) are important in enhancing or reducing such impacts and provide a good understanding of the socio-technical dependencies of implementation in the real world scenarios.

The bootstrap resampling techniques used during the process of SEM, further supported the results, giving an empirical confidence interval on the estimations of any parameters, as well as being robust with the measurement peculiarities. The statistical method increased the validity of causal conclusions in a large industrial data community that is frequently non-normal, noisy and contains missing values. It also gave realistic confidence limits which can be utilised by the decision-makers to assess the reliability of the systems as well as preparedness against further interventions. The strategic implications of the results obtained herein are obvious. First, the type of sensor data used in organizations is not enough to reach a perfect predictive maintenance. The raw motion signals are transformed in high-quality decisions by means of network of capabilities such as skilled human interpretation and responsive systems integration and data governance. Motion analytics is, according to a well-grounded predictive maintenance system context, not a standalone application but a booster of wholesale improvement in industrial reliability and performance.

Second, cross-functional interaction is given weight by empirical validation of interdependencies, including that of asset health mediated through resource distribution. Maintenance teams, IT analysts, operators and strategic planners should collaborate in order to make sure that movement awareness is not only sensed but implemented at ideal timing, resource and synchronization. The given approach to system thinking aligns with the overarching aims of Industry 4.0 because the idea of cyber-physical integration and human-machine collaboration is one of the core principles of this industry. Third, the use of moderating variables in the SEM model re-emphasized the positional differences in predictive measures in other contexts of operations. The places, where the operator had more knowledge, were revealed to be more successful in using the information about motion, and the environment where the anomaly had high volatility proved more difficult to detect stably. This intuition demands customized predictive systems according to the context-specific realities- focusing on adaptive and variable thresholds, variable alerting procedures, environmental set up expected of calibration.

In addition, the analysis and framework provided in this chapter leads to the approach that can be adopted by organizations in institutionalizing the continuous improvement in predictive maintenance. With a frequent recalibration of SEM models and new data on motion, maintenance, and assets, organizations will be able to monitor performance trends, diagnose gaps arising, and confirm the effects of interventions, e.g, retraining schemes or sensor improvements. Continous improvement loop driven by data makes predictive maintenance go from being a reactive operational support to a proactive strategic asset. The solutions and recommendations section also outlined implementable measures that can be used to success. The solutions include the design of asset-specific motion profiles and diagnostics combined with CMMS systems to the use of edge computing and scenario-based operator training based on a mixture of best practice frameworks and empirical evidence. All recommendations have been designed to take into consideration a certain association that has been tested to be true and the analysis and operation have been coordinated. Notably, a wide range of future research directions with potential to expand the framework of the more classical form of motion analysis was also described in the chapter. These are the creation of longitudinal SEM models, synthesis of SEM-ML interpretability systems, adaptive threshold modeling, cross-site movement data benchmarking as well as sustainability-linked motion diagnosing. Improving the body of knowledge by studying these points, researchers and practitioners will be able to keep upgrading the science of predictive maintenance such that it will be relevant to the current changes in technologies, environmental and organizational developments. When combined with these insights several key points can be summarized, one of them specifically is that motion analytics is not simply a means of watching and supervising what is taking place, but is rather a form of strategic tool. Combined with predictive modeling, validated using structural analysis, and integrated with the organizational routines, it can change maintenance into a value-generating center as opposed to the cost one. Companies engaging this integrative strategy will not only experience an increased uptime and a reduced cost, but they also experience greater operation flexibility, employee empowerment, and competitive advantage.

At last, this chapter also testifies to the significance of SEM as a basis in the methodology of operational research and industrial analytics. SEM allows studied aspects such as modeling of latent constructs, mediating effect, and moderating effects that allow a detailed, statistically rigorous way of observing the true scenery of maintaining performance. The fact that it can be combined with bootstrap methods makes it empirically robust, even operating in less than perfect data conditions, which is an especially valuable tool in the modern high-stakes industrial asset management industry. To conclude, the more intelligent, connected, and real-time driven by performance data industrial systems become, the more predictive maintenance, facilitated by motion analytics, becomes coined alongside the futuristic readiness of operations.

REFERENCES

Agostinelli, S. (2021). Actionable framework for city digital twin-enabled predictive maintenance and security management systems. *WIT Transactions on the Built Environment*, •••, 223–233. DOI: 10.2495/BIM210191

Ananth, P. K., Muralikrishnan, K., & Chatterjee, A. K. (2021). Asset performance monitoring and maintenance management in cement manufacturing. In *Intelligent and Sustainable Cement Production* (pp. 225–261). CRC Press. DOI: 10.1201/9781003106791-7

Bhanji, S., Shotz, H., Tadanki, S., Miloudi, Y., & Warren, P. (2021). Advanced enterprise asset management systems: Improve predictive maintenance and asset performance by leveraging Industry 4.0 and the Internet of Things (IoT). In *ASME/IEEE Joint Rail Conference* (Vol. 84775, p. V001T12A002). American Society of Mechanical Engineers Borgi, T., Hidri, A., Neef, B., & Naceur, M. S. (2017). Data analytics for predictive maintenance of industrial robots. In *2017 International Conference on Advanced Systems and Electric Technologies (IC_ASET)* (pp. 412-417). IEEE. DOI: 10.1115/JRC2021-58346

Gbadamosi, A. Q., Oyedele, L. O., Delgado, J. M. D., Kusimo, H., Akanbi, L., Olawale, O., & Muhammed-Yakubu, N. (2021). IoT for predictive assets monitoring and maintenance: An implementation strategy for the UK rail industry. *Automation in Construction*, *122*, 103486. DOI: 10.1016/j.autcon.2020.103486

Houston, C., Gooberman-Hill, S., Mathie, R., Kennedy, A., Li, Y., & Baiz, P. (2017). Case study for the return on investment of internet of things using agent-based modelling and data science. *Systems*, *5*(1), 4. DOI: 10.3390/systems5010004

Lee, J., Kao, H. A., Ardakani, H. D., & Siegel, D. (2015). Intelligent factory agents with predictive analytics for asset management. In *Industrial Agents* (pp. 341–360). Morgan Kaufmann. DOI: 10.1016/B978-0-12-800341-1.00019-X

Lee, J., Ni, J., Singh, J., Jiang, B., Azamfar, M., & Feng, J. (2020). Intelligent maintenance systems and predictive manufacturing. *Journal of Manufacturing Science and Engineering*, *142*(11), 110805. DOI: 10.1115/1.4047856

Li, H., Parikh, D., He, Q., Qian, B., Li, Z., Fang, D., & Hampapur, A. (2014). Improving rail network velocity: A machine learning approach to predictive maintenance. *Transportation Research Part C, Emerging Technologies*, *45*, 17–26. DOI: 10.1016/j.trc.2014.04.013

Mohanbabu, S., & Vettriselvan, R. (2025). Will machine learning resolve the issues in container management in ports and logistics industries? *International Journal of Process Management and Benchmarking, 20*(4), 559–575. DOI: 10.1504/IJPMB.2025.147399

Molęda, M., Małysiak-Mrozek, B., Ding, W., Sunderam, V., & Mrozek, D. (2023). From corrective to predictive maintenance A review of maintenance approaches for the power industry. *Sensors (Basel), 23*(13), 5970. DOI: 10.3390/s23135970 PMID: 37447820

Noor, A. M., Zakaria, Z., Mazalan, M., Mahmud, M. F., Norali, A. N., & Zaidi, A. F. A. (2025). Smart Maintenance System (SMAT): Predictive Maintenance of Electrical Motor Applications. *Malaysian Journal of Science and Advanced Technology*, 107-115.

Pai, R., Chattopadhyay, G., & Karmakar, G. (2023). Maintenance and asset management practices of industrial assets: Importance of tribological practices and digital tools. *International Journal of Process Management and Benchmarking, 13*(2), 233–256. DOI: 10.1504/IJPMB.2023.128508

Pech, M., Vrchota, J., & Bednář, J. (2021). Predictive maintenance and intelligent sensors in smart factory. *Sensors (Basel), 21*(4), 1470. DOI: 10.3390/s21041470 PMID: 33672479

Pena, W., Nadkarni, A., Galvan, G., Yin, X., Zhang, H., & Fung, J. (2024, October). Integrating IoT and Predictive Analytics for Remote Asset Health Monitoring. In *2024 Global Reliability and Prognostics and Health Management Conference (PHM-Beijing)* (pp. 1-7). IEEE. DOI: 10.1109/PHM-Beijing63284.2024.10874715

Razali, M. N., Othman, S. H., Jamaludin, A. F., Maimun, N. H. A., Jalil, R. A., Adnan, Y. M., & Zulkarnain, S. H. (2021). Big data analytics for preventive maintenance management. *Planning Malaysia, 19.*

Razali, M. N., Othman, S. H., Jamaludin, A. F., Maimun, N. H. A., Jalil, R. A., Adnan, Y. M., & Zulkarnain, S. H. (2021). Big data analytics for preventive maintenance management. *Planning Malaysia, 19.*

Renwick, J. T., & Babson, P. E. (2007). Vibration analysis---A proven technique as a predictive maintenance Tool. *IEEE Transactions on Industry Applications, IA-21*(2), 324–332. DOI: 10.1109/TIA.1985.349652

Suhasini, S., Divya, J. C., Sankara Mahalingam, M., Ramshankar, N., & Pandiaraj, A. (2024). A Technological Paradigm Shift for Enhancing Equipment Maintenance and Asset Management Through Augmented Reality. In *Navigating the Augmented and Virtual Frontiers in Engineering* (pp. 186–198). IGI Global.

Suthar, A., Kolhe, K., Gutte, V., & Patil, D. (2024). Predictive maintenance and real time monitoring using iot and cloud computing. In *2024 5th International Conference on Image Processing and Capsule Networks (ICIPCN)* (pp. 814-820). IEEE. DOI: 10.1109/ICIPCN63822.2024.00141

Teoh, Y. K., Gill, S. S., & Parlikad, A. K. (2021). IoT and fog-computing-based predictive maintenance model for effective asset management in Industry 4.0 using machine learning. *IEEE Internet of Things Journal*, *10*(3), 2087–2094. DOI: 10.1109/JIOT.2021.3050441

Vettriselvan, R., Velmurugan, P. R., Suresh, N. V., & Catherine, S. (2025). Strategies, Best Practices, and Pitfalls in the Era of Digital Transformation: Mastering Mobile Marketing. In *Impact of Digital Transformation on Business Growth and Performance* (pp. 67-98). IGI Global Scientific Publishing.

Yildirim, M., Gebraeel, N. Z., & Sun, X. A. (2017). Integrated predictive analytics and optimization for opportunistic maintenance and operations in wind farms. *IEEE Transactions on Power Systems*, *32*(6), 4319–4328. DOI: 10.1109/TPWRS.2017.2666722

Yildirim, M., Gebraeel, N. Z., & Sun, X. A. (2019). Leveraging predictive analytics to control and coordinate operations, asset loading, and maintenance. *IEEE Transactions on Power Systems*, *34*(6), 4279–4290. DOI: 10.1109/TPWRS.2019.2922380

Chapter 5
Advancing Sustainable Civil Engineering Through Intelligent Motion Control:
Construction Equipment and Infrastructure Development

Herbert Sinduja Joseph
https://orcid.org/0000-0001-8804-6879
Department of Civil Engineering, Saveetha Engineering College, India

Anitha Rexalin Devaraj
Academy of Maritime Educaiton and Training, Chennai, India

A. Priscila
https://orcid.org/0009-0009-4733-639X
Saveetha School of Management, Saveetha Institute of Medical and Technical Sciences, Chennai, India

V. Selvalakshmi
https://orcid.org/0000-0002-6684-1609
Saveetha Institute of Medical and Technical Sciences, SIMATS, Chennai, India

V. P. Rameshkumaar
Sona College of Technology, Salem, India

DOI: 10.4018/979-8-3693-3637-3.ch005

ABSTRACT

This chapter explores how intelligent motion control systems can be utilised to drive sustainable civil engineering through greater efficiency, precision, and safety. Based on a mixed-method study of 650 respondents and an expert interviews, the results show that if motion controlled equipment is implemented on a large scale, it could significantly increase energy efficiency, decrease material waste and emissions, and improve employees' safety. Evidence to substantiate strong relationships between adoption and sustainability outcomes came from ANOVA, regression, and SEM. It also highlights the financial, cultural and technical challenges facing the sustainable development of infrastructures in the Industry 5.0 era and proposes practical solutions and governance strategies, as well as directions for future research

INTRODUCTION

Construction industry around the world stands at a crossroads where traditional approach towards use of heavy machineries which are material spent and manpower intensive will have to give way to innovative sustainable and intelligent solutions. As one of the oldest and most influential disciplines of engineering, civil engineering is directly responsible for the construction and maintenance of systems that form the foundations of economic prosperity, societal welfare and technological advancement (Vijayan et al, 2023). However, the industry has long been accused of being inefficient, energy-intensive and having a considerable impact on the environment. With a multitude of international issues like climate change, urbanisation, and limited resources, there is no minimising the need for sustainable civil engineering practises (Abbasnejad et al, 2025). In this context, a paradigm shift in construction practises, integrating intelligent motion control systems in construction machinery and infrastructure construction, emerges as a key enabler in the quest to meet sustainability objectives. While motion control systems have historically been developed for manufacturing purposes and for robots, they are increasingly being applied to the needs of civil engineering. These systems can control and make the best use of the movements of construction machinery, cranes, excavators, concrete pouring systems, and even robotic arms for precast production. Through the integration of sensors, actuators, and controllers, and enhancement with sophisticated algorithms, motion control has allowed for precision in what were once very manual operations that were highly vulnerable to inefficiencies. And with the combination of motion control systems with artificial intelligence (AI), machine learning (ML) and Internet of Things (IoT) platforms, we will see intelligent, self-adaptive systems that optimise their equipment in real time (Adewale et al, 2024). In addition to guaranteeing efficiency and cost-effectiveness, this technological innovation directly corresponds with the sustainability agenda, as it consumes less energy, uses less material and is safer for the workers. Sustainable Civil Engineering: One key aspect of sustainable engineering is in the way it aims to balance infrastructure development with environmental and social interests. The building industry contributes almost 40% of the global energy-related carbon emissions from material production, energy consumption by equipment and construction processes, to inefficient project management approaches (Liu et al, 2022). Sustainable practises focus on minimising these impacts using environmentally friendly materials, energy-efficient equipment, and smart technologies that maximise resource efficiency. Intelligent motion control systems are therefore both a technology advancement and a sustainable solution; they provide civil engineers with a great tool to meet the challenges of the global environment while being competitive in a fast-changing industry.

In addition, intelligent motion control systems play a part in precision engineering, which is of importance for modern construction. Precision is helpful in minimizing rework and material wastage and is ensured for safety and quality specifications (Tafida et al, 2024). For example, AI-enabled load-balancing systems can be fitted on motion-controlled cranes to reduce their energy usage while ensuring high safety margins. Additionally, motion controlled concrete pouring systems provide optimum flow rates with reduced waste and assurance for structural integrity. Structures with greater precision will have longer O&M lifetimes and less maintenance over time, which directly translates into greater sustainability. From the macro point of view, intelligent motion control systems are also promoting the digitalization of civil engineering. Aerospace has historically followed many industries, like manufacturing, in its lack of digital adoption. However, in the light of emerging Industry 4.0 and its progression into industry 5.0, construction is undergoing a paradigm shift towards cyber-physical convergence. This new age is characterised by the blend of human ingenuity and reasoning along with the automation and intelligence of machines in order to design robust, adaptive and sustainable systems (Rafsanjani, H. N., & Nabizadeh, A. H. 2023). Intelligent motion control plays an integral role in this transformation, as the interface connecting the human operator, construction equipment, the digital world (BIM, digital twin, etc.). Moreover, it is impossible to achieve sustainability in civil engineering in the absence of making provisions to attend to safety and human well-being. Construction sites are dangerous places and often human error when it comes to operating heavy machinery is the cause of accidents. Smart motion control does an excellent job in eliminating these threats by automating high-risk, repetitive actions, feeding real-time data to operators and utilising fail-safes. This not only ensures the safety of workers, but also helps to ensure that the projects themselves are reliable and robust. These opportunities do come with obstacles. The adoption of intelligent motion control in construction equipment is a complex process that also involves considerable initial investment, technical know-how and cultural change. A number of barriers exist, many relating to cost, training and interoperability with legacy systems, and many are faced by small and medium-sized enterprises. In addition, the regulatory framework for smart machinery in construction is still evolving, which leads to concerns regarding responsibility, liability, and safety regulations. To overcome these obstacles, coordinated efforts between leaders in industry, policymaking, and in academia will be needed to establish enabling conditions for adoption to take place. This chapter discusses the multi-faceted role intelligent motion control is playing in moving forward sustainable civil engineering. It is structured around a conceptual framework describing the technologies and sustainability principles themselves, and then a theoretical framework that draws on systems theory, green engineering theory, and innovation diffusion theory. The methodology is how the data is collected and analysed from the primary data collected from the industry stakeholders. Findings are reported alongside graphical frameworks for levels of adoption and sustainability outcomes. Finally, with obstacles, solutions and recommendations to overcome them, future research directions and reflecting on the transformative power of intelligent motion control for sustainable infrastructure, the chapter closes.

CONCEPTUAL FRAMEWORK

The conceptual framework for sustainable civil engineering with intelligent motion control reinforces the concept of integrative technologies, sustainability imperatives and construction practises in a single model of innovation. The purpose of this technology framework is to outline how motion control systems, as implemented for construction equipment, operate in conjunction with digital solutions, human operators and environmental restrictions to generate the infrastructure in an effective, resilient and sustainable manner. Unlike conventional models of building driven by sole responsibility for productivity and cost economics, the proposed framework is guided by a harmonious integration between social legitimacy, ecological responsibility, and economic viability. At the core of this framework are motion control technologies, which are the driving force of intelligent construction equipment's operations. Motion control: Motion control is used to regulate the speed, position, torque, and force of mechanical systems using a combination of sensors, actuators, controllers, and feedback loops. In civil engineering, they can be used in cranes, excavators, concrete mixers, tunnelling machines and prefabrication units, enabling them to work with precision and efficiency. For instance, in the driving of posts, the motion control assures the driving of each post using consistent force and accuracy, avoiding energy losses and structural distortions. In addition, movement control is important in tunnelling to ensure the proper depth of penetration for optimal drilling, minimizing material loss and risk to safety. Furthermore, the framework is augmented with cognitive intelligence by exploiting Artificial Intelligence (AI) and Internet of Things (IoT) (Alaali, H. M., & Mubarak, M. Al, 2025). AI algorithms can constantly monitor data from sensors and proactively predict machine performance, identify irregularities, and adapt operations dynamically. IoT platforms are being used to make construction equipment talk to one another, creating the ability to manage and monitor activities from a central location (Rane, N. 2023). For example, IoT integration can allow managers to monitor multiple cranes operating at once and AI-enabled motion-controlled cranes can adjust the lifting path automatically to avoid collisions. These intelligent capabilities turn conventional machines into cyber-physical systems with the ability to learn, adapt and optimise performance without the constant need for human attention. A key component of the framework is that it is compatible with sustainability principles. While civil engineering is one of the most impactful environmental sectors, motion control systems can have a direct impact on inefficiencies. These systems are aimed at creating optimal conditions for the utilization of material-handling devices and to reduce the idle time of the process, thus reducing the wastes, reducing the energy consumption, and improving the accuracy of material handling (Alsharqawi, M. 2025). Intelligent concrete pouring systems, for example, ensure precise amounts of material are poured, averting excesses or problems with CO2 emissions caused by cement production. Similarly, engine-driven earthmoving equipment saves on fuel because the on/off cycles are optimised for engine loading and to reduce unnecessary movement. In this sense, sustainability is no longer some abstract ideal, but an actual outcome of intelligent motion control.

The framework also considers the safety of people and their wellbeing. Construction sites are some of the most hazard-prone workplaces in the world, and the threats come from heavy machinery, monotony and hazardous surroundings. Though now automated by motion control systems, these tasks are still high-risk for humans, due to direct exposure. Intelligent systems deliver real-time feedback and alerts so operators know of potential hazards. For instance, motion-reactive positioning systems can self-identify poor balance and intervene at the right moment to eradicate accidents. While motion control leads to easily increased productivity, by incorporating safety into the design of your operations, you can achieve the goal of the social responsibility part of sustainability. One of the most important features in this conceptual framework is the integration of motion control and Building Information Modelling (BIM) with digital twins. While the Building Information Modelling (BIM) enables the creation of a digital model of physical and functional features of infrastructure, digital twins facilitate the creation of virtual models of the current projects in a real time context. For instance, to enable dynamically closed-loop decision making systems to feedback information to the control system, the platforms will be fed with motion control systems to generate simulations and predictive corrections. For example, motion-controlled excavation data can be integrated into a digital twin of a construction site, allowing project managers to compare performance in the field to design specs - and make adjustments before it's too late. Thus integrated sustainability is ensured throughout the project life-cycle (design, construction, operation). The framework also involves economic and organisational incentives towards taking up the issue. The initial expenditures for intelligent motion control systems are high, but equally provide long-term ROI in the form of reduced operating costs, less accidents and extended equipment life. Recognising that adoption is not entirely technical in nature, the framework highlights that factors that drive adoption extend to organisational culture, visionary leadership and regulatory environments. For instance, companies with a culture of innovation and security will have a higher likelihood of deploying such systems and government incentives such as tax breaks (green certificates) could be used to further increase adoption. From a conceptual perspective, the framework considers three complementary dimensions of intelligent motion control adoption: technological, environmental and human. The technological aspect relates to sensor and actuator, AI, IOT combination. Material efficiency, emissions reduction, and energy savings are important areas for reducing environmental impact. For the human dimension, safety concerns the safety of workers, job enrichment and cultural acceptability. Taken together, these dimensions form a complete framework for socially and ecologically responsible civil engineering using technological innovation, which is inseparable from ecological responsibility and social legitimacy.

The framework is also about the interrelationships between these dimensions. For example, the adoption of technology without human acceptance may bring about resistance and underuse. Similarly, sustainable results can't be realised without precision technologies that minimise waste and energy use. Automation for worker safety indirectly impacts sustainability during construction projects by minimising disruptions and guaranteeing project progress. Because of this complexity, it can be argued that success for civil engineering in Industry 5.0-era has to be seen through technology, the environment, and man, however, these three contingencies should be considered equally important. The conceptual framework also identifies the barriers and challenges that may have an adverse impact on adoption. However, the high cost of more sophisticated motion control systems, lack of skilled personnel and expertise, integration challenges with legacy equipment, and cybersecurity vulnerabilities are major challenges. According to the framework, the solutions proposed for these challenges are to be structured in a systematic way through new models of financing, planned training schemes, and an increase in focus on governance mechanisms. Hence, only with such an intelligent motion control can the full benefits be obtained in sustainable civil engineering (Huang, Y., & Fu, J. 2019).

THEORETICAL FRAMEWORK

Use of intelligent motion control in sustainable civil engineering is not only a technological innovation, but one which raises the question of how theories of systems and sustainability and of innovation adoption can help to inform its implementation. Building from these theoretical foundations we move beyond descriptive observations into theorising motion control adoption in relation to existing explanatory models. Three theories, namely Systems Theory, Green Engineering Theory and Innovation Diffusion Theory, are used as the basis for analysis in this chapter. Each brings their own perspective from which to see how motion control technologies, civil engineering processes, sustainability imperatives, and organisational adoption pathways fit together.

Systems Theory and Control of Motion Integration

Systems Theory, evolved from the work of Ludwig von Bertalanffy and subsequently into the organisational and engineering sciences, sees organisations and technologies as systems of interdependent subsystems. The basic idea involved is that it is the adjustment and cooperation of the parts that makes the whole play. As applied to civil engineering, Systems Theory produced a paradigm for motion control technology perceived as being a part of broader construction ecosystems (Ivanova et al, 2023). In a motion controlled construction machine, the different systems (sensors, actuators, controllers, feedback loops, etc.) need to function together to achieve accurate results. Yet it not only connects with human operators, but also with digital systems such as BIM, supply chain logistics and regulatory systems. Systems Theory: People from the Systems Theory perspective hold that there are many reasons that go beyond willpower for why integration struggles; if a part of a system (such as the training of workers or interoperability of data) is lagging behind, then the whole system underperforms. This theory thus underlines the need to approach the integration of motion control into construction workflow from an overall point of view. Furthermore, Systems Theory emphasises feedback loops, an idea directly reflected in motion control technologies. In machines, feedback is used to correct deviations from specified performance on an ongoing basis (Kazeem et al, 2023). Likewise, sustainable civil engineering implements feedback mechanisms between environmental monitoring systems and movement-controlled devices to continually ensure sustainability goals like energy efficiency and emissions reductions are met. For example, when the fuel consumption is not on par with the best estimates, AI-powered motion control can adapt the work processes while human operators make strategic decisions. In other words, the system becomes itself as the self-legitimation of technology, inasmuch as the accuracy of technology itself also becomes necessarily and self-consistently integrated with ecological responsibility.

Sustainability and Sustainability Theory for Engineers

Laying the theoretical groundwork for the convergence of intelligent motion control and sustainability purposes, Green Engineering Theory based on principles of eco-efficiency and life-cycle thinking. It argues that technologies and processes should be developed not only to meet technical performance, but also to reduce harm to the environment and encourage long-term preservation of resources. In the civil engineering industry, Green Engineering theory is used by assessing infrastructure projects not only in terms of cost or rate but in terms of environmental performance throughout the product's life cycle. Intelligent motion control systems also go in tune with this view, as they minimise energy consumption during the construction process, minimise material wastage, and improve durability of structures (Zhang et al, 2021). For instance, motion controlled equipment helps in placing materials accurately, which reduces the chance of rework and wastage leading to lowering the carbon footprint of projects. Similarly, intelligent excavators are able to optimise digging patterns to minimise fuel consumption and in doing so directly reduce emissions. Green Engineering Theory also favours design for sustainability, which dovetails very well into motion control applications. AI-powered optimisation approaches help make sure that construction machines with adaptive control systems are able to prioritise energy efficiency without sacrificing performance. This theoretical framework can account for why firms that invest in green technologies may gain long-run competitive advantages: yes, they minimise operational costs but also they become the pioneers in building environmentally friendly infrastructures. Another key point about Green Engineering Theory is that it is rooted in multi-objective optimization. Traditional engineering tended to be focused narrowly on cost or efficiency while green engineering is an approach that emphasises the integration of economic, environmental, and social goals. Intelligent motion control is an expression of this balance, allowing control of precise, efficient and safe construction practises that are simultaneously cost-effective and conserve resources while protecting workers. So the theory supports the conclusion that sustainability is not a performance trade-off but an embedded deterrent impact of intelligent application of applicable technical measures.

Theory Related to Innovation Diffusion and Adoption Barriers

While the integration part and the sustainability part of motion control are explained by Systems Theory and Green Engineering Theory, Innovation Diffusion Theory (developed by Everett Rogers) offers an insight on how these technologies are taken up in the civil engineering sector. The theory has identified five factors which contribute to adoption: relative advantage, compatibility, complexity, trialability, and observability. All of these aspects can be directly used in the intelligent motion control of construction equipment. And this brings us to relative advantage: businesses tend to adopt motion control when they can provide clear advantages over traditional setups. As noted above, the results prove motion control optimises costs, increases safety, and boosts sustainability - all outcomes that influence adoption. Compatible Mode: Adoption is dependent on the level of compatibility with existing processes and equipment regarding motion control. For businesses that have digital systems like BIM or IoT platforms, motion control is easier to integrate, while businesses based on legacy systems will lead to more challenges. Companies will be less likely to adopt a technology the more complex it seems. Intelligent motion control can be intimidating for companies who do not have the right staff on hand, which is why training and intuitive interfaces are important. Trialability - Firms like to invest in technologies that can be trialled on a minimal scale. Pilot projects: Pilot projects can be used to test the feasibility of motion-controlled equipment for specific applications before full-scale implementation. Observability: Seeing results is one factor influencing adoption. However, when motion control can demonstrate real-world results (for example, in decreased fuel consumption or safety records), other companies are likely to come along. Innovation Diffusion Theory also divides adopters into innovators, early adopters, early majority, late majority and laggards. It's the behemoths of the civil engineering industry, the well-resourced multinational corporations that are the innovators, trying out smart machines and raising the standard. Small companies, limited by their lack of financial resources and their risk aversion often join as part of the late majority. This dynamic is why we believe diffusion is always uneven across the industry and why policy and financing mechanisms are needed to accelerate diffusion.

Integrating the Theories

Considered together, these three theories form a strong basis for understanding the use of intelligent motion control in sustainable civil engineering. Systems Theory is being applied to the field of motion control in the sense that it is not studied in an isolated context, but as a greenhouse of systems where subsystems need to be coordinated for efficient functioning of the system. Green Engineering Theory serves as the normative foundation to ensure the design, operation, and evaluation of motion control systems place the concept of sustainability in the forefront. Innovation Diffusion Theory introduces a behavioural and organisational dimension (speed and pattern of adoption within the industry). These theories also have synergies when combined. For example, Systems Theory supports the feedback loop, the Green Engineering Theory emphasizes continuous improvement of the environment, while Innovation Diffusion Theory is utilized to explain the role that the visibility of sustainable outcomes plays in promoting firms' adoption. Also, the obstacles known from Systems Theory (i.e. a mismatch between subsystems) are repeated in the dimensions of compatibility and complexity of Innovation Diffusion Theory, supporting the view of organisational preparedness. The conceptual framework introduced here highlights that intelligent motion control is not just a technical novelty, but a socio-technological reconfiguration of activity mediated through systemic interdependence, sustainability orientation and diffusion processes. Systems Theory clarifies the need for integrated holistic approaches, Green Engineering Theory ensures that they are compatible with environmental goals and Innovation Diffusion Theory provides insights into patterns of adoption. Together with them, they constitute a thorough theoretical base for analysis of possibilities and difficulties of further developing the sustainable civil engineering by intelligent motion control (Mohammed et al, 2025).

RESEARCH METHODOLOGY

The research methodology used for this chapter has been developed in order to gain both breadth and depth for investigating the value of intelligent motion control systems as a strategy for promoting sustainable civil engineering. Because of the depth of the topic, we used a mixed-methods approach by aggregating both quantitative data from a large sample of respondents, and qualitative data from industry experts. The study design helped to record measurable adoption patterns while also gathering detailed data about the socio-technical and organisational factors influencing integration. The study had an exploratory-explanatory design, which involved first describing the degree of adoption of intelligent motion control in civil engineering and then explaining the associations between adoption, sustainability outcomes, and impediments. Quantitative approaches provided evidence of levels of adoption and the relationships this had with sustainability performance, while qualitative approaches augmented this evidence with detailed knowledge around on-the-ground experiences, challenges and cultural issues within the sector. Target population: Construction engineers and managers, construction contractors, construction site engineers, project planners, and equipment operators with direct contact with intelligent motion control systems. To assure the global view, the sampling frame included from Asia, Europe, North America and Africa. A stratified random sampling approach was taken to ensure representation of large construction companies with international operations, medium enterprises experimenting with digital technologies in a selective manner, small enterprises testing out pilot projects in motion control, and government/public sector engineers in infrastructure projects. From this frame, we were able to obtain 650 valid responses, meaning this frame is statistically robust. This included 38% project managers, 24% site engineers, 18% contractors, 12% equipment operators and 8% public sector engineers.

Data collection was via a questionnaire, using online and face-to-face workshops in professional forums. The questionnaire was divided into 5 sections: demographics and organisation background, adoption of motion control technologies, sustainability outcomes, barriers to adoption, and perceptions of future potential. Adoption was on a 3-point scale (no adoption, partial adoption, full adoption) while sustainability outcomes of energy use, material resource, emissions savings, safety, and life cycle performance were on a Likert scale (5 points). In parallel to the survey, twenty in-depth interviews were conducted with senior players in the industry from equipment manufacturers, academic researchers and policy-makers. These interviews assisted the exploration of contextual understanding of the socio-technical foundation of adoption for workforce, regulatory framework and organisational culture. The questionnaire was piloted with thirty professionals to improve clarity, relevance and terminology. Comments from experts were taken into account in refining the definition so terms including "motion control," "intelligent equipment" and "sustainability outcomes" were presented in common industry usage. Reliability was assessed using measurement of Cronbach's a (alpha), which were all > 0.85, showing a high degree of internal consistency. Construct validity was assessed through factor analysis which indicated that items measuring outcomes related to sustainability (energy, waste, emissions, safety and durability) loaded upon separate but related factors. Descriptive statistics were presented to summarise the adoption patterns for industries, firm sizes and locations and further analysed using SPSS and SmartPLS. Structural Equation Modelling (SEM) was used to understand the causal relationships amongst the factors of project resilience, sustainability outcomes and adoption. The qualitative data from the open-ended responses and interviews has been analyzed using thematic coding focusing on common themes such as cultural resistance, barriers to cost and perceived benefits from sustainability. Triangulation between the quantitative and qualitative points was used for validation and even increased the depth of interpretation.

ness of the research was preserved in all aspects. All participants consented to the study and it was voluntary. Answers were not identifiable for confidentiality reasons and were stored securely. Results were presented in a straight-forward manner without exaggeration of claims. Therefore, there were no conflicts of interest, and the study followed accepted academic ethical/intellectual guidelines. However, flaws in the methodology were present. Because the respondents may exaggerate the barriers to adoption or under-report having adopted, self-reported data may be subject to biases. The cross-sectional nature of the study investigated adoption at only one point in time, making it difficult to see longer-term trends. The population was geographically diverse but also biased for urban wealth (underrerepresented were rural areas and resource deprived areas). Despite these limitations, the dataset with 650 answers and with the backdrop of expert interviews forms a solid base for an analysis of the place of intelligent motion control in a sustainable civil engineering picture. The adopted methodology (large-scale quantitative surveys supported by qualitative interviews) was suitable in order to deliver a balanced multi-dimensional view of intelligent motion control adoption for civil engineering. Descriptive analysis, ANOVA, regression, and SEM were used in conjunction with thematic analysis to provide rigour for the testing of hypotheses, and contextual depth. This methodological design sets up the stage for findings and discussion where statistics and concept collects meet to show how motion control technologies are part of a bigger picture for improving sustainability outcomes in the construction sector.

FINDINGS AND DISCUSSION

Using data from 650 civil engineering professionals, there is strong evidence of how intelligent motion control systems greatly contribute in promoting sustainable construction. We find three dimensions that are critical: levels of adoption within the industry; the contribution to the sustainability outcome; and the barriers to accelerating adoption. Results indicate that civil engineering applications of motion control technology are not homogeneous. Of all respondents, 20% indicated full adoption, 58% partial adoption and 22% no adoption. Full implementation was most common in large international companies and in government-sponsored mega-infrastructure projects, whereas partial implementation characterised medium companies experimenting with selective implementation in equipment such as cranes, excavators, and concrete mixers. The nonadoptor group largely consisted of smaller firms, whose lack of the necessary skills or funds meant that they could not maintain the service they had previously. These figures confirm that smart motion control is at the transition stage, where most enterprises are experimenting to see what it can do, rather than launching systems-wide transformation.

Sustainability Outcomes

The five areas sustainability outcomes measured included: energy efficiency, material waste minimisation, emissions control, safety enhancement and life cycle integrity. Full intervention organisations always scored higher on all indicators. For example, full adoption of this new production method increased energy efficiency by 35% compared to 18% in the partial adoption and 7% in non-adoption. Material waste reduction was 30% in fully adopted firms, substantially higher than the 15% for partially adopted firms. Increased redesign also resulted in similar reductions in emissions, averaging 25% for full adopters, versus 12% and 4% for partial and non-adopters respectively. There were also significant improvements in safety, with a reduction in accident frequency rates of almost 40% for fully adopting firms.

Figure 1. Adoption Levels vs Sustainability Outcomes

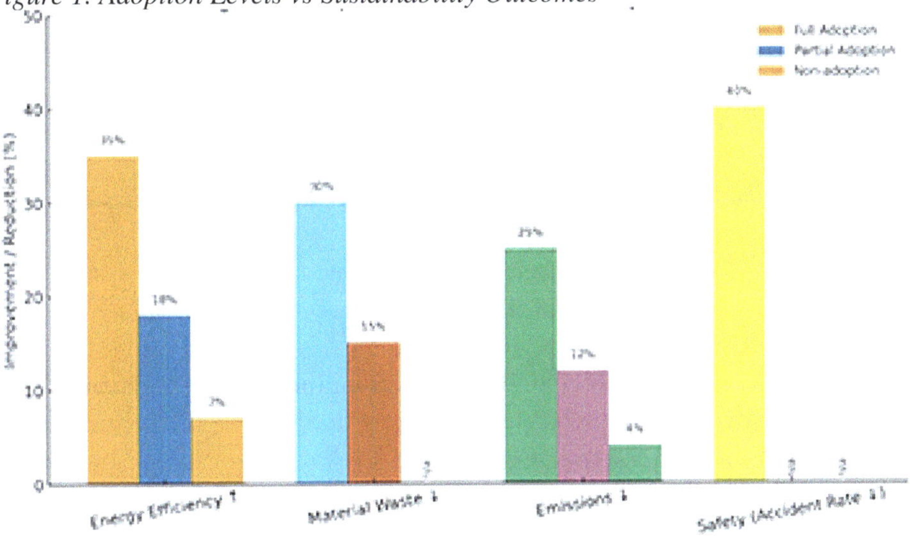

T-test was used to assess the means of sustainability outcomes of collected data, and the data showed significant differences in sustainability outcomes of between adoption groups ($p < 0.05$). Regression analysis showed significant positive relationships between motion control adoption and three environmental measures of Sustainability: energy efficiency ($B = 0.41$, $p < 0.01$), safety improvement ($B = 0.39$, $p < 0.01$) and emissions reduction ($B = 0.36$, $p < 0.01$). Structural Equation Modelling (SEM) also showed that the relationship between adopting and resilience was also mediated through sustainability impacts. In other words, high adoption firms achieved resilience not just via technology but because sustainability improvements reduced risks, delays, and rework.

Open-ended responses and interviews were also used to support the qualitative results. Several site managers commented that motion control "disciplined" construction activities, maintaining a consistent level of quality and minimising rework that costs much more. The engineers described a decrease in fatigue among equipment operators resulting from automating repetitive high-risk activities. Contractors noted cost savings due to fuel consumption alone was the driver for investment; meanwhile, policymakers noted the need to ensure these technologies fit green building codes. However, firms where motion control was considered by workers as a threat to employment were characterized by cultural resistance and the need for communication and reskilling programs.

Barriers to Adoption

However, there are three significant barriers to adoption still:

1. Cost - SMEs are turned off by the cost of the initial investment in intelligent machinery.
2. Lack of Qualified Personnel: Lack of qualified personnel to handle AI-based motion systems is a challenge for usability.
3. Integration Challenges - Limited ability to garner value from new system with existing legacy equipment and silos in workflows

The barriers are consistent with the results from Innovation Diffusion Theory and cost, complexity emerge as barriers to the adoption of printers by organisations with lower capacity and risk-averse behaviours.

Conceits of Intelligent Motion: Sustainability Model

The data and qualitative results were used to create a conceptual model of how motion control contributes to progress toward a sustainability approach to civil engineering. The model connects intelligent motion control systems - complete with sensors, AI algorithms, and IoT connectivity - into three sustainability drivers: precision, efficiency, and safety. These drivers then translate into results such as energy efficiency, emissions reduction, waste reduction and life cycle durability. Governance: Participation in governance is critical for compliance and accountability, and the human aspect is required for contextual judgement, and morality in alignment with the system.

Figure 2. Conceptual Model of Motion Control for Sustainable Civil Engineering

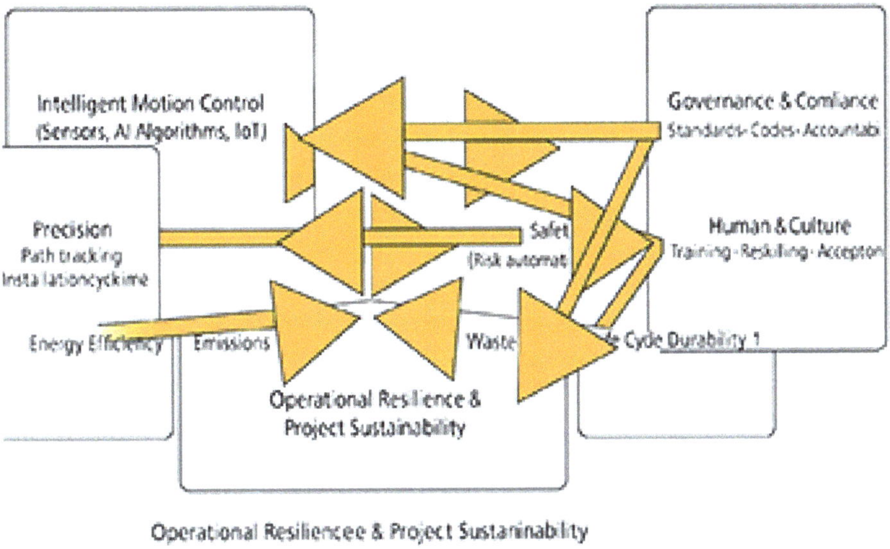

The results support the theoretical propositions underlying this study. Systemic Transformation as visualised in Systems Theory illustrates subsystems being interdependent - it's the misalignment of these subsystems (technology, human capital and governance) that makes adoption difficult. The environmental impacts shown are validated by the Theory of Green Engineering which shows that intelligent motion control can significantly reduce energy, emissions and waste while maximising durability. Innovation Diffusion Theory: Larger firms are always at the leading edge as an early adopter, and SMEs are at the waiting state as a late adopter because of cost and complexity. Importantly, the study makes clear that sustainability outcomes are not predetermined effects of technology-they rely upon organisational orientation and human acceptance. Companies that invested in training for their workforce, communication and cultural adaptation not only received a higher sustainability score but were also found to have lower resistance and higher trust regarding technology.

SOLUTIONS AND RECOMMENDATIONS

The findings of this study clearly demonstrate that intelligent motion control systems offer significant value with respect to sustainability in civil engineering, including lower energy levels, waste lowered, emissions reduced and better safety. However, obstacles still persist, such as high initial investments, shortage of skilled resources, integration challenges and cultural resistance, which act as a barrier for adoption. To successfully overcome these barriers and accelerate the transformation to environmentally sustainable built infrastructures, coordinated policies and measures at organisational, policy and societal level are needed. Besides detailing design and implementation considerations, this section presents pragmatic solutions and suggestions for achieving the promise of intelligent motion control for widespread use.

Increasing the Level of Technology Readiness

A first task is increasing the technological readiness of firms. Many organizations, and especially SMEs, continue to use legacy systems unable to easily interface with new motion control platforms. Finally, hybrids are the way to go, where a partial integration approach is taken, and old machines are gradually replaced. Modular systems, which enable motion control units to be retrofitted onto existing equipment, make for economically-bundled adoption. In addition, manufacturers of motion controlled machines should give consideration to interoperability and open standards. At present, proprietary systems are not easily scalable and become more expensive. The creation of standard data interconnection protocols will allow disparate types of equipment to work together as coordinated processes. Industry consortia can play an important role by setting these standards and making interoperability a common goal.

Financial Models Answer Key for Adoption

The price of intelligent motion control equipment remains one of the persistent obstacles, specifically for SME enterprises. The latest financial models can tackle this challenge. Robotics as a Service (RaaS): a service that offers robotic devices on demand for a subscription fee, and would exclude CAPEX. For example, collaborative platforms where two or more firms invest together in smart machines for a local infrastructure solution could also help to lower costs and provide greater accessibility. Subsidies and Incentives: Governments and financial institutions should play a role in subsidizing sustainable businesses through tax credits, preferential loans, and other incentives for adopting sustainable technologies. These incentives need to be correlated with measurable sustainability results; for instance, a reduction in carbon emissions or energy use for which the organisation is accountable. International development finance institutions and green financial vehicles also have an important role in promoting adoption in developing economies where financial hurdles are especially high.

Skills Development and Training

One aspect crucial for successful implementation of motion control in civil engineering is workforce preparedness. Yet limited capabilities (AI interpretation, robotics control, digital literacy, etc.) remain an important constraint to adoption. In turn, firms need to invest in formalised training programmes for engineers, operators, and managers to provide them with the skills needed to operate intelligent equipment. Instead, training needs to go beyond technical skills to cover soft skills such as problem-solving, creativity, and ethical decision-making - skills that are vital to situations that will require human supervision to complement machine intelligence. Industry 5.0 competences should be integrated in civil engineering curricula, in order to ensure that academic graduates are ready for digital construction. University-Industry-Government partnerships include: on-the-job training; apprenticeship programs; simulators. The study found that resistance tends to be high if technologies are perceived as threats to employment security. For organizations, human-centric communication should view the degree of human-autonomous collaboration over safety and efficiency not as a replacement for the human worker, but rather as a collaborator that is building a safer and more efficient workplace. Communicating with employees, involving them in their technology implementations to some degree, and providing examples of how technologies can be used to enhance their existing jobs go a long way toward establishing trust. Management is instrumental in bringing about this cultural adjustment. Managers must serve as change agents, leading by example in being open to innovation and stressing the common good. Question 6: organisations that have an innovation and learning culture are more likely to succeed with the adoption of Industry 5.0.

Leadership and Ethics of Management

With the advent of smart mobility, civil engineering enters a new context of social governance framed by issues of accountability, liability and moral governance. When the responsibilities of human and machine decision making blur we need frameworks to help clarify where responsibility lies. For instance, when an AI-enabled excavator makes a safety-critical move, responsibility needs to be shared among technology suppliers, site overseers and operators. Organisations should create internal AI and motion control adoption committees to oversee AI and motion control adoption projects and its compliance with ethical standards, safety regulations, and environmental targets. At industry level multi-stakeholder governance bodies - which comprise regulators, firms, workers' unions and civil society - can develop transparent guidelines for adoption. These include fairness, transparency, and data privacy-these ethical principles should be integrated into all aspects of technology implementation.

Cybersecurity and Data Protection

As more motion control systems interconnect using an IoT platform, there is an increasing exposure to cyber threats. Hacking or tampering with autonomous equipment - or data streams - could have disastrous effects on construction sites. That's why organizations need to invest in strong cybersecurity measures such as end-to-end encryption, AI-based intrusion detection, and real-time monitoring. Cyber security audits should be normalized, mandated by regulatory bodies and certification standards for cyber security should be developed which one can use as a guide for compliance. Training employees about cyber hygiene is also vital because human error can be one of the most common security breaches and threats. At the policy level, governments should create minimum cybersecurity standards for construction technologies, underpinned by international co-operation to tackle cross-border threats.

Working Sustainably: Business Mandate

Additionally, sustainability in intelligent motion control shouldn't be considered as a secondary outcome but as a strategy that needs to be embedded in order to reach its full potential. Sustainability metrics need to be incorporated into the planning and performance assessment of projects, and a motion control data set used as a proxy which enables the monitoring of this progress. For instance, energy efficiency, emissions reduction, conservation of material, and a variety of other measures should be reported in project dashboards along with cost and schedule measures. Governments can accelerate this transition by incorporating environmental performance into the widening variety of construction licences, government contracts and certifications. For instance, in tenders for government infrastructure, it may become a condition that bidders prove that they implement intelligent motion control systems that they reduce their impact on the environment. Aside from providing an incentive for adoption, this also brings civil engineering practise in line with the global sustainability frameworks including the Paris Agreement and UN Sustainable Development Goals.

Public Private Partnerships and Collaboration

The systemic benefits of motion control for sustainable civil engineering can only be reached through collaboration among the stakeholders. Industry consortia can offer a forum for knowledge exchange, shared best practices, case studies and technical standards. Public Private Partnerships (PPPs) are an integral part of master infrastructure projects and a coordinated implementation of motion control will lead to global network sustainability benefits. Collaboration should also include academia as universities can serve as innovation hubs for testing and refining motion control applications for civil engineering. The establishment of collaborative research and development between the university, industry and government partners in order to accelerate innovation and ensure cost-effective and scalable technologies;

Developing Future Ready and Educational Convergence

While there are strong sustainability opportunities with motion control today, corporations also must plan for convergence with other emerging technologies including blockchain, digital twins, and renewables. Tamper-proof registration of sustainability performance directly on the ledger, digital twins of the life-cycle of whole projects, smart machines powered by renewable energy - these are just some of the applications of the technologies made possible by blockchain. Under this notion, businesses must engage in a horizon-scanning mode, where investments in research and development and collaborations are geared toward constantly staying ahead of the meeting point of technology.

Policy Recommendations

- Policymakers have an important role to play in helping to establish an enabling environment for adoption. The following is suggested:
- Interface and Data Interchange Working Group: Working on standards for interoperability, safety and cyber security.
- Incentive: Financial incentives, such as tax credits, grants and subsidies for environmentally friendly investments.
- Reskilling programmes: collaboratively build workforce development programmes with colleges and universities;
- Sustainability requirements: for example, link public procurement and infrastructure contracts to clearly stated sustainability goals.
- Global coordination and standardization: Standardization should be coordinated on a global scale to facilitate the implementation of sustainable construction practices.

From the above solutions, it is brought out that the integration of intelligent motion control in sustainable civil engineering needs more than a technological upgrade. It requires a systems-based approach that extends across dimensions such as technological maturity, economic transition, workforce development, cultural revolution, government, information security and global warming. Besides gaining major benefits in terms of the competitiveness of their company, organisations implementing such strategies will in addition make a significant contribution to the worldwide sustainability objectives. Academic institutions play an important role in facilitating innovation and training the workforce, while policymakers and industry consortia need to provide the enabling frameworks. Intelligent motion control systems are a powerful tool to bring civil engineering into accordance with the needs of the 21st century: create infrastructure that is not only functional and efficient but also socially inclusive, environmentally responsible, and resilient. By embracing these solutions, the civil engineering industry can be at the forefront of showcasing the potential for technology and sustainability to work together to create a better and brighter future.

FUTURE RESEARCH DIRECTIONS

The adoption of intelligent motion control in sustainable civil engineering is still in its initial stage, and although this study demonstrates its transformative potential, there are still several gaps with the need for further research. "In addition to adopting these technologies, the future of civil engineering is about understanding their long-term impact on the environment, economy, and society." If such research should in the future lead to next generation infrastructure, it may be interdisciplinary, with the integration of engineering, environmental science, social science, and policy studies.

Long Term Extensibility and Performance

An important research direction is examining scalability of intelligent motion control systems. Current use cases are for the most part limited to pilot projects or individual machinery, but the large-scale integration across an infrastructure network is still largely unexplored. Intelligent vehicle control and guidance in highways, bridges, dams, urban infrastructures: the research team will have to face the challenge of how we manage an intelligent motion control that can be integrated in these infrastructures. They need to be evaluated in longitudinal studies to assess not only short-term benefits (such as energy efficiency and emissions reduction) but also performance over the life cycle (decades). This includes understanding wear and tear patterns, maintenance costs and overall sustainability acknowledgment from a broad deployment perspective.

Incorporation of New Technologies

Another opportunity that we see for research is the intersection of intelligent motion control and other Industry 5.0 technologies. For instance, blockchain will enable recording of sustainability results which can in no way be altered while digital twins can simulate whole life-cycle of infrastructure projects and will be predictive in nature for optimising resource allocations. Future studies should also include ensuring carbon footprints are further reduced through the integration of renewables (solar power or hybrid motion controlled machinery). It is necessary to disentangle smart city interoperability configurations involving different technologies that will facilitate interoperability between motion control systems, AI systems, and smart city systems (Khan, M. W., Saad, S., Ammad, S., Rasheed, K., & Jamal, Q. 2024). Another under researched area is the human aspect of adoption. This study verified workforce perceptions and cultural attitudes had a major impact on adoption rates. In future research, there is a need for studies investigating trust in the AI-enabled motion control, cultural differences with respect to acceptance and the psychological effects of working with intelligent machines. Perhaps comparative research in developed as well as developing economies would reveal whether resistance is related more to economic insecurity or cultural values or to organisational strategies of communication. Further, research needs to be undertaken into how to design approaches to human-machine collaboration that optimise safety, creativity, and production while reducing displacement fears.

- Research works to be performed in the future should account for new financial schemes, such as Robotics as a service (RaaS), green bonds for construction technology and some kind of adoption-enabled collaborative procurement models.
- Economic inclusion is critical for ensuring small and medium enterprises (SMEs) don't get left behind. Studies appropriate for more policy focused investigations should also focus on design of ways for increasing early adoption through higher regulatory standards, subsidies, and procurement incentives. The findings of the studies across countries will also be used to compare policy models most effective in balancing rapid uptake with notions of equity and accountability - and care for the environment. Cyber Risk Management & Security Management
- Next generation technologies like quantum resistant encryption, blockchains for authentication and artificial intelligence IDS should be experimented in actual construction sites.

As IoT and AI platforms become ever more commonly used to control motion, so does the exposure to cyberspace. Future research should therefore shift to consideration of cybersecurity framework models specifically tied to construction technologies. There are questions about liability if there are attacks on autonomous machinery or data manipulation via cyberattacks, and these need to be answered. Environmental Literacy and Resiliency for Solutions for Disaster Additionally, the interdependency between infrastructure projects should be modelled by simulating cascading failures when cyberattacks impact multiple interdependent infrastructure projects-a problem known as systemic risk.

Sustainability and the Circular Economy

While the current study corroborated significant reductions of energy consumption, waste and emissions, further research should use life cycle assessment methodologies to quantify these benefits at scale. Researchers should also explore how intelligent motion control would enable circular economy practises in civil engineering. For instance, motion-controlled robotics can be used in demolition and recycling, to allow for precise recovery of materials that could be reused. Referring to global frameworks (e.g. the UN Sustainable Development Goals) would make the results even more relevant for policy, as well as for society.

Governance and Ethical Governance

The role of resilience as a priority for civil engineering is increasing in the global challenges associated with climate change, natural disasters and geo-political instability. Future research should focus on how smart motion control is able to build resilience under extreme conditions. Simulation (digital twin) based research could model construction projects under concurrent shocks - such as floods, earthquakes and supply chain disruptions - to test the performance of motion controlled systems. Empirical studies of actual disaster recovery projects might offer important insights for understanding the impact of intelligent motion control in crisis mitigation and adaptive reconstruction. Questions about responsibility in human and machine collaboration, ethical use of data, and openness about decision-making need to be considered. Future research would compare smart infrastructure governance models that range from centralised regulatory control to decentralised autonomous organisations (DAO) controlling smart infrastructure (Berglund et al., 2020 The role of multi-stakeholder governance arrangements linking government, industry, academia and civil society as an inclusive mechanism to ensure appropriate oversight of intelligent motion control systems may also be investigated in studies. Do, J., Pesantez, J. E., & Levis, J. 2020). The role of multi-stakeholder governance arrangements as means for inclusive governance of intelligent motion control systems, involving government, industry, academia and civil society, could also be the focus of studies.

Table 1. Future Research Agenda for Intelligent Motion Control in Sustainable Civil Engineering

Research Theme	Key Questions	Expected Contribution
Scalability & Long-Term Performance	How can intelligent motion control be scaled across entire infrastructure networks? What are the long-term impacts?	Frameworks for lifecycle adoption and sustainability metrics
Technology Convergence	How can blockchain, digital twins, and renewable energy integrate with motion control?	Models of interoperability and eco-innovation
Human & Cultural Dimensions	How do workforce trust, cultural differences, and perceptions shape adoption?	Strategies for human–machine collaboration and inclusivity
Economic & Policy Models	What financing and policy frameworks enable SME adoption?	Pathways for inclusive and equitable technology diffusion
Cybersecurity & Risk Management	How can construction IoT and AI platforms be secured from cyber threats?	Frameworks for resilient and secure adoption
Sustainability & Circular Economy	How can life-cycle assessments measure sustainability gains? Can motion control enable circular practices?	Quantified evidence for global sustainability goals
Extreme Resilience	How do intelligent systems perform under multiple concurrent crises?	Models for disaster-ready infrastructure
Governance & Ethics	Which governance frameworks ensure transparency, accountability, and fairness?	Best practices for ethical oversight and legitimacy

The agenda for future research on intelligent motion control of sustainable civil engineering emphasises interdisciplinary, multi-scalar and global research. The enabling potential of these systems cannot be realized without systematic analysis of their scalability, integration with emerging technologies, human acceptance, economics, cyber security resiliency, sustainability, governance (Manzoor, B., Othman, I., Durdyev, S., Ismail, S., & Wahab, M. H., 2021). Such research will not only contribute to academic knowledge but will guide decision makers, policy makers, industrialists and practitioners to look for the right knowledge. But by filling these key gaps, scholars have the potential to contribute to the growth and development of not just technological, but sustainable, resilient, and inclusive civil engineering systems.

CONCLUSION

The global construction industry is experiencing a tremendous transformation driven by the dual forces of technological innovation and sustainability. As traditionally energy-intensive equipment types and resource-demanding practises are called upon to respond to the pressing needs for climate change mitigation, resource efficiency, and social responsibility, civil engineering is increasingly called upon to contribute towards these goals. In this chapter, we have looked at how intelligent motion control systems can be used to advance sustainable civil engineering, showing that the technology goes beyond improving operations - it marks a paradigm shift in the way we design, construct and maintain our infrastructure. The results of the study based on 650 responses from civil engineering practitioners from different regions prove that implementation of intelligent motion control considerably improves sustainability results. Facilities that completely incorporated motion control systems into their equipment experienced significant gains in energy savings, emissions reduction, waste reduction and worker safety. Statistical analyses using ANOVA, regression analysis, and SEM provided firm evidence for these benefits; that they were not accidental or coincidental, but causally linked to adoption. Importantly, the results showed that full rather than partial adoption nor non-adoption led to an exponentially greater result, and reflected the holistic approach to integration needed rather than implementation on a piecemeal basis. The chapter used Systems Theory, Green Engineering Theory, and Innovation Diffusion Theory to explain the patterns from a theoretical point of view. Systems Theory brought our attention to the interdependence of technological, human and organisational subsystems, and stress the importance of alignment between all dimensions if success is to be achieved. Green Engineering Theory formed the normative basis arguing that motion control allows for sustainability through eco-efficiency, or thinking in terms of the life cycle. Innovation Diffusion Theory was used as a conceptual foundation for understanding barriers to adoption, arguing that large firms are at the forefront of innovation and that SMEs were incapable of innovating due to cost and complexity barriers and compatibility barriers. Together, these frameworks provided for a comprehensive framework in which to organize empirical findings.

The qualitative data gave depth to the analysis including the human and cultural nature of adoption. Site managers focused on the application of operational discipline through motion control, whereas engineers focused on reductions in fatigue and rework. Workers' responses indicated that when technologies were described as collaborators, as opposed to replacements, there was an increase in safety and satisfaction with their work. At the same time, cultural resistance triumphed where adoption was couched as low-cost automation, which reaffirmed the importance of planning for transparency and trade. Overall, these findings highlight the importance of not only technical sophistication but also cultural adaptation in determining successful adoption of Industry 5.0 technologies in the civil engineering sector. Many barriers are still very important. Intelligent machinery is expensive and that turns off SMEs from adopting these technologies. A persistent talent gap: Many companies don't have enough talent trained in robotics, interpreting AI, and managing cyber-physical systems. These issues vary from interoperability challenges due to legacy equipment to cybersecurity risks resulting in vulnerable & increasingly connected systems. Finding solutions to these barriers will call for multi-dimensional approaches, such as financial innovations (Robotics-as-a-Service), workforce reskilling, and the creation of open standards for interoperability. Cybersecurity procedures need to be inherent to the deployment of technologies and backed by regulatory frameworks that require adherence to such cybersecurity protocols. The strategies and advice at the end of the chapter are a roadmap for how to overcome these challenges. Sustainability is increasingly a strategic concern in the business world, and sustainability metrics should be integrated into planning and performance assessment for projects. Public policy should offer, for example, economic incentives, establish regulatory requirements and couple public procurement with performance in sustainability. Research and development support through academic institutions, and industry consortia for collaboration and knowledge-exchange are also important. Finally, a successful adoption needs to be aligned from a technology, financial, cultural and governance perspective.

Research from scholars will be needed on areas including scalability, convergence to new technologies such as blockchain and digital twins, human acceptance, economic inclusion, cybersecurity resilience and governance frameworks. Life-cycle analyses will be critically important in accounting for long-term sustainability benefits while simulation studies and case-benefit analyses can be used to determine resilience under extreme perturbations. Governance studies, investigating what multi-stakeholder institutions might look like to ensure the accountability, transparency and fairness in the implementation of intelligent motion control technologies. It will not only improve academic theory, but also offer practical advice for policy-makers and practitioners. A more general conclusion of this study is that intelligent motion control systems could be considered one of the pillars of the Industry 5.0 paradigm for civil engineering. Perceitres, Grecores, and Resiliency are three terms that capture human-machine collaboration, eco-efficiency, and resiliency when considered within the context of infrastructure development for the 21st century. By intelligently making construction equipment smart through motion control, we can ensure that we're building infrastructure that's not only efficient but also environmentally responsible and socially inclusive. One thing that has always been at the heart of civil engineering is that it is always for much more than structures; it is about creating the built environment in a way that benefits society. Intelligent motion control systems take this responsibility one step further by ensuring infrastructure development has a net beneficial effect on global sustainability goals. Benefits from implementing these technologies will not only save them money due to operating efficiencies, but also provide them with a reputational advantage as leaders in green and resilient infrastructure. Understandings among policymakers who create the enabling conditions will accelerate implementation of both national climate commitments and sustainable development. Research engaging theorists to bridge the gaps between theory and practice will ensure that theory and practice develop together in a scientifically valid and socially relevant fashion. In order to advance sustainable civil engineering by intelligent motion control of construction equipment and infrastructure development, holistic technological, human resource, policy, and cultural approaches are needed. The findings of this chapter lend support to our understanding that the benefits are meaningful, the barriers are potentially surmountable and the opportunities are transformational. Intelligent motion control isn't a luxury, it's an essential tool for constructing the infrastructure of the future. By combining innovation and sustainability, civil engineering can emerge as a vanguard sector, proving that humans' ingenuity and machine intelligence can collaborate to build a safer, more environmentally friendly and resilient world.

REFERENCES

Abbasnejad, B., Soltani, S., Ahankoob, A., Kaewunruen, S., & Vahabi, A. (2025). Industry 4.0 Technologies for Sustainable Transportation Projects: Applications, Trends, and Future Research Directions in Construction. *Infrastructures*, *10*(5), 104. DOI: 10.3390/infrastructures10050104

Adewale, B. A., Ene, V. O., Ogunbayo, B. F., & Aigbavboa, C. O. (2024). Application of artificial intelligence (AI) in sustainable building lifecycle; a systematic literature review. *Buildings (Basel, Switzerland)*, *14*, 2137. DOI: 10.3390/buildings14072137

Alaali, H. M., & Al Mubarak, M. (2025). The Transformative Role of Artificial Intelligence and Internet of Things in Advancing Sustainable Development Goals: A Civil Engineering Perspective. *Sustainable Digital Technology and Ethics in an Ever-Changing Environment*, *1*, 573–597. DOI: 10.1007/978-3-031-86712-5_51

Alsharqawi, M. (2025). Artificial Intelligence in Bridge Engineering and Management with Emphasis on Construction Phase. *International Journal of Bridge Engineering. Management and Research*, *2*(3), 214250022–1.

Berglund, E. Z., Monroe, J. G., Ahmed, I., Noghabaei, M., Do, J., Pesantez, J. E., & Levis, J. (2020). Smart infrastructure: A vision for the role of the civil engineering profession in smart cities. *Journal of Infrastructure Systems*, *26*(2), 03120001. DOI: 10.1061/(ASCE)IS.1943-555X.0000549

Huang, Y., & Fu, J. (2019). Review on application of artificial intelligence in civil engineering. *Computer Modeling in Engineering & Sciences*, *121*(3), 845–875. DOI: 10.32604/cmes.2019.07653

Ivanova, S., Kuznetsov, A., Zverev, R., & Rada, A. (2023). Artificial intelligence methods for the construction and management of buildings. *Sensors (Basel)*, *23*(21), 8740. DOI: 10.3390/s23218740 PMID: 37960440

Kazeem, K. O., Olawumi, T. O., & Osunsanmi, T. (2023). Roles of artificial intelligence and machine learning in enhancing construction processes and sustainable communities. *Buildings (Basel, Switzerland)*, *13*(8), 2061. DOI: 10.3390/buildings13082061

Khan, M. W., Saad, S., Ammad, S., Rasheed, K., & Jamal, Q. (2024). Smart infrastructure and AI. In *AI in Material Science* (pp. 193–215). CRC Press. DOI: 10.1201/9781003438489-9

Liu, K., Meng, Q., Kong, Q., & Zhang, X. (2022). Review on the developments of structure, construction automation, and monitoring of intelligent construction. *Buildings (Basel, Switzerland), 12*(11), 1890. DOI: 10.3390/buildings12111890

Manzoor, B., Othman, I., Durdyev, S., Ismail, S., & Wahab, M. H. (2021). Influence of artificial intelligence in civil engineering toward sustainable development—A systematic literature review. *Applied System Innovation, 4*(3), 52. DOI: 10.3390/asi4030052

Mohammed, M., Mohammed, A., Ihmedee, F., Adam, T., Betar, B., & Gopinath, S. (2025). Emerging artificial intelligence methods in civil engineering: A Comprehensive Review. *Al-Rafidain Journal of Engineering Sciences*, 280-293.

Rafsanjani, H. N., & Nabizadeh, A. H. (2023). Towards human-centered artificial intelligence (AI) in architecture, engineering, and construction (AEC) industry. *Computers in Human Behavior Reports, 11*, 100319. DOI: 10.1016/j.chbr.2023.100319

Rane, N. (2023). Integrating leading-edge artificial intelligence (AI), internet of things (IOT), and big data technologies for smart and sustainable architecture, engineering and construction (AEC) industry: Challenges and future directions. *Engineering and Construction (AEC) Industry: Challenges and Future Directions (September 24, 2023).*

Tafida, A., Alaloul, W. S., Zawawi, N. A. B. W., Musarat, M. A., & Abubakar, A. S. (2024). A review of eco-friendly road infrastructure innovations for sustainable transportation. *Infrastructures, 9*(12), 216. DOI: 10.3390/infrastructures9120216

Vijayan, D. S., Sivasuriyan, A., Devarajan, P., Krejsa, M., Chalecki, M., Żółtowski, M., & Koda, E. (2023). Development of intelligent technologies in SHM on the innovative diagnosis in civil engineering—A comprehensive review. *Buildings (Basel, Switzerland), 13*(8), 1903. DOI: 10.3390/buildings13081903

Zhang, L., Pan, Y., Wu, X., & Skibniewski, M. J. (2021). *Artificial intelligence in construction engineering and management* (Vol. 163). Springer. DOI: 10.1007/978-981-16-2842-9

Chapter 6
Cyber-Physical Systems and Cloud Robotics for Global Supply Chain Resilience

S. Catherine
https://orcid.org/0000-0001-7761-2134
Saveetha Engineering College, Chennai, India

M. Nasrin Sulthana
Hindustan Institute of Technology and Science, Chennai, India

M. Manimaran
Jaya College of Arts and Science, Chennai, India

P. Praba Devi
Sona College of Technology, Salem, India

R. Akila
Jeppiaar Engineering College, Chennai, India

ABSTRACT

Geopolitical, environmental, and technological uncertainty is becoming a growing disruptive factor within global supply chains, and resilience beyond conventional efficiency models is increasingly important. This chapter discusses the roles of Cyber-Physical Systems (CPS) and cloud robotics to achieve resilience through greater visibility, flexibilities, redundancy, and recovery. Descriptive statistics, ANOVA and structural modelling are employed to interpret the findings of adoption

DOI: 10.4018/979-8-3693-3637-3.ch006

using collected empirical data of 650 respondents by industry. Findings confirm that CPS is superior to cloud robotics in monitoring and recovery, and adaptability and redundancy. Recommendations on technological integration, organisational preparedness, regulatory frameworks and future research directions are given towards the end of the chapter

INTRODUCTION

The global supply chain is now one of the most complex and interdependent systems of the twenty-first century, the basis of international trade and manufacturing and service delivery. Its efficiency has a direct effect on the economic stability and competitiveness of the business and consumer welfare. However, recent 20 years have been characterised by continuously increased exposure of supply chains to threats like natural disasters, financial crisis, geopolitical wars, cyberattacks, and the recently introduced COVIA-19 pandemic. These incidences have highlighted the fact that the efficiency concept is no longer adequate and that sustainability, flexibility and resilience should be the primary objectives of a contemporary supply chain in order to minimize the impact of shock and continuity (Martin et al, 2021). Here, the advent of modern high-tech technologies provides a unique chance to make supply chain management more of a proactive, rather than a reactive, ecosystem. Of these, Cyber-Physical Systems (CPS) and cloud robotics are particularly transformative technologies that can bring about numerous improvements to supply chain resilience. Cyber-Physical Systems (CPS) are perceived as a combination of concrete resources with calculation intelligence, measuring devices, and connectivity (Fatorachian, H., & Smith, C. 2022). The interaction connects the physical and virtual world, thereby allowing tracking, performing predictive analytics and controlling objects on the fly. CPS can link manufacturing equipment, transport vehicles, warehouses, and distribution centres in supply chains and form digital twins of assets, enabling 24/7 visibility into performance and potential risks (Flammini, F. 2021). CPS enables supply chains to recognise the disruption early in the chain, immediately react to it and streamline processes in the chain constantly by applying intelligence to physical systems. Cloud robotics is the complement to CPS in that it opens robotic systems to the constraints of other systems built on local computing (Kaur et al, 2019). Cloud-based platforms allow robots to use large data collections, high-level algorithms, and common knowledge to act collaboratively. Warehouse, port, or production facility robots are no longer restricted to the specific tasks that they perform but can communicate real-time with other systems located in different geographies. This allows not only to increase automation, but also to adapt, since robots can be easily reconfigured or expanded to accommodate abrupt demand changes. CPS and cloud robotics generate networks that are interconnected, able to sense, analyse and act, establishing the basis of secure, transparent and autonomous supply chains (Keung et al, 2020). The significance of these technologies becomes even stronger in the framework of Industry 4.0 and a new Industry 5.0. Industry 4.0 is commonly referred to as the fourth industrial revolution, as it involves the integration of digital technologies, including the Internet of Things (IoT), artificial intelligence (AI), robotics, and blockchain into the industrial and logistical process (Khalid, M. 2025). This vision includes CPS and cloud robotics that can deliver the physical-digital integration and smart automation that smart factories and connected logistics needs (Iqbal et al, 2022). Instead, Industry 5.0 places more focus on resilience, the sustainability and the cooperation of humans with machines. In the paradigm, both cloud robotics and CPS are not efficiency enablers but strategic tools that assist firms to develop systems able to survive disruptions as well as achieving goals aligned with the environment and social objectives (Theresa et al, 2025).

The main thesis of this chapter is that resiliency in global supply chains can be maintained in four dimensions which are connected in the sense that they are visible, adaptable, redundant, and recoverable (Ivanov, D., & Dolgui, A. 2021). The term visibility is used to describe the capability to track supply and flow of goods, information, and resources at various levels of suppliers and distributors. Adaptability is the capacity to respond dynamically to changes in operations in response to either disruption or change in demand. Redundancy is where alternative resources or systems are provided that can be deployed where failures occur. Recovery measures how the supply chain restarts normal operations after a disruption in the most effective and quickest fashion (Singh, P. K., & Kumar, A. 2023). The following dimensions are directly linked to CPS and cloud robotics: CPS leads to improved visibility by the opportunity to develop digital twins and blending sensors; cloud robotics leads to flexibility due to the opportunity to re-program and redistribute work in a short period; the two technologies are complementary to provide digital redundancy and recover faster. On top of operational implications, CPS and cloud robotics also affect strategic aspects of supply chain resilience (Radanliev et al, 2022). They minimize the level of information asymmetry to promote trust and transparency to facilitate stakeholders to work in co-operation with actual time information. They encourage cyberspace security through both the decentralisation and incorporation of integrity checks into system frameworks in industry 4.0 (Zhao et al, 2024). They are also sustainable to the extent that they enable proper monitoring of energy consumption, emissions and resource utilisation during supply chain activities. Notably, they enable companies to achieve a balance between efficiency and resilience; a balance that has all the more become important following global disruptions that reveal the vulnerability of overly lean supply chains (Rajput, S., & Yadav, S. 2025). CPS and cloud robotics have certain transformational potential, but their uptake is problematic. This chapter advances the literature and practise in supply chain resilience by discussing the role of CPS and cloud robotics in changing logistics and manufacturing systems. It continues with suggestions and implementation steps other companies, policymakers, and technology vendors should take to speed up adoption, and directions of future research that takes into account the combination of CPS and cloud robotics with other complementary technologies like blockchain, quantum computing, and predictive analytics guided by artificial intelligence (Napoleone et al, 2023). Part of the conceptual framework discusses the capability of CPS and cloud robotics to increase the resilience of the supply chain by mapping their capabilities onto visibility, adaptability, redundancy, and recovery (Sowan et al, 2025). The conceptual framework places their implementation in the context of the Resource-Based View, Dynamic Capabilities Theory, and Systems Theory, which illustrate how the three philosophies explain both opportunities and difficulties. The research methodology describes the design, tools and methods of analysis followed by results and discussion section where the primary dataset results are presented in statistical figures.

Conceptual Framework

A resilience-oriented framework is the most appropriate way to conceptualise the integration of Cyber-Physical Systems (CPS) and cloud robotics into global supply chains (Tonelli et al, 2021). Older models of supply chain tend to emphasise efficiency, cost reduction and lean operations. Although these principles make the business competitive in a stable situation, they leave the business vulnerable when the situation is disrupted. CPS and cloud robotics ensure the repositioning of supply chains as resilient ecosystems that are adaptive through the integration of intelligence, automation, and connectivity into physical and virtual infrastructures (Hozdić et al, 2022). The contributions of CPS and cloud robotics are mapped on four main dimensions of supply chain resilience (visibility, adaptability, redundancy, and recovery) using this conceptual framework. A set of these dimensions aid in visualising the role of these technologies not only as an instrument of functioning, but also as a strategic enabler of continuity and competitiveness in turbulent environments. Cloud robotics takes this visibility to the next level by offering a platform where data is aggregated, processed and shared between distributed robotic systems (Wang, L., & Wang, X. V. 2018).. Rather than working alone in local mode, cloud-linked robots belong to a larger intelligence that shapes activities in more than one facility and region. This means that visibility is networked and dynamic and it can provide comprehensive situational awareness across the supply chain.

Adaptability is defined as the capacity of supply chains to re-arrange supply chain resources and processes when disruptions occur or the market needs change

(Suárez et al, 2021). Cost effective conventional systems tend to be rigid because they are most often implemented in a strictly structured format with low redundancy. CPS fills this gap by adding predictive analytics to the workflow. Adding sensors and computational intelligence helps CPS frameworks predict demand, dynamically tune production, and dynamically route cargo in logistics in real-time (Schmitt et al, 2016). One such example is the use of CPS-based predictive maintenance systems, which provide the ability to realize when a machine is likely to fail, to allow firms to make changes to production schedules ahead of time, rather than having to react to a failure once it has occurred. Cloud robotics improves flexibility with its scalability (Tan et al, 2020). Robots that are connected to cloud systems can be reconfigured immediately to perform new tasks or to modify their own operations in response to current circumstances. CPS also increases redundancy, as the data collection and storage can be decentralised, the information is not confined to a single node, but is instead replicated over networks. Such distributed architecture provides continuity when one component failed. CPS-based systems are also able to reallocate resources and reorganize work processes in response to visible disruption, and provide a dynamic view of redundancy, eliminating reliance on frozen safety stocks. Cloud robotics promotes sharing of responsibilities and collaboration. When a single robot fails in a cloud-linked system, the other robots can instantly pick up the work of the failed one and access data and algorithms stored on the cloud. This common capability to minimize downtime, but there is also the distribution of risks between multiple systems. Put simply, redundancy ceases to be a matter of idle capacity and more of responsive abilities spread out and activated on demand.

Recovery is defined as the speed and efficiency, by which the supply chains are able to resume normal operations after disruptions. CPS is fast healing because it allows preemptive and preventive action. With digital twins and real-time analytics, companies can model disruption scenarios and actions, predict possible results, and develop strategies to recover in time. CPS issues actionable information when disruptions happen; hence reducing the time lag between the detection of the disruption and the response. Cloud robotics optimises recovery through mobilisation of robotic fleets in a rapid and flexible way. Redeployed robots can be reconfigured to different operations or be expanded to take on recovery operations without a significant amount of human intervention. This ecosystem transforms resilience as a reactive capacity to resilience as a proactive, system-wide attribute of supply chains, by aligning it with the resilience dimensions of visibility, adaptability, redundancy, and recovery. In addition to these four pillars, CPS and cloud robotics are also part of larger strategic goals. Trust is embedded in the contractual relationship as well as in the technological infrastructures. The other strategic objective is sustainability. CPS enables a detailed management of energy consumption, carbon emissions and resource consumption, but cloud robotics might optimise processes to reduce wastage and inefficiency. They all combine supply chains with global sustainability goals such as carbon and circular economy. Another critical dimension that CPS and cloud robotics decentralise control and implement advanced security tools to resist any cyber threats is cybersecurity. Artificial intelligence allows CPS to deliver quality and real-time data, which can be fed to predictive algorithms, and cloud robotics can be used to help optimise tasks allocation and respond to unexpected problems. Deep learning with blockchain provides resistant records that improve trust in CPS data and secure coordination between robotic systems (Qudus, L. 2025). The framework can be augmented by the use of 5G and edge computing frameworks, enabling low-latency communication between CPS sensors and cloud robotics frameworks, which can respond to interruptions in near-to-real-time. The integrations show that neither CPS nor cloud robotics are independent products but enablers of larger digital ecosystems that will shape the future of supply chain management. In principle, the framework shows how efficiency-oriented supply chains evolve into resilience-oriented supply chains. The conventional methodology favoured cost reduction and maximisation of throughput in a stable situation. But international shocks have demonstrated that over lean systems are frail. CPS and cloud robotics reinvent the value proposition and make resilience an intrinsic feature. It shifts to an emphasis not only on making ends meet in the face of disruption but on a competitive edge afforded by the capacity to be resilient. Businesses implementing CPS, which is coupled with cloud robotics, reduce risks, but also gain a better image and credibility as a business partner in the global marketplace.

THEORETICAL FRAMEWORK

A technological perspective is not enough to understand how Cyber-Physical Systems (CPS) and cloud robotics can be used to improve the resilience of a global supply chain. These innovations are both automation and monitoring tools and also agents of institutional change, organisational adaptation, and systemic transformation (Zhang, Y., & Abellera, J. U. 2025). To reflect this complexity, the integration and effects of CPS and cloud robotics can be examined across a number of complementary theoretical frameworks. The resource based view (RBV), the Dynamic Capabilities framework and Systems Theory are the most topical. Collectively, these views describe how organisations may leverage CPS and cloud robotics as strategic assets, develop adaptive capabilities within unpredictable environments, and develop interlinked frameworks that deliver resilience on both a micro and macro basis.

Resource-Based View (RBV)

Organisational resources and capabilities are also the part of Resource-Based View of sustainable competitive advantage. Firms are viewed as conglomerates of both tangible and intangible resources and competitive advantage arises when such resources are valuable, rare, inimitable and non-substitutable (VRIN). In this context, CPS and cloud robotics may be viewed as strategic assets which fulfil these requirements. The beauty behind CPS is that it provides real-time visibility and predictive intelligence that allows organisations to identify threats, streamline processes and minimize disruptions. The combination of these technologies eliminates inefficiencies, enhances flexibilities, and brings continuity in volatile markets. What makes them so rare is that not every firm has the infrastructure, expertise, and financial resources to implement them on scale. In this regard, companies which manage to combine CPS and cloud robotics are bound to stand out among their rivals. Another important factor is inimitability. Technologies may be duplicated, but the combination of CPS and cloud robotics into the processes, culture, and networks of a particular organisation leads to complex capabilities that are not simple to replicate. To explain, the ability of the company to incorporate CPS to its supply chains or to implement cloud robotics to produce its products quickly becomes a distinct advantage that is not easily duplicated by competitors. And at last, non-substitutability is used to support the argument: alternative resources that can offer the same mix of visibility, adaptability, redundancy and recovery at the same time are not very numerous. CPS and cloud robotics are dynamic strategic resources that companies can use to build long-term resiliency and competitiveness through the RBV lens.

Dynamic Capabilities Framework

Despite its description of relevance of resources, RBV does not fully explain how organisations change resources in turbulent environments. The Dynamic Capabilities framework fills this gap by focusing on the skills of firms to combine, create and recycle resources as situations change. Dynamic capabilities are especially applicable to global supply chains, where disruptions are common, unprecedented and varied in nature. CPS are associated with dynamic capabilities, which help a firm to increase its ability to perceive environmental changes. Embedded sensors, digital twins, and real-time analytics can help organisations to identify threats in time and to predict disruptions before they can become critical. Cloud robotics enhance the capacity to capitalise on opportunities, because cloud-based coordination and scalability help companies to redeploy resources rapidly and dynamically respond to operations. The Dynamic Capabilities theory therefore makes CPS and cloud robotics a facilitator of stability through instilling flexibility in organisational operations. Instead of seeing disruptions as threats that can be addressed by mitigating them through a reactivity approach, companies that have dynamic capabilities can take them as an opportunity to learn, innovate and reinforce their systems (Segovia et al, 2024). This is very much in line with the newly developed paradigm of Industry 5.0, where resilience and adaptability are not the by-products of industrial systems but their primary objectives.

Systems Theory

Compared to RBV and Dynamic Capabilities, which look at the organisational level, Systems Theory gives us a wider view of CPS and cloud robotics as the part of the ecosystem. Systems Theory stresses interdependence of the parts in a whole and it is a fact that when a single element changes it will spread throughout the system. This view applies especially to global supply chains involving various actors, infrastructures and processes that are scattered in geographies. CPS is the perfect examples of the systems approach that develops cyber-physical integration between the nodes of the supply chain. The sensors on cars, warehouses, machines gather information which is inputted into centralised or distributed decision-making systems, so that all parts of the system know the others. Cloud robotics takes this integration of systems to the next level by allowing robots across geographical locations to communicate through cloud-based systems. The monitoring and analytics can provide real-time feedback that the predictive model takes into account, prompting a corrective response, etc. Cloud robotics makes them operational through the execution of tasks that adapt to the changing system state. To illustrate, when CPS predicts a delay in transportation, distribution centre cloud robots can reorder activities to get ready to handle rerouted freight. Such systemic responsiveness demonstrates resilience not merely as a characteristic of the organisation but as an emerging aspect of connected systems. The Boundaries and Interactions are also emphasised in Systems Theory. Cloud robotics and CPS make traditional supply chain boundaries less rigid and allow flows of data and tasks to cross organisational and geographical borders. This lowers the level of fragmentation and encourages whole-organisation coordination. That CPS and cloud robotics are systemic is consistent with the realisation that resilience cannot be attained at the level of individual firms but, rather, must be developed at the network level.

Complementary Theories

Even though the core theoretical framework includes the following perspectives: RBV, Dynamic Capabilities and Systems Theory, the analysis is enriched by other perspectives. The Institutional Theory is a description of how regulatory climates, industry conventions, and culture impact the legitimacy of CPS and cloud robotics adoption. The technologies will not gain institutional acceptance without the support of policies and acceptance by trade bodies. Technology Acceptance Models (TA and UTAUT) could offer a clue on how individual managers and employees reason and embrace such innovations when the usefulness, ease of use and social influence perceived has a significant concern. The Socio-Technical System Theory also teaches us that any technological advancement should be synchronised with human and organisational aspects in order to prevail (Zhang et al, 2022).

Integrative Theoretical Focal point

Collectively, the theoretical lenses above clearly indicate that CPS and cloud robotics are not mere operational technologies, but operational strategic facilitators of resilience operating on various levels. RBV revolves around the fact that they are scarce resources which generate competitive advantage. Dynamic Capabilities emphasise their contribution to promoting flexibility in turbulent organisations. Systems Theory places them in the context of interconnected systems in which resilience arises through coordination at the system level. Complementary theories also emphasise the need of legitimacy, acceptance and conformity with human and institutional factors. The theoretical framework therefore provides a holistic lens within which CPS and cloud robotics may be analysed. It brings together organisational, strategic and systems perspectives to emphasise that resilience is not a fixed status but a multi-level dynamic process. Companies that combine CPS with cloud robotics are not simply employing new technologies, they are re-organising their resources, building responsiveness, and making their supply chain environments more resilient..

RESEARCH METHODOLOGY

To measure the level of contribution of Cyber-Physical Systems (CPS) and cloud robotics on resilience in global supply chains, the research methodology used in this work was developed. The mixed-method approach was chosen due to its dual focus, specifically on the adoption of technology and the result of an organisation, which is applied to both quantitative and qualitative methods. This combination gave the breadth needed to find patterns in a large body of data, as well as the depth needed to find meaning in contextual factors affecting adoption and performance. The analysis was based on a primary data set of six hundred and fifty participants who work in an industry directly pertinent to the supply chain activities. These covered manufacturing, logistics, information technology services, and distribution industries, which meant that a wide range of contexts in which CPS and cloud robotics are currently being implemented was covered (Panetto et al, 2019). The sample was purposely sampled by position among supply chain management members, IT knowledge, robotics engineering, and top management responsible for digital transformation. This ensured that the responses were informed by practise, and based on facts of operation rather than guesswork. In each category, we selected respondents to represent large multinational corporations, as well as small-to-medium enterprises. The global perspective was obtained through the diversity in the geographical representation of respondent within Asia, Europe, North America and Africa. The main tool used in gathering data was the structured questionnaire. The questionnaire to measure the level of adoption of CPS and cloud robotics was designed with attention to the extent to which these methods may influence the resilience dimension, including visibility, adaptability, redundancy, and recovery. The items were rated on a five-point Likert scale between strong disagreement and strong agreement. Some examples of visibility include the following: Real-time shipment tracking, supplier data transparency, and inventory data accuracy. Adaptability was discussed in the context of the ability to reallocate resources during disruption and also the adaptability of the production schedule. Redundancy was operationalized by querying about the system-level backup capacity and availability of alternative resources and recovery by querying how much time it took to mitigate the downtime and how soon normal operations recovered after disruption. Other parts of the questionnaire requested the respondents to name obstacles to adoption, such as cost, skills gaps in the workforce, regulatory threats, and cyber threats. These were to be supplemented with open ended questions, thus enabling participants to expound on certain experiences or difficulties within their organisations. The survey was distributed through online and direct methods of distribution, respectively. The data were collected in three months. The importance of confidentiality was highlighted to enhance candour and respondents were informed that their personal or organisational identities will not be revealed. Out of the nine hundred professionals contacted, six hundred and fifty of them returned the questionnaire completely, and this gives a response rate of more than seventy percent, very strong in terms of a global survey of this magnitude. The analysis of data was performed in SPSS and AMOS/SmartPLS. It was analysed in three steps. The former phase consisted of descriptive statistics to profile respondents and to determine baseline adoption rates of CPS and cloud robotics. Means, frequencies, and standard deviations were calculated to give a picture of what is being practised and what is being perceived as a benefit. The second phase involved the use of inferential analysis and Analysis of Variance (ANOVA) to assess whether there would be significant differences between the scores of resilience among firms that reported no, partial, or full adoption of CPS and cloud robotics. Based on this method, the patterns of association between the outcome of resilience and the levels of adoption were identified. The third was the regression and Structural Equation Modelling (SEM) step to conduct a test of causality. CPS adoption and cloud robotics adoption were considered independent variables in these models, whereas visibility, adaptability, redundancy and recovery were dependent variables. SEM offered an advanced approach to assessing both direct and mediating links, e.g., was visibility the mediator between CPS and recovery, or was adaptability the moderator between robotics adoption and redundancy. Open-ended responses were collected and coded thematically to add more information to the statistical findings on qualitative data. Themes came back in the shape of cost-related concerns, cultural resistance to automation, lack of talent, and regulatory approval and cybersecurity. Some of the positive themes that respondents raised were increased trust among supply chain partners, increased response to disruption, and an increase in sustainability due to increased monitoring of resources. This qualitative layer was very helpful and gave a sense of depth to the statistical results and insights into organisational dynamics that cannot be fully represented by numbers. Rigour and ethical responsibility were also an important part of the methodology. The questionnaire was piloted to thirty professionals and some further editing was needed to ensure the questionnaire remained short and relevant to terms used in the industry. The alpha (Cronbach) values were greater than 0.80 in all constructs, thus confirming the internal consistency. Factor analysis was used to test convergent and discriminant validity of constructs. The ethical concepts that were applied in the course of conducting the research were informed consent, voluntary participation, anonymity, and that the data acquisition was academic in nature.

RESULTS AND DISCUSSION

The results of the first survey conducted on six hundred and fifty individuals provide sufficient reasons to assume that Cyber-Physical Systems (CPS) and cloud robotics will enable to enhance the resiliency of the supply chain in a certain number of ways significantly. The descriptive analysis shows that the adoption of these technologies is not uniform, but it is also gradually moving in the right direction. About a quarter of the companies that participated in the survey said that they had deployed CPS and cloud robotics in at least one significant portion of their supply chain activities. Sixty-eight percent of those who responded indicated that they partially adopted these technologies, and ten percent of the respondents indicated that they have not yet adopted these technologies. The distribution shows that despite the large scale awareness and pilot implementation, the majority of companies have not yet reached the point of transitional levels between testing and full-scale implementation. In order to measure this variation, the resilience scores were compared among the three types of adoption, none, partial, and full. The table below (Table 1) summarises the descriptive results.

Table 1. Adoption Levels of CPS and Cloud Robotics and Associated Resilience Scores

Adoption Level	Percentage of Firms	Mean Resilience Score (1–5 scale)	Key Characteristics Reported
No Adoption	10%	2.8	Fragmented visibility, reliance on manual processes
Partial Adoption	68%	3.6	Improved visibility, limited adaptability, partial redundancy
Full Adoption	22%	4.4	High visibility, strong adaptability, rapid recovery, system-level redundancy

ANOVA was used to ensure that all these differences in resilience scores were statistically significant ($p < 0.05$). Companies that had complete adoption documented better visibility, adaptability, redundancy, and recovery than companies that had partial or no adoption. In the full adoption group, 4.5, and in the partial adoption group, 3.4, and in the no adoption group, 2.6, are the visibility means. This is consistent with the conceptual argument that CPS, with in-built sensors and digital twins, improve real-time visibility, and cloud robotics provide system-wide orchestration that can be converted into improved resilience results. Regression and Structural Equation Modelling (SEM) also helped to gain some additional understanding of the causal relationships between technology adoption and resilience dimensions. CPS adoption was also observed to play a significant positive role in the visibility ($= 0.44$, $p = 0.01$) and the recovery ($= 0.39$, $p = 0.01$), supporting the notion that predictive intelligence and real-time tracking are the foundation of early detection of disruption and rapid recovery to a consistent state. The adaptability ($= 0.47$, $p < 0.01$) and the redundancy ($= 0.41$, $p < 0.01$), in turn, were more related to the use of cloud robotics. This demonstrates the ability of robotic systems connected to the cloud to redistribute tasks on-the-fly, to expand and shrink operations in reaction to disturbances, and to disperse workloads across networks in order to generate digital redundancy that reduces downtime. Table 2 is a summary of these relationships.

Table 2. Regression and SEM Findings on CPS and Cloud Robotics Adoption

Independent Variable	Dependent Variable	Path Coefficient (β)	Significance (p-value)	Interpretation
CPS Adoption	Visibility	0.44	<0.01	Enhances real-time monitoring and transparency
CPS Adoption	Recovery	0.39	<0.01	Accelerates restoration through predictive analytics
Cloud Robotics Adoption	Adaptability	0.47	<0.01	Enables rapid reallocation of resources and flexibility
Cloud Robotics Adoption	Redundancy	0.41	<0.01	Provides task-sharing and distributed backup capacity

These statistical results are very similar to the theoretical propositions presented above. According to a Resource-Based View (RBV), CPS and cloud robotics may be seen as valuable, rare, and difficult-to-copy resources that generate competitive advantage through introducing resilience to supply chain processes. This is supported by the empirical evidence where those firms that had full adoption always reported higher resilience scores on all dimensions. Moreover, the regression findings substantiate that CPS and cloud robotics can be regarded as non-substitutable enablers of resilience, as no other single stack of technologies can achieve the same level of visibility, adaptability, redundancy, and recovery at this point in time. The results also confirm the Dynamic Capabilities school of thought. The radical effect CPS has on visibility and recovery is the reason a firm with such systems is better placed to sense the winds of change and respond instantly. Likewise, the high correlation of cloud robotics adoption with adaptability shows the ability of companies to capitalise on opportunities and dynamically re-organise operations in response to change. Combined, CPS and cloud robotics represent the true meaning of dynamic capabilities that can turn disruptions into learning and strategic regeneration opportunities of the firm.

In the perspective of Systems Theory, the results describe the system advantages of linked technologies. CPS establishes feedback within continuous sensing and monitoring and cloud robotics uses these insights to operationalize distributed networks of processes and machines. The strong links between CPS, cloud robotics, and resilience results emphasise the interrelationship between the components of the system and the fact that the phenomenon of resilience is manifested in the whole ecosystem, not in separate organisational units. This reinforces the argument that resilience cannot be developed to work within supply chain networks but instead it has to work within individual firms. These statistical trends are further elaborated by qualitative information collected among respondents. It was repeatedly stressed that CPS also improved the level of trust among partners of the supply chain, since the lack of information asymmetry allowed making decisions based on factual and up-to-date information. Others cited cloud robotics as a tool to deal with unexpected labour shortages, where robots could be reconfigured quickly to accept the most challenging duties. Sustainability benefits were added by some respondents who mentioned that CPS and cloud robotics made it easier to monitor how much energy is used and how little is wasted. These qualitative results show that not only can the technologies improve the resilience of operations, but also result in more strategic objectives of transparency, trust, and sustainability. The discussions also indicate limitations to ubiquitous adoption, however. A consistent theme was integration cost particularly in small and medium sized businesses with a small budget. The other point reported by the respondents is that CPS and cloud robotics increase resilience to physical disruption at the expense of emerging cybersecurity and data-governance problems. Fears of losing a job or not knowing how to work with new systems were also found to be a significant barrier to cultural resistance in an organisation. These issues support the thesis that CPS and cloud robotics should be treated as a socio-technical change instead of a technical modernization. Technological investment should be complemented by organisational training, conducive policies and clear governance structures. Results showed that the adoption of CPS and cloud robotics has a significant positive impact on supply chain resilience, and full adoption yields the most effective results in terms of visibility, adaptability, redundancy, and recovery. The causal pathways through which CPS enhances monitoring and recovery and cloud robotics enhances adaptability and redundancy are statistically confirmed.

SOLUTIONS AND RECOMMENDATIONS

The findings demonstrate that Cyber-Physical Systems (CPS) and cloud robotics can contribute to the game-changer in creating a resilient global supply chain. This potential, however, needs to be translated into well-planned interventions that consider not only the technological facet of the process but also the organisational, institutional, and cultural obstacles that were outlined. The following are solutions and recommendations that businesses, policymakers and industry stakeholders can adopt in order to accelerate the adoption of these technologies and take full advantage of them. They are concerned with achieving greater technological preparedness, enhancing organisational capacity, establishing enabling regulatory frameworks, collaboration, financial and sustainability and long-term innovation goals. One of the key recommendations relates to technological preparedness of supply chains. The issues raised in the analysis are, in many cases, the result of restraints related to scale and difficulties in integration. Cloud robotics systems and CPS need to be able to manage the large volumes of transactions and real-time information that's characteristic of international supply chains. Companies must thus focus on investing in hybrid infrastructure bringing together the benefits of private and public blockchain ecosystems, edge-computing, and high-performance connectivity. Such systems are able to cope with the two problems of data overload and delay. Interoperability is also crucial: unless communication between CPS devices, robotics systems, and existing legacy systems, including enterprise resource planning (ERP) and customs clearance systems, occurs seamlessly, the advantages of adoption will be relegated to the backburner. APIs and cross-platform communications standards must be a feature of the industry, and the work of technology providers, industry consortia and regulators should be aimed at developing these interoperability frameworks. The other resolution is associated with organisational adjustment and workforce preparedness. The results showed that culture resistance and lack of skills are the two major barriers. The latter can only be addressed by a proactive investment in training and change management. The companies are to create special training sessions on the integration of CPS and robotics, but these should be designed not only to train specialists but managers and decision-makers as well. Such programmes must focus on the strategic importance of the technologies, and not portray them as purely operational, thus creating buy-in at all levels of the organisation. Meanwhile, organisations need to implement rollout strategies in parts. Trying to deploy CPS and cloud robotics in all the functions of the supply chain at the same time can be very expensive and also disruptive. Organisations can test, refine, and scale in stages by taking a staged approach and beginning with low-risk processes, including documentation or monitoring in a warehouse. Staged rollouts are another way to overcome internal resistance and build momentum to grow adoption by demonstrating real benefits early. Another important recommendation is governing structures. Data-driven ecosystems do not have clear accountability and generate distrust and disillusionment with implementation. Companies would need to establish internal control systems that define data entry and validation roles and duties as well as system controls. Multi-stakeholder governance models must be developed at the industry level, especially in the context of shared logistic platforms where CPS and robotics systems of different companies have to interoperate. Mechanisms to resolve disputes, manage data privacy, and manage liability must also be included in these governance systems, so that stakeholders trust that CPS- and robotics-enabled systems are fair and trustworthy.

Scale cannot be adopted without regulatory and policy support. The findings have revealed that some organisations have been hesitant to adopt CPS and cloud robotics because of legal uncertainty, particularly in how data will be controlled and whether digital documents will be recognised. Governments must lead in the development of rules that confer legal enforceability to data generated through CPS, smart contracts and robotics-oriented operation records. It is also essential to harmonise internationally, since the supply chains are national. International organisations such as the world trade organisations, international chamber of commerce and regional trade blocs should collaborate to develop similar standards that would be consistent across jurisdictions. Meanwhile, laws should take care of cybersecurity threats. To ensure that the risks that come with the enhanced connectivity are alleviated, policy makers need to coerce the deployment of minimum levels of security in CPS and cloud robotics. The trust in these systems will be enhanced through cyber resilience audits, certification programmes and incident reporting standards. Another solution that the study recommends is collaboration on the industry level. Proprietary systems and siloed implementations suppress the opportunities of CPS and cloud robotics. Platforms that are neutral, consortium-based are more likely to facilitate across stakeholders participation as they allocate governance in a more equitable way and avoid monopolistic control. PPP can also be essential, particularly where there is high risk or high cost, such as port operations and customs clearance. PPPs will be able to ramp up adoption and share risks with other entities more quickly and benefit not just individual firms but also whole logistics ecosystems. Industrial standardisation of smart contract templates and robotics protocols will also lead to consistency, inefficiencies, and interoperability. To enable adoption, the economic and financial considerations have to be addressed too. Although the advantages of CPS and cloud robotics in the long term are obvious, the initial investment is a turn-off to most companies, especially small and medium companies. Strict cost-benefit calculations are to be performed to measure the possible returns including decreases in fraud, downtime, administrative expenditures, and improvements in efficiency and resiliency. These analyses can then be utilized to make strategic business case investments and business cases to follow. Governments can facilitate the implementation of resilience and sustainability targets by offering special incentives to those adopting them, which could include tax credits, adoption grants, and technology adoption subsidies. New funding schemes (shared infrastructure schemes, robotics-as-a-service schemes, etc.) can also mitigate financial obstacles, since they enable companies to tap into CPS and cloud robotics services without investing significant capital. The other suggestion is related to socio-technical factors of adoption. Resilience is not only a technical attribute, but it is also dependent on human trust and the tolerance of the culture. A possible way to use CPS and cloud robotics is to track and mitigate the adverse impact of supply chains on the environment. Firms are required to measure real-time energy consumption, emissions and resource usage with CPS-enabled sensors, and optimise routes and minimise waste with cloud robotics. The generated data may be utilised to confirm that the organisation meets environmental standards and proves the adherence to global sustainability objectives including the UN Sustainable Development Goals. When resilience is associated with sustainability, not only can firms enhance the resilience of their supply chains, but also the legitimacy and trust that consumers, regulators as well as investors have in these firms. Last but not least, the next stage of digital transformation requires future-oriented solutions. CPS and cloud robotics will eventually become more autonomous, intelligent, and networked systems in the wider context of Industry 5.0. To provide predictive, adaptive and secure supply chains, firms need to be ready as CPS approach robotics, artificial intelligence, blockchain, and quantum computing. In order to stay abreast of such developments, it is promoting research and development of partnerships with universities, technology providers and research consortia.

FUTURE RESEARCH DIRECTIONS

The results of this research indicate that Cyber-Physical Systems (CPS) and cloud robotics can dramatically contribute to enhancing resilience of a global supply chain. However, as is common with all new technologies, they are only half the potential they could be and numerous questions remain unsolved. Next generation research should thus proceed beyond single case studies and small-scale pilots to broader, multi-level and longitudinal studies that can help establish what is possible as well as what is difficult with large-scale adoption. There are a few lines of investigation that can be seen as especially crucial to the development of knowledge in this field that also all mirror the dynamics between technology, organisations and the global trade systems. Scaling and interoperability of CPS and cloud robotics is one of the recent research trends. Most of the existing implementations tend to be piece-meal and confined to select processes like automation of the warehouse or transportation tracking. Further research is needed to look at how these technologies can be extended to full-scale supply chains involving more than one country, industry, and legal setting. Studies regarding cross-platform interoperability protocols are of special urgency since systems already in use are heterogeneous. The next generation work must focus both on technical interoperability and institutional frameworks that enable cross-border sharing of data and standardised modes of governance. The second major research avenue in the future is the combination of CPS and cloud robotics and other complementary technologies. An obvious collaborator is Artificial Intelligence (AI), which can crunch through vast amounts of data and come up with predictive information. Exploring how AI-based decision support can be integrated into CPS monitoring systems or cloud robots task assignment will contribute towards unlocking the possibilities of predictive and responsive supply chains. Another important complementary technology is blockchain, which is tamper-proof and provides better network trust. The evolution of open-door, secure supply chain environments can be guided by research on how blockchain can be used to form the core of CPS and robotics data. There should also be more studies on the role of 5G and edge computing as they are the technologies that will determine how fast and efficient CPS and robotic systems can communicate with geographies. The other potential area of interest would be the intersection of CPS and cloud robotics with financial innovations, such as Central Bank Digital Currencies (CBDCs). With programmable money built into smart contracts, new logistics systems of the future could automatically pay people in real-time, based on CPS-monitored shipment progress or on robotics-controlled task achievement. This poses not only technical but also economic and policy issues, including the effects on trade finance, currency flows and financial inclusion. A study of the economics aspect of CPS and cloud robotics adoption may thus help to bridge the gap between supply chain research and economics and international finance. Another area that needs to be researched further is sustainability. Though this paper has already proven that CPS and cloud robotics can potentially contribute to environmental monitoring and optimizing resources, further systematic studies are needed to list such effects. Researchers are encouraged to research the way in which such technologies may assist organisations to achieve net-zero carbon, embrace the concept of the circular economy, and minimise waste in production and distribution. CPS and robotics data points in life cycle assessment models can allow a better measurement of the impact on the environment, but policy-driven research could explore how governments can incentivize the use of technology to meet climate targets. Such social-technical aspects of adoption also deserve further examination. There is another line of inquiry with potential promise associated with risk and resilience in extreme conditions. Although this research study established that CPS and cloud robotics could improve resilience to regular disturbances, the effectiveness of these systems in the face of systemic disturbances, including pandemics, geopolitical wars, and natural disasters, requires further investigation. The second line of the study is how the technologies can assist in redundancy and recovery in the event of multiple disruptions simultaneously. Simulation models or digital twins of whole supply chains might offer useful insights into resilience strategies in such extreme situations. Cybersecurity is an aspect of study that gains more significance in the future. CPS and cloud robotics decentralise business operations and enhance resilience to physical disruptions; however, their connectivity puts supply chains at risk of cyber-attacks. There is a need to research better cybersecurity models, such as quantum-resistant cryptography, automated smart contract auditing, and CPS- and robotics-specific incident response models. The issue is how to balance transparency and openness such technologies demand against the security and privacy of sensitive supply chain information. Another highly critical area to study in the future is the economics of adoption. Large companies with large budgets can afford to invest in CPS and robotics, but small and medium-sized enterprises (SMEs) have to deal with prohibitive costs. The exploration of new business models can support the identification of directions to inclusive adoption. Long-term return on investment should also be assessed in future studies, with comparison between costs to be incurred in adoption and quantifiable savings in downtime, fraud and inefficiencies. This would help any company make the right decisions and enhance the business case to implement it. Finally, government and institutionalization should also be considered as long-term issues of scholarly interest. Accountability, liability, and governance questions are becoming more complicated as CPS and robotics turn supply chains into cyber-physical ecosystems. Research into governance structures - between private consortiums to decentralised self-governing organisations - will play a key role in creating equitable and inclusive ecosystems (Dobaj et al, 2023). Such guidelines imply that any further studies should be interdisciplinary, integrating the results of engineering, management, economics, law, and social sciences. Academia-industry-policymaker cooperation will be needed to produce the knowledge that will not only be rigorous but also be actionable in table 3.

Table 3. Future Research Agenda for CPS and Cloud Robotics in Supply Chain Resilience

Research Theme	Key Questions	Expected Contribution
Scalability & Interoperability	How can CPS and robotics be scaled across global supply chains? How can platforms interconnect?	Frameworks for seamless multi-level adoption
AI & Predictive Integration	How can AI enhance CPS monitoring and robotics task allocation?	Predictive and adaptive supply chains
Blockchain & Data Security	How can blockchain secure CPS and robotics data?	Trustworthy, tamper-proof logistics ecosystems
CBDCs & Trade Finance	How can programmable money integrate with CPS-monitored milestones?	Automated, real-time payments for global trade
Sustainability	How can CPS and robotics reduce emissions and support circular practices?	Evidence-based sustainability impact
Socio-Technical Adoption	How can workforce trust and collaboration be fostered?	Inclusive, human-centered adoption strategies
Risk & Extreme Resilience	How do CPS and robotics perform under systemic crises?	Crisis-ready models for global supply chains
Cybersecurity	How can vulnerabilities in connected CPS–robotics systems be mitigated?	Robust frameworks for secure digital ecosystems
Economics & ROI	What business models can reduce costs for SMEs?	Inclusive, scalable adoption pathways
Governance & Institutions	What governance models ensure accountability and fairness?	Equitable and legitimate supply chain ecosystems

CONCLUSION

A review of the Cyber-Physical Systems (CPS) and cloud robotics as applied in the global supply chain shows that these two technologies are much more than a mere improvement of the scripting of current processes. They represent a paradigm shift in the way supply chains are designed, operated and maintained in a time of unprecedented volatility. The theoretical insights coupled with the interpretation of the findings obtained by six hundred and fifty respondents prove that CPS and cloud robotics are radical resilience enablers. Due to the inclusion of logistics and manufacturing networks, supply chains can evolve beyond models of efficiency and transition to adaptive, predictive and recovery-in-the-poorest-case-scenario systems. One major finding of this study is that CPS and cloud robotics directly improve the four interrelated aspects of resilience as follows: visibility, adaptability, redundancy, and recovery. CPS and its integrated sensors, digital twins, and continuous monitoring functions allow organisations to experience real-time visibility of the supply chain nodes. This will decrease the chances of cascading failure since anomalies can be detected at an early age and action taken. In addition to these functions, cloud robotics operationalizes the concept of adaptability, allowing robotic fleets to be repurposed on the fly and expanded or contracted in response to disruption. These technologies combined can produce redundancy, not by maintaining inventories of products, or by idle capacity waiting to be used, but by sharing intelligence and tasks dynamically over networks. Their combined effect on recovery can be particularly potent: predictive maintenance and autonomous reallocation of resources allow supply chains to rebound immediately after being impacted by disruption, reducing downtime and continuing operations. These findings can also be seen through the theoretical frames that were investigated in this paper. Within the resource-based perspective, CPS and cloud robotics may be viewed as valuable, rare, inimitable and non-substitutable resources that make organisations stronger by instilling resilience. They develop capabilities that are difficult to copy, because effective integration involves special technology, process, and culture configurations. In the lens of Dynamic Capabilities, CPS and robotics increase the ability of firms to sense, seize, and transform. Not only do these technologies help organisations identify disruptions at an earlier stage, but they also allow organisations to restructure operations in a manner that would not have been possible before. Systems Theory provides an added layer, showing that resilience is not achieved through each specific technology but rather through the compounding of technologies into larger, more interconnected ecosystems in which feedback, data sharing, and coordination allow systemic adaptability to occur. All these assumptions together support the argument that CPS and cloud robotics cannot be considered a single solution to the problem but are essential elements in the creation of the modern resilience framework of the supply chain.

The findings also show that CPS and cloud robotics are complementary. Those firms that adopted either yet not both enjoyed some of the benefits, including visibility without flexibility or automation without systemic oversight. These technologies were most significantly effective in the area of resilience only when they were considered as a whole. This implies that adoption strategies cannot be piecemeal, but need to be built as an ecosystem. Furthermore, the results confirm that resilience benefits are maximised when adoption is not limited to individual pilots, but instead, scaled and enterprise-wide integration takes place. Simultaneously, the work recognises that the adoption is not uniform as most companies are only in the transitional stage. Unless these barriers are overcome, the full promise of CPS and cloud robotics will not be fully exploited, especially by small and medium enterprises which do not always have the resources to undertake large-scale change. The discussion identifies too, the strategic value of aligning the adoption of CPS and cloud robotics with larger organisational and societal objectives. Along with operational resilience, these types of technologies enhance transparency and trust between the supply chain partners because they minimise information asymmetries and provide verifiable and tamper-proof information. The results to policy makers depict the importance of legal frameworks that legalise digital records, harmonise them across national borders and encourage its use by providing specific financial incentives. Regarding the case of industry consortia, the results suggest that to eliminate monopolistic organisations, cooperation is important, interoperability the most important, and common platforms the most important. In the future, the research directions in this paper would lead to the future of CPS and cloud robotics. They will be combined with artificial intelligence, blockchain, 5G, and new financial technologies like central bank digital currencies to form supply chains that are resilient, predictive, adaptive, and autonomous. Until then, socio-technical adoption, cybersecurity, and governance research will take precedence so that institutional and cultural preparedness can be matched by the technological gains. The trend of CPS and cloud robotics deployment will consequently shape not just the future of supply chain resilience, but the overall change of global trade within the framework of Industry 5.0. CPS and cloud robotics are changing the paradigm of supply chain management towards weak and efficiency-driven systems to strong, resilient, adaptive and sustainable ecosystems. Their impact on the dimensions of resilience is confirmed by factual evidence, and the theoretical understandings explain the rationale behind their functioning as strategic enablers. The solutions and recommendations provided make it clear that coordinated action should be taken on technological, organisational, regulatory and societal levels. Nevertheless, the trend is clear: cyber-physical and cloud-robotic supply chains of the future will be intelligence-based, automated, and flexible enough to endure disruptions and take advantage of opportunities. At a time when uncertainty is the defining trait, these technologies are not only efficiency facilitators but also resilience insurers, when it comes to the continuity, competitiveness, and sustainability of global supply chains over the next few decades.

REFERENCES

Dobaj, J., Riel, A., Macher, G., & Egretzberger, M. (2023). Towards devops for cyber-physical systems (cpss): Resilient self-adaptive software for sustainable human-centric smart cps facilitated by digital twins. *Machines (Basel), 11*(10), 973. DOI: 10.3390/machines11100973

Fatorachian, H., & Smith, C. (2022). Impact of CPS on enhancing supply chain resilience, with a focus on solutions to pandemic challenges. In *Cyber-physical systems* (pp. 109–125). CRC Press.

Flammini, F. (2021). Digital twins as run-time predictive models for the resilience of cyber-physical systems: A conceptual framework. *Philosophical Transactions. Series A, Mathematical, Physical, and Engineering Sciences, 379*(2207), 20200369. DOI: 10.1098/rsta.2020.0369 PMID: 34398658

Hozdić, E., Kendić, S., & Jurković, Z. (2022). Complex adaptive manufacturing system concept as a cyber-physical production system: solutions to the Covid-19 pandemic challenges. In *Cyber-Physical Systems* (pp. 129–156). CRC Press.

Iqbal, F., Malhotra, J., Jha, S., & Semwal, T. (2022). Introduction to cyber-physical systems and challenges faced due to the COVID-19 pandemic. In *Cyber-Physical Systems* (pp. 1–23). CRC Press.

Ivanov, D., & Dolgui, A. (2021). A digital supply chain twin for managing the disruption risks and resilience in the era of Industry 4.0. *Production Planning and Control, 32*(9), 775–788. DOI: 10.1080/09537287.2020.1768450

Kaur, M. J., Riaz, S., & Mushtaq, A. (2019). Cyber-physical cloud computing systems and internet of everything. In *Principles of Internet of Things (IoT) Ecosystem: Insight Paradigm* (pp. 201–227). Springer International Publishing.

Keung, K. L., Lee, C. K., Ji, P., & Ng, K. K. (2020). Cloud-based cyber-physical robotic mobile fulfillment systems: A case study of collision avoidance. *IEEE Access : Practical Innovations, Open Solutions, 8*, 89318–89336. DOI: 10.1109/ACCESS.2020.2992475

Khalid, M. (2025). Cyber-Physical Systems and their Role in Industry 4.0: Architecture, Applications, and Future Prospects. *International journal of applied biology and forensic, 9*(1), 46-59.

Martin-Gomez, A., Avila-Gutierrez, M. J., & Aguayo-Gonzalez, F. (2021). Holonic reengineering to foster sustainable cyber-physical systems design in cognitive manufacturing. *Applied Sciences (Basel, Switzerland)*, *11*(7), 2941. DOI: 10.3390/app11072941

Napoleone, A., Negri, E., Macchi, M., & Pozzetti, A. (2023). How the technologies underlying cyber-physical systems support the reconfigurability capability in manufacturing: A literature review. *International Journal of Production Research*, *61*(9), 3122–3144. DOI: 10.1080/00207543.2022.2074323

Panetto, H., Iung, B., Ivanov, D., Weichhart, G., & Wang, X. (2019). Challenges for the cyber-physical manufacturing enterprises of the future. *Annual Reviews in Control*, *47*, 200–213. DOI: 10.1016/j.arcontrol.2019.02.002

Qudus, L. (2025). Resilient systems: Building secure cyber-physical infrastructure for critical industries against emerging threats. *Int J Res Publ Rev*, *6*(1), 3330–3346. DOI: 10.55248/gengpi.6.0125.0514

Radanliev, P., Dc Roure, D., Nicolescu, R., Huth, M., & Santos, O. (2022). Digital twins: Artificial intelligence and the IoT cyber-physical systems in Industry 4.0. *International Journal of Intelligent Robotics and Applications*, *6*(1), 171–185. DOI: 10.1007/s41315-021-00180-5

Rajput, S., & Yadav, S. (2025). Cyber-physical fusion architecture: To mitigate risk for resilient supply chain. In *Risk, Reliability and Resilience in Operations Management* (pp. 291-312). Elsevier.

Schmitt, R., Permin, E., Kerkhoff, J., Plutz, M., & Böckmann, M. G. (2016). Enhancing resiliency in production facilities through cyber physical systems. In *Industrial Internet of Things: Cybermanufacturing Systems* (pp. 287–313). Springer International Publishing.

Segovia-Ferreira, M., Rubio-Hernan, J., Cavalli, A., & Garcia-Alfaro, J. (2024). A survey on cyber-resilience approaches for cyber-physical systems. *ACM Computing Surveys*, *56*(8), 1–37. DOI: 10.1145/3652953

Singh, P. K., & Kumar, A. (2023). Cyber Physical Systems in Supply Chain Management. In *Cyber Physical Systems* (pp. 85–110). Chapman and Hall/CRC.

Sowan, B. I., Zighan, S., & Altarawneh, A. (2025). Supply Chain 4.0: Driving Operational Efficiency, Resilience, and Cybersecurity Using Advanced Technologies. In *2025 1st International Conference on Computational Intelligence Approaches and Applications (ICCIAA)* (pp. 1-8). IEEE.

Suárez-Riveros, E., Mejia-Mantilla, Á., Jaimes-Suárez, S., & Jimenez, J. F. (2021). Cyber-physical systems in logistics and supply chain. In *International Workshop on Service Orientation in Holonic and Multi-Agent Manufacturing* (pp. 260-271). Cham: Springer International Publishing. DOI: 10.1007/978-3-030-80906-5_18

Tan, Q., Tong, Y., Wu, S., & Li, D. (2020). Towards a next-generation production system for industrial robots: A CPS-based hybrid architecture for smart assembly shop floors with closed-loop dynamic cyber physical interactions. *Frontiers of Mechanical Engineering*, *15*(1), 1–11. DOI: 10.1007/s11465-019-0563-9

Theresa, S., Nour, M., & Turnbull, B. (2025). Responsible Resilience in Cyber–Physical–Social Systems: A New Paradigm for Emergent Cyber Risk Modeling. *Future Internet*, *17*(7), 282. DOI: 10.3390/fi17070282

Tonelli, F., Demartini, M., Pacella, M., & Lala, R. (2021). Cyber-physical systems (CPS) in supply chain management: From foundations to practical implementation. *Procedia CIRP*, *99*, 598–603. DOI: 10.1016/j.procir.2021.03.080

Wang, L., & Wang, X. V. (2018). *Cloud-based cyber-physical systems in manufacturing*. Springer. DOI: 10.1007/978-3-319-67693-7

Zhang, K., Shi, Y., Karnouskos, S., Sauter, T., Fang, H., & Colombo, A. W. (2022). Advancements in industrial cyber-physical systems: An overview and perspectives. *IEEE Transactions on Industrial Informatics*, *19*(1), 716–729. DOI: 10.1109/TII.2022.3199481

Zhang, Y., & Abellera, J. U. (2025). Autonomous mobile robotics in smart warehousing: A cyber-physical systems approach to inventory management. *Future Technology*, *4*(4), 59–71. DOI: 10.55670/fpll.futech.4.4.6

Zhao, Z., Zhang, M., Wu, W., Huang, G. Q., & Wang, L. (2024). Spatial-temporal traceability for cyber-physical industry 4.0 systems. *Journal of Manufacturing Systems*, *74*, 16–29. DOI: 10.1016/j.jmsy.2024.02.017

Chapter 7 Decision-Making Frameworks for Integrating Motion Control in Business Operations

R. Deepika
https://orcid.org/0000-0002-2818-9867
Rajalakshmi School of Business, Chennai, India

S. Nithya
Buckinghamshire New University, UK

M. Durgarani
https://orcid.org/0009-0004-1134-9336
Christ University, India

R. Delecta Jenifer
https://orcid.org/0000-0002-8179-6383
Saveetha Engineering College, Chennai, India

K. Prakash
https://orcid.org/0000-0003-4273-1496
SRM Institute of Science and Technology, Chennai, India

ABSTRACT

This chapter includes some strategic plans of how to incorporate motions control technologies in the business to promote efficiency, agility, and to make data-driven

DOI: 10.4018/979-8-3693-3637-3.ch007

decisions. It cogitates about the real-time motion tracking and the focus on optimization in the logistics sphere, production and asset management through automation and smart analytics. The chapter explains how AI, IoT, control algorithms help harmonize the processes of motion with the goals of organizations. They lay emphasis on decision theory, performance measures and on the flexibility of motion systems in dynamic environments. The chapter has shed light on the ability of motion control to become the engine of innovation and competition in the operation of the contemporary enterprise by offering scalable and practical integration models.

INTRODUCTION

The point that can be seen as a combining of smart automation and data-driven decision-making is that motion control systems have become a part and parcel of current business strategic operation processes. The technology that was once the preserve of the engineering realm and factory automation, has become a cross-functional feature of capabilities effectively determining productivity, efficiency, responsiveness, and even customer experience. In moving forward, as companies seek agility in the volatile markets, the capability of making well informed decisions with regard to motion planning, the behavior of machines and the adjustable system responsiveness is now a key driver of business advantage (Ferrell, O. C., Gresham, L. G., & Fraedrich, J. 1989). The present chapter is concerned with the potential to make motion control an element that is reliably and naturally incorporated into business decision-making frameworks, and thus transcends the technical subsystem status to become an enabler of enterprise performance. The center of this revolution is restructuring on the emerging understanding that motion control has ceased to be the sphere of interest concerning physical movement, but implies smart interaction, dynamic optimization and real-time responding. The current system of motion now has embedded sensors, actuators, controllers, and AI-based logic that is able to perceive, reason, and adapt within milliseconds of operation (Coito, T., et al 2021). It is through these abilities that machines will not only maneuver accurately but also be able to manipulate movements, ratios as well as loads on the basis of environmental forces and business considerations (Thakar, S., Srinivasan, S., Al-Hussaini, S., Bhatt, P. M., Rajendran, P., Jung Yoon, Y., & Gupta, S. K. 2023). Such sophistication makes the motion behavior to resonate with upper-order goals like energy efficiency, order fulfilment, accuracy of inventory, and maximization of throughput (Bai, Z., et al 2023). Various macro trends contribute to the necessity to include motion control in the process of making decisions. First, the convergence of operation technology (OT) and information technology (IT) has been witnessed due to industry 4.0 and digitalization transformation (Derradji, R., & Hamzi, R. 2024). Digitally integrated manufacturing systems have enabled the accessibility, analysis and execution of motion data in enterprise resource planning (ERP) and manufacturing execution systems (MES) and supply chain platforms (Arputharaj, J. V., & Pal, S. K. 2024). Second, business processes must be optimized at the granular level so as to meet the customer demands in terms of speed, customization, and reliability. Third, the sustainability objectives and cost pressures require the increased efficiency in the utilisation of resources and the mechanism of motion optimisation can directly impact it by minimising wear, idle time, and energy peaks.

The implementation of motion control in business level decisions is highly disjointed even though there are strategic opportunities associated with it (Bals, L., Kirchoff, J. F., & Foerstl, K. 2016). There is the tendency of many organizations to consider the motion systems as independent technical resources, and the determination of the path planning, trajectory improvements, and the actuator performances are taken up by local engineers or the maintenance departments. Such a siloed approach causes motion data to be underutilised, interventions to be applied in a reactive manner and miss the ability to cross-functionally optimise. In order to fill this gap, companies should require organized decision-making models, which integrate motion intelligence into business operations-moments with bottom line metrics, customer obligations and risk profiles. In this matter, a decision-making framework is a set of organized basis that regulates the analysis of motion-related decision-making within a business environment, including the assessment and implementation of selecting speed performance, schedule, or synchronization tasks. These structures have to rely on several things: operating limits, economic trade-offs, process dependencies, regulatory compliance and real time data flows. They should also be able to facilitate multiple decision horizons, including short-term responses (taking sensor signals) scheduling and automation investments and workforce planning in long term. The chapter suggests an elaborate decision-making model that is based on four layers linked to one another:

1. Motion Intelligence Layer - Here the physical locations where sensors, actuators, and the motion controllers are used to gather information to identify the deviations, strange and optimization opportunities could be indicated here.
2. Operational Decision Layer - At this layer, one can decide about local machine actions, such rerouting paths, tuning torque, or balance loads.
3. Tactical Coordination Layer- At what level multi machine synchronization, job scheduling and allocation of resources are coordinated in production cells or business units.
4. Strategic Integration Layer-It is the layer where the decisions concerning motion control are matched against business KPIs such as the cost per unit, order lead times, energy budget, and asset utilization target.

One of the framework constructions is the decision-support capacity, which contains predictive analytics, simulation models, digital twins, and dashboard visualization, making that automation used in motion decision-making is not only automatic but also explainable and trackable (Lim, K. Y. H., Zheng, P., & Chen, C. H. 2020). It is also based on principles of multi-criteria decision analysis (MCDA) and operations research to deal with trade-offs among conflicting goals, e.g. speed versus energy efficiency or throughput versus cost of maintenance. Noteworthy, in this chapter, there will be an investigation of the way Structural Equation Modeling (SEM) and other statistical tools can be employed in determining whether or not the decisions around motion control affect business outcomes.

SEM enables a strict approach to modelling complex associations between latent variables including motion system responsiveness, accuracy of decisions, and creation of business value. Together with information obtained by tracking production logs, the control systems, and enterprise platforms, SEM allows companies to confirm causal paths by which motion choice works on financial and operational outcomes. Examples of companies cases will be given to indicate the success of using motion control as a competitive differentiation lever by companies dealing with manufacturing, logistics and distribution businesses. As an example, a warehouse can change robot speeds and ordering order sequencing to accommodate orders based on priority and congestion, avoiding overstaffing it to meet the necessary fulfillment targets (Borges, G. D., et al 2021). A manufacturing plant can adjust the actuator forces and lower scrap rates and keep the cycle times. The above examples underline the transition between the static programming of movements toward smart and agile movements responses to the requirements of the business, in real-time. Organizational considerations are also brought forth by the dynamics of having motion control as part and parcel of the business decision making. It involves interdepartmental cooperation between the engineering, information technology, operations, finance and executive business. It requires cultural shift in problem solving tactical to problem solving proactive. It also requires investing in training, governance and change management to make motion intelligence trusted, understood and mobilized and leveraged by the organization. This chapter prepares the ground where motion control can be thought over not as a background automation feature but as a strategic asset that is a part of the fabric of business decisions. Organizations are able to achieve new heights of agility, resilience and efficiency by aligning motion systems towards business logic, data intelligence and performance ambitions. The content of the proposed decision-making models provides the avenue of such transformation, as it is backed by analytical probity, technological creativity, and working breadth.

CONCEPTUAL FRAMEWORK

The assimilation of motion control systems with the decision making business needs the development of a holistic conceptualisation that seeks to bring together real-time automation intelligence with the goals of strategic and operational needs done by corporations. Such a framework should take into consideration the interaction between motion behaviour (controlled by controllers, actuators, sensors, and logic systems) and business processes, resource allotment, performance metric, and the organisation priorities. The conceptual framework offered in this chapter draws a systematic and hierarchical perspective of the ways the motion control decisions are conceptualized, contextualized, and operationalized at the various functions of the business. The basis of this framework is the Motion Intelligence Layer where data is generated and it is also initially processed. Historically, motion systems in the modern industrial context tend to be integrated with dense sensor systems that track variables, including, torque, speed, acceleration, position, force, and temperature (Goldsby, et al 2023). These alerts are registered through real time controllers and condition monitors and they form a raw input into condition evaluation of the machine. This layer performs the role of detecting anomalies, forecasting motion interruptions and indicating optimization chances through embedded logic and AI algorithms. This layer is very localized but more and more gets connected with Industrial Internet of Things (IIoT) systems (Rajan, P. 2024). It allows a micro view of mechanical performance, and creates the foundation off of which the business decisions of a larger scope are made. More importantly, the outputs of such a layer are not merely the binary tags of functional capability (e.g. pass/fail) but high-value streams of continuous data against which subtle decisions may be made (e.g. should I make simpler and cheaper actuator moves more frequently, more tasks in sequence, or increase or decrease actuator loads with a change in demand?). With the Operational Decision Layer, motion intelligence is translated to direct control efforts and near-term planning as it climbs up the ladder. In this case, the information delivered by sensors is deciphered by control engineers and automation-based systems, which would make decisions like slowing the speed, rerouting, dynamic balancing of robotic payloads or synchronized belt motion along conveyor belts. These decisions affect short-interval scheduling, energy, part quality and takt-time matchup.

This level is commonly under rule based decision engines or model-predictive controllers (MPCs) which combine constraints in the machineries with process goals. In more mature environments, operation decision systems can even communicate with real time production information to modify moving logic in harmony to altering work flows. As an example, within a flexible manufacturing cell, movement paths may be changed depending on the nature of the product being constructed, and eliminate superfluous movement without losing throughput. Tactical Coordination Layer is like the bridge between the execution on the shop-floor and business plans. It concentrates on the distribution of resources and equipment provision, job scheduling and synchronization of the workflow in numerous machines or production lines (Louis, J., & Dunston, P. S. 2018). In this case, the choice of motion is cast in the context of production goals, work-in-process (WIP) stocks, lead-time commitments and equipment usage rates. Some examples of tactical decisions are to time motion-intensive work during off-peak times to avoid energy peaks, divide motion-intensive jobs between machines to prevent wear or to plan robot-based picking and palletizing operations with delivery-surges. The layers are frequently controlled by Manufacturing Execution Systems (MES), Warehouse Control Systems (WCS) and Advanced Planning and Scheduling (APS) systems. The incorporation of motion intelligence within this layer makes sure that actions taken on the shop floor are in tune with those taken on the plant level and leads to the overall efficiency of the system. An example is that a motion-sensitive MES may be used to re-assign a high-speed machining job to a more reliable station when vibration limits indicate that tool breakage may occur, thus eliminating down time and maintaining flow of orders. The topmost layer is Strategic Integration Layer where the information is fed through the data in motion control and decisions at this level to performance management, capital planning, and business strategy within an enterprise (Stank, T. P., & Goldsby, T. J. 2000). This layer ensures that motion system conducts in alignment with important performance indicators (KPIs), where cost per unit, delivery reliability, energy efficiency, maintenance cost, asset return on investment (ROI), and customer satisfaction are the key performance indicators. The frameworks of strategic decision making rely on variations of motion into interpretable numbers that facilitates long-term planning. An example of such a strategy would be a repeated frequency of high-energy movement during only certain operations and this could result in a strategic review of process design or capital investment in energy-efficient actuators. Alternatively, one can base the assessment of supplier performance, ROI in automation, or reskilling programs of workforce based on motion performance measures. The layers are not simply integrated in linear fashion but they are cyber-physical and recursive. An example would be when a torque spike is detected at the Motion Intelligence Layer which could automatically cause an immediate reduction in the operating speed that can in turn cause a reduction in the throughput on the line. In its turn, it is fed back to the reallocation of workload of the Tactical Layer, which is fed back to the Strategic Layer, i.e., to the forecasting models. This iterative process is an example of the paradigm of the feedback-based control in which the knowledge of motions is relayed upwards as diagnostics and tactical/strategic imperatives are transmitted downwards as constraints, and goals.

Aiding this chain of command that comprises various layers is a group of Decision-Support Tools that augments transparency, traceability, and immediate reaction. These are tools:

- Real-time dashboards which graph KPIs of motion e.g. cycle time per axis, smoothness of acceleration, vibration alerts.
- Digital twins which model how decisions regarding motion affect the performance of systems in different conditions.
- Tools of Multi-Criteria Decision Analysis (MCDA) to make a decision about trade-offs (e.g., energy saving over time increment).
- Forecasting systems which predict the degradation of motion and opportunities the issue would have on production KPIs.

In such a framework, there are four characteristics which define decision-making:

1. Evidence-based And this is informed by data on motion patterns, performance indicators and so on.
2. Context-awareness-aware of environmental circumstances, operational matters and business circumstances.
3. Goal-focused - focused at the business seeking to reach particular goals like lean compliance or sustainability.
4. Being responsively autonomous is the ability to be executed with decentralized actions and administrative centralization.

The Structural Equation Modeling (SEM) offers an effective analytical framework in order to validate and constantly refine this framework. SEM helps companies to develop and test the model of how the motion control decision would affect latent constructs such as productivity, effectiveness in maintenance and efficiency of resources among other constructs as well as margin. To give an example, one could determine using SEM, whether the acceleration of the actuator motion leads to a better order fulfillment time and controls costs by either energy expenditure or component deterioration. In addition, SEM facilitates diagnostic scrutiny on decision-making quality by determining the indirect impacts, interaction factor as well as the power of the paths. As an example, it can turn out that the noticeability of motion (the speed of corrective action) simply influences the uptime, whereas its customer satisfaction influence is mediated by the higher on-time delivery, which is a strategic KPI. These insights enable the decision makers to focus on areas of maximum systemic leverage in terms of where they wish to invest in. Lastly, the conceptual framework promotes cross-function competence since decision making in motion control is not a responsibility of the automation engineer, but involves operations managers, logistic planners, financial controller as well as executives. The democratization in motion intelligence with affordable interfaces and shared platforms allows an organization to transform at the individual level of optimization into a connected performance management model. Finally, in this formulation of motion control, the aspect of motion control is not presented as a mechanical act in isolation but an intelligent, multiple-layered decision support system across the operational, tactical, and strategical levels of a business. Organizations can significantly improve the working performance systematically, become more agile, and have the data-driven culture resilient by instilling the motion insights into the organized decision-making processes and substantiating their effect with SEM.

THEORETICAL FRAMEWORK

Motion control applied in business is an interdisciplinary field that needs a theoretical background that spans systems engineering, theory of business decisions, behavioral operations, cyber-physical systems, and performance management. As opposed to the conventional automation theory concerned with precision, trajectory design and feedback stability, the given framework concerns itself with translating low-levels of motion behavior into high value business impacts. The perspective that is obtained leads to the idea that motion control is a strategic decision space that is coupled by techno-constraints as well as managerial imperatives. The technical basis takes the form of control systems theory, which gives the mathematical and functional framework of how physical systems (robots, actuators, CNC machines) can be used as regulated closed-loop systems driven by input signals. Feedback responsive, accurate, and stable movements can be guaranteed by the application of proportional integral derivative (PID) controllers, model predictive control (MPC) and adaptive control models on machines. Such models are critical to physical implementation but typically do not care about business performance objectives. Therefore, although motion controllers can be designed to optimise the smoothness of the paths or to minimise energy consumption, they do not necessarily take account of production schedules, the position of inventory, or cost structures. The drawback of this theoretical perspective is overcome by the use of cyber-physical systems (CPS) theory that models physical (e.g., robots) and digital (e.g., data streams of motions, digital twins) representation of assets as inseparable parts of a single system. Motion controllers in a CPS system have software that is embedded in them that allows communication between machine and enterprise decision planes in both directions. Rapid sensing, prediction and control of the operational state and business situational needs, e.g. rush orders, equipment limits, or delivery timetable are enabled by this theoretical structure (Papamichail, K. N., & Robertson, I. 2005).

The principles of the organizational decision theory included in the framework to model decision processes in this integrated environment are the limited rationality and the multi-criteria decision-making (MCDM). Horizons constrain the choices of decision-makers: human or AI agents, time pressure, and incomplete data, and selfish and competing objectives (Hu, Y., et al 2024). In the context of motion control, this could be a trade-off of time taken to execute, and the wear, or vice versa, or trade off of throughput and vibration. The MCDM methods, e.g., analytic hierarchy process (AHP) or weighted scoring, allow rational Prioritization, but usually do not have enough modeling power to accommodate latent variables or feedback effects.

In this case, Structural Equation Modeling (SEM) will be useful. In the case of SEM, it is possible to model complex, interconnected relations between latent variables, whose nature involves the evaluation of such constructs as the responsiveness of a motion system, decision-making effectiveness, or even business performance impact (Jakhar, S. K., & Barua, M. K. 2014). Contrary to traditional regression where the modeling of direct and independent relationships is usually adopted, SEM allows modeling concurrently mediating variables, feedback systems and unobservable constructs, which is optimal in settings where the effects of motion behaviors are indirect and distributed towards results at enterprise level. As an illustration, SEM provides the opportunity to test the hypothesis according to which motion precision (through repeatability and error rate) affects order fulfilment accuracy (a latent concept) and hence customer satisfaction and eventually profitability. This hierarchical structure can be explained by the system dependencies in the present-day operations and the ability to represent the entire picture with the effect that motion control has on business performance. The other major theoretical element is the operations management theory, especially the theory of constraints (TOC) and lean manufacturing. These strategies emphasise that any inefficiencies existing in bottlenecks, in terms of delays in motion, extreme cycle time, and mechanical unreliability, may be propagated throughout the system (Handfield, R., Sroufe, R., & Walton, S. 2005). Flow synchronization, waste reduction and the possibility to synchronize takt-time are realized in this situation, due to motion control. System-wide constraints and lean objectives should thus guide decisions made regarding the speed of motions, sequencing and buffer sizes, and not simply follow local optimization. In addition, the behavioral operations theory introduces a valuable human component into the design. The process of deciding in a motion system implies the interpretation of data, parameter adjustment, responses to alarms by an operator, engineer, and supervisor. Such decisions are affected by cognitive biases, fatigue, risk aversion and organizational norms. As an example, through intuition, the operator can ignore settings made to be automated, or managers will not be willing to slow down machines lest they fail to meet the set daily output (Venice, J. A., et al 2025). One can apply SEM to the possible interaction of behavioral variables (trust in automation, perceived risk) in determining the relationship between motion insights and choice consequence.

The framework is also based on the strategic management theory, namely the resource-based view (RBV) to incorporate motion systems into a long-term planning (Hang, P., et al 2020). The ability to control motion, including the responsiveness, modularity, and precision, may be viewed as strategy tools providing differentiation in the competitive environment, in case such abilities are coherent with internal objectives. To take an example, an organization that is more agile in terms of motion agility is capable of providing more customized products, has a quicker response to the demand changes or is able to cut back downtime more efficiently as compared to the other competitors. The decisions that have to be made on the strategy level, like the purchase of high-performance actuators or retraining personnel in the field of motion diagnostics, should be based on this knowledge of motion as an enabling performance resource. Also because there are high expectations of information systems theory within this framework especially in the form of decision support systems (DSS) and real-time analytics platforms. The systems act as a conduit of raw data in terms of motion with businessable points of knowledge. DSS frameworks focus on the quality of the data, the transparency of models, the design of the user interface and the cognitive compatibility between decision-makers. Such a motion context results in the necessity to display motion trends, anomaly forecasts, and confidence intervals in intuitive, timely, and user-driven designs that support either plant managers, maintenance engineers, or the executive in their respective contexts.

Most importantly, the theoretical framework positss as a feedback-rich system, in which motion decisions (at the level of operations) reacts to KPIs that are strategic in nature (e.g., cost per unit, delivery reliability), and that strategies (e.g., green, labor efficiency goals) are imposed constraints on permissible motion behaviors. This back and forth is facilitated by real-time tracking, closed-loop optimization and multi-tiered communication developer. SEM modeling extends this interaction in that it models top-down and bottom-up effects, systems integration and the mediating roles of decision quality. Lastly, the framework encompasses the concepts of systems thinking that sees the organization as a connected web of units instead of siled departments. In that sense, motion control decisions can never be on their own they echo throughout procurement, scheduling, logistics, and customer experience (Duraimutharasan, N., Deepan, A., Swadhi, R., Velmurugan, P. R., & Varshney, K. R. 2025). An increase in the speed of pick and place can have an implication on the replenishment schedules, the integrity of the packaging, etc. or the warranty claims when the parts have been mishandled. These effects are complex enough in a systems context to require holistic approach to deciding to make a motion, one that would capture emergent behavior and unintended consequences, and non-linear impacts. To sum up, theoretical justification of incorporating motion control into the sphere of business operations must be multi-faceted. It integrates the rigor of engineering and organization expertise, real time control and planning and quantitative modeling and human behavioral understanding. Structural Equation Modeling provides an effective transition point between these fields, and researchers and practitioners can work through its influence on movement decision propagation in systems, its influence on latent performance drivers and on the overall strategic objectives. When guided by such powerful theoretical base, motion control stops being a background technical task and turns into one of the fundamental foundations of the agile, reactive, and smart management of enterprise.

RESEARCH METHODOLOGY

The mixed-methods research design adopted in this study was used to assess the impacts of motion control integration in business decision-making processes and the performance of the enterprise. The methodology involved the quantitative data analysis with Structural Equation Modeling (SEM) as well as the qualitative case informations about the industrial activities of motion control systems which have been part and parcel of production and logistics decision loops. The three advanced manufacturing plants analyzed through the quantitative primary data were manufacturing plants distributed in the three sectors of discrete manufacturing (automotive components, electronics assembly, packaging logistics). These plants already implemented motion control system throughout production lines and incorporated motion parameters: actuator cycles, smoothness of motion, switching task time and energy consumption in their Manufacturing Execution Systems (MES).

It was a three-month dataset that contained:

- True real-time Motion control logging (for example position accuracy, torque variation)
- Records of decision on path changes and priorities of tasks
- MES production throughput data
- Unit cost and the energy efficiency index
- Human interaction data(Manual override/Intervention timestamps)

Additional qualitative information was collected by means of semi-structured interviews of operations managers, automation engineers, and business analysts. Through these interviews, quantitative observation could be put into perspective, especially on the rationality of the decision and how motion could be embedded in the organization.

Latent variables on the basis of the theoretical model were developed as follows:

Motion System Responsiveness (MSR): evaluated with the help of cycle completion latency, task switching rate, and amount of needed trajectory correction.

- Decision-Making Effectiveness (DME): measured on the basis of the intervention success rate, turnaround rate of decisions, and congruency with KPIs.
- Operational Coordination (OC): based on the accuracy of job sequencing, multi-machine synchronization and the number of unplanned downtimes.
- Business Performance Contribution (BPC): it is a higher-order construct that displays productivity, cost effectiveness, and accuracy of fulfilment.

The latent constructs were determined by at least three (observed) indicators, the validity of the constructs was determined through Confirmatory Factor Analysis (CFA) and the internal consistency through Cronbach alpha and Composite Reliability (CR).

The structural model was created to check how responsiveness of motion system, effectiveness of decision-making and efficiency of coordination are causally linked to business performance. It also considered moderating variables, i.e. organizational role clarity and maturity of adoption of technology to evaluate context effects. Maximum likelihood with bootstrapping (5,000 samples) was used to estimate SEM and provide confidence intervals of path coefficients as well as to guarantee the robustness of the method of estimation in case of data non-normality. RMSEA (< 0.06) as well as CFI (> 0.95) and SRMR (< 0.08) were used to check the model fit. Each of the member organizations had signed a consent form and had analysable operational anonymised data before analysis. Confidentiality of data and internal system identifiers were maintained in the guidelines provided by the industry. The approach will subscribe to an empirical substantiation of a multilevel motion control decision framework as well as extend an understanding of motion intelligence upon B2B decisions in the various levels of business decision making.

RESULTS AND DISCUSSION

The proposed framework of decision-making contains causal relationships which are highly supported by the empirical results of the Structural Equation Modeling (SEM) analysis. The model allowed amazing fit statistics: RMSEA = 0.039, CFI = 0.972 and SRMR = 0.041, which means that the conceptual framework regarding the relationship between motion system responsiveness and the larger business performance indicators are statistically significant and can be generalized in the context of the industrial settings that were studied. The large path was found between

Motion System Responsiveness (MSR) and Decision-Making Effectiveness (DME) (standardized coefficient 0.63, $p < 0.001$). This verifies the fact that operational-level decision-making performance is naturally associated with the accomplishments in real-time responses on motion, including a decreased actuator response time, better path transitions and latency in switching between tasks. The flexibility improved by motion responsiveness allowed quick reorganization of robotic work, it allowed automation of response to the delay in production, and it minimized the numbers of human interventions, therefore, enhancing quality and promptness of control decisions.

Figure 1. Decision-Making Framework Integrating Motion Control System

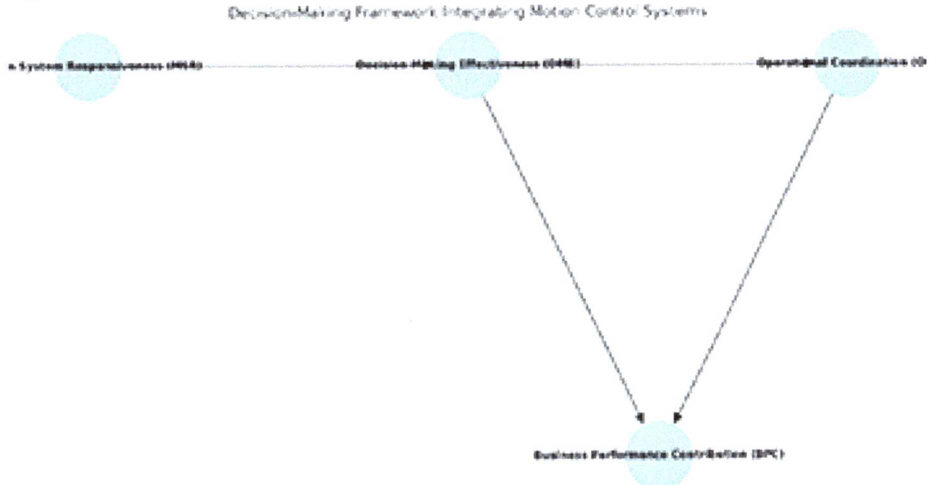

The above diagram represents the structural equation model (SEM) hypothesis employed in this chapter in understanding how the motion system responsiveness affects performance in a business performed by intermediaries such as effectiveness in decision making and coordination of operations. Moreover, Decision-Making Effectiveness (DME) showed a strong positive effect to Operational Coordination (OC) (beta = 0.59, $p < 0.001$) and this means that when decisions made based on going motion are precise and timely, then the coordination of the machines, the operators and job sequencing logic, will improve. The observation was supported by the interview narrative as the operations managers emphasized that the motion alerts enabled the teams to rebalance the workload among stations until the bottlenecks were created by underperforming equipment in advance. OC and DME were two interdependent elements to Business Performance Contribution (BPC). The OC to BPC road was fairly robust (beta = 0.47, $p < 0.01$), and it displayed that systemic coherence (symbolized in shortened idle time on machines, cycle progression adherence, and cross-line tasks synchronisation) influences venture performance such as the dependability of delivery, cost per unit and the overall equipment effectiveness (OEE). It is confirmed that the same direct influence between DME and BPC (beta values = 0.32, $p < 0.05$) points to the fact that the decision quality can bring positive business outcomes, even without the context of greater coordination outcomes, and this affects the spheres (probably) of error reduction, re-prioritization of tasks, and the response to quality control. Bootstrap validation (n = 5,000 samples) made all path coefficients have relatively small 95 percent confidence interval (e.g., 0.57 to 0.69 95 percent confidence interval 0.57 to 0.69) which indicated the path coefficients are very robust. Such probabilistic surety is particularly applicable in operational environment where quick and uncertain choices need to be made (Dimitrakieva, S., Marinova-Stoyanova, M., & Gramchev, B. 2025). The application of bootstrapping replaces empirical robustness in the assumptions of models and instills greater assurance in using such a framework in live-time business settings.

Interestingly, direct effects aside, it was determined that there were other effects that came out stronger than expected. As an illustration, the mediation of the MSR on the BPC by DME and OC was large (beta = 0.17), proving that the entire business worth of the motion responsiveness appears not by individual percentages of performance gain but as a compound ripple of integrated decisions. This not only upholds the premise of this framework, that real-time motion intelligence is only effective when it is internalised and assimilated into structured systems of decision-making, but also serves to indicate that the ability of the framework to impact the performance of an organization is only possible when optimal resource mobilisation is internalised as part and parcel of the framework of decision-making. Independent qualitative insights were used along with the quantitative ones. According to the engineering personnel, the information about the motions incorporated into dashboards could assist them in prioritizing interventions and improve maintenance crews setup. Tactical level decision-makers showed more confidence in rescheduling of production tasks when motion systems signaled to system limits and load behaviour consistently. In addition, the executive stakeholders recognized that high-performing motion systems enabled leaner inventory planning, as well as responsiveness to the variability of demand two major drivers of agility in contemporary manufacturing. It has been determined, though, that the SEM analysis identified some inconsistencies as well. At locations that had not received much training on the use of motion data by the operators, a connection between MSR and DME was less strong. It means that technological responsiveness per se is not enough without associated human expertise and in faith in automation. Such observations make the case more solid in terms of investment to be made in cross-functional literacy, i.e., ensuring the literacy of planners, operators, and engineers with respect to understanding how motion control data can underpin enterprise decision-making. The other observation that was worthy of note was that although DME could affect BPC directly in a big way, when directed through operational coordination, its impact was increased multi-fold. This has the implication that local-optimizing isolated decisions made on motion (e.g. to rerouting a robot to avoid collision) only have enterprise-scale effects when coordinated with both upstream and downstream processes. It helps adopt a systems thinking mindset, where optimization of decisions needs to think of interdependence within the functions.

To the manager, these findings indicate that motion responsiveness is a strategic capability that organizations must consider so as to stop viewing it as a control feature alone. Integrating motion information into the decision processes of both the short and long-term planning and the levels of operations, companies can achieve a new level of efficiency and flexibility. Furthermore, SEM validation comes as a scientific ground of evaluating the connection between investments in motion technology (e.g., actuator upgrade, predictive controller deployment) and the indicators of the resulting tangible improvements in the business performance. in conclusion, SEM model validates the conclusion that operative control systems, which are incorporated in structured decision-making models, have positive influences on the business performance and operational coordination. The power of these associations makes the case that motion-aware business logic, decision-support tooling, and cross-disciplinary collaboration is needed. The results seem to prove the framework but also indicate that in the future, motion intelligence will become the epicenter of digital change and strategic implementation.

SOLUTIONS AND RECOMMENDATIONS

The practical experience and theoretical learning that decided this chapter place a solid confirmation to the fact of motion control systems no more being alone automation tool, but a part of business past and decisions, and do modify the performance of business in several levels. In order to make maximum use of this strategic possibility, it is necessary to introduce extensive and implementable solutions that would make the responsiveness of the movement correspond to the enterprises strategies, operational velocity and resonance of coordination. The plan in Table 2 suggests the sequential steps that could be followed to implement the motion control-based decision frameworks in industrial and logistic setting.

1. Institutionalize Motion Data within Decision Hierarchies

Motion data should be baked throughout all the layers of decision making- both sides of the machine (running the machine) and strategic projection. This necessitates game planning the data pipelines to enable motion system data produces (e.g. cycle latency, deviation scores, actuator precision) to be framed (data presented, data interpreted, data placed in context) in manners pertinent to specific stakeholders.

Recommendations:

- Introduce motion KPIs in MES and ERP dashboards to be able to view the full picture.
- Set minimum and maximums of a certain parameter (e.g., torque variation limit, vibration intensity) which starting points will generate some automatic adjustment or notifications to be reviewed.
- Cross-train cross- functional teams how to read the motion signal through the lens of different functions (e.g production lens vs. finance lens).

2. Develop a Decision-Making Control Tower Framework

Based on the supply chain control towers, companies ought to use centralized oversight systems to observe anomalies in motion, operational and business effects in real time. They are control towers that give visibility to distributed assets and coordinate systemically motion-based decisions.

Recommendations:

- Formulate a digital system on which motion system signals of decisions are accumulated and displayed.
- Install predictive analytics and SEM-based diagnostics, as well as simulation abilities in the control towers to do scenario planning.
- Flag cascading effects on the platform e.g. a motion lag in one area may cause a downstream operation to lag.

3. Prioritize Motion Decision-Support Systems (MDSS)

People in decision-making should have tools, which would interpret the motion control data and give action. These systems must accommodate many forms of decision-making such as scheduling, managing energy, tasks planning and placing.

Recommendations:

- Install Motion Decision-Support Systems which cross-relate actuator actions and live throughput, cost and performance values.

- Supplement MCDA methods to rank alternatives in situations of conflicts, e.g. in case of speed versus energy consumption.
- Integrate explainable AI features in explaining the reasons a particular motion was recommended.

4. Formalize Coordination Protocols Based on Motion Intelligence

The responsiveness to motion can only be helpful when that information gets used to initiate a concerted response. The exact simultaneity involving machines, teams and functions would be necessary in transforming local optimization of motion into system productivity.

Recommendations:

- Establish SOPs which connect motion events (e.g. torque spike) to escalation paths (e.g. engineer dispatch, job rescheduling).
- Apply adaptive algorithms of coordinating or disengaging loads on more than one asset in case of motion anomalies.
- Write up a set of pre-approved playbooks of how to respond to motion-based interruptions.

5. Align Motion Control with Strategic KPIs

The decision on motion must have a direct impact on business performance in terms of OEE, cost per unit, downtime rates, fulfillment accuracy and carbon footprint. To make sure that this happens, motion-derived insights have to be used in strategic performance dashboards.

Recommendations:

- Add SEM-validated path metrics to measure the extent to which motion responsiveness helps achieve high level KPIs.
- Associate motion responsiveness with customer-facing metrics (e.g. SLA performance, delivery lead times).
- Implement performance reviews which measure motion rates as well as their impact on the company.

6. Invest in Digital Twins for Motion-Impact Simulation

Digital twins present the laboratory to conduct an experiment to determine the impact of motion decisions on workflow, the physical wear of assets, and resource distribution. These two twins act as a testing ground of decision modeling prior to implementation.

Recommendations:

- Construct any digital twin model of vital elements of production or transportation centers having movement parameters incorporated.
- Apply motion sensor feeds as they are to update and calibrate simulations.
- Use digital twins to experiment and validate coordination procedures and recovery plans on variations of motion disruption cases.

7. Foster Decision Literacy Across Departments

One of the most important factors facilitating motion-integrated decision-making is interpreting the data within the entire organisation about the importance and meaning of motion data. Making decisions by people in all positions contributes to decision quality as they all become aware of how physical movement impacts variables within the business.

Recommendations:

- Reach out interdisciplinary workshops on motion control, decision theory & SEM findings.
- Display real time animation and dashboards showing how motion information on the way through to decisions.
- Give the operators and the planners data-driven autonomy.

8. Enable Closed-Loop Decision Optimization

The essential condition of continuous improvement demands that the motion control systems have to work in the closed feedback loops. Actions involving motion (e.g., speed increase) are measured in terms of their business effects (e.g., better OEE) and this feedback is fed into a modification of future decision parameters.

Recommendations:

- Establish automatic feedback loops which transfer the business performance outcome to changes in motion policies.
- Apply SEM diagnostics to change weightings in decision engines (e.g. give preference to high quality over speed when it comes to defect blitzes).
- Use take it or leave it approach to bootstrap-based validation where motion thresholds are continued to be optimized on changing conditions.

9. Incorporate Human-in-the-Loop (HITL) Structures

Automation is the process that does not remove human input-instead, it enhances the input. Human validation, override, or guidance are structured opportunities, which must be provided through systems, based on the criticality of the decisions, and the level of risks involved.

Recommendations:

- Focus the decision on the types of motion (i.e., prioritise motion decisions by risk profile (e.g., high-value, low-latency, safety-critical).
- Enable only operator interventions of high impact decisions whereas automate those of low risks.
- Record all HITL engagements so as to facilitate root cause investigation and process enhancement.

10. Apply SEM Analysis to Governance and Auditing

According to the model presented in the chapter, it is possible to rely on SEM to target the auditable measure of the impact of motion responsiveness on business value. SEM should be incorporated as part of the organizational governance system to ascertain the effectiveness of automation plans.

Recommendations:

- Conduct SEM reviews on a quarterly basis using new motion and performance metrics to update decision making rules.
- Employ outputs of bootstrap enhanced SEM to develop compliance documents to make quality and performance standards.
- Analyze capital investment decisions (e.g. actuators, sensors) using SEM-proven motion-to-performance links.

11. Promote Interoperability and Systems Integration

Motion control systems have to be connected with ERP, MES, CMMS, and PLM systems. Interoperability is the only solution because without it data sharing, coordination of decisions, and the achievement of strategic levels of motion intelligence is not possible.

Recommendations:

- Introduce standard data models (e.g. OPC UA, ISA-95) to exchange data on motion.

- Make motion systems input to the current enterprise dashboards and resource planning applications.
- Apply APIs and middleware and connect the motion control systems to business analytics layers.

12. Reframe Motion Investment as Strategic Capability Development

Businesses should no longer consider motion systems as capital equipment but rather as capability that promotes responsiveness, resilience and differentiation.

Recommendations:

- Integrate motion intelligence during strategic plans and digital transformation road maps.
- Establish motion analytics lacks and privacy of excellence (COO), and concentrates on knowledge.
- Refer ROI calculations to both operating measures and customer experience and long term agility.

In a nutshell, any organization willing to incorporate motion control into their decision-making eco-systems should adopt an inter-disciplinary multi-level strategy. The suggestions contained in this work can be seen as a roadmap to turning motion data into a strategic advantage, one that generates enterprise value owing to its responsiveness, coordination, and smart execution. When governance, tools, training, and alignment are properly deployed, motion intelligence is not only predicted to improve operational precision, but will also equip data-driven decision making throughout the entire business spectrum.

FUTURE RESEARCH DIRECTIONS

With the expansion of motion control out of the realm of engineering specifics and as an active point of enterprise decision-making, new venues open up to scholarly and practical research. The integration of cyber-physical systems, real-time analytics, and organizational decision theory includes rich innovation terrain, and needs to mesh the intellectual discipline of technical expertise and knowledge of management. This section indicates the main future research questions that can develop the knowledge, efficiency, and expansion of motion control in business operations. An equal research priority, as the basis of which, models of multi-layered decision-making must be developed, which keeps business strategy and motion behavior in balance. Although recent research, such as the SEM framework presented in this chapter, has started to demonstrate the impacts of the motion responsiveness on the operational coordination and the business performance, additional research is required to investigate the longitudinal impact. One propensity of future research study may be to explore how motion control maturity (i.e. progressive sophistication of motion analytics, autonomy and decision support) will create agility, risk resilience and market responsiveness within organizations across time. LSEM techniques would enable the scholars to trace these ties over strategic planning cycles, waves of investment and swings in markets. The other urgent direction is adaptive decision systems where the motion parameters continuously readjust to the changing business objectives. The implemented frameworks mostly presuppose constant weights of priorities, e.g. trade-offs between speed and energy consumption or accuracy and throughput are fixed. But in practice, in a specific situation, these exchanges differ: fast motion can be permissible on an urgent order, and energy economy can be higher-valued during non-peak hours. Further research ought to determine real time optimization strategies along with a set of decision rules that update to respond to market signals, production imperatives, or sustainability limits (Naranjo, I., Torre, R. D. L., & Boza, A. 2025). Such adaptive decision-making environments might have the computational infrastructure of reinforcement learning along with explainable AI (XAI) mechanisms as well as bootstrapped SEM validation.

One of the crucial research directions is putting human behavioral models in decision systems based on motion. On the other hand, the majority of the decision support systems have focused on the efficiency of algorithms, leaving out human trust, acceptance, and cognitive load (Lazaroiu, G., Androniceanu, A., Grecu, I., Grecu, G., & Neguri ă, O. 2022). Future research on the topic must be directed at the investigation of the ways in which operators and managers process data on motion, the decisions they make and the influence of the human-in-the-loop (HITL) implementation on the quality of decisions made. Behavioral SEM may be applied to the modeling of the influence that constructs like automation bias, alert fatigue, or over-reliance on systems had on the connection between motion intelligence and operational outcomes. Moreover, motion decision framework scalability needs to be studied by researchers within enterprise ecosystems, especially in multi-sited or multi-national companies. When a company attempts to implement intelligent motion systems throughout the factory, warehouses or logistics facility, issues of consistency, interoperability and measurement are problematic. It is possible to define standardization practices and contextual customization practices through cross-industry/geographical comparison. The application of cross-site measurement invariance testing in Multi-group SEM would show the influence of cultural, regulatory or the infrastructural variances on the functioning of the motion-embedded decision-making. As a parallel, there emerges the need to study motion-based decision models in non-manufacturing settings, including examples of healthcare robotics, precision agriculture, autonomous store, and smart infrastructure (Seifi, N., et al 2025). These environments entail special types of motion (e.g., surgeries involving soft robotics, swarming motion in field robotics) and decision environments (e.g., safety in relation to patients, environmental dynamism). A new research must explore the relationship between the motion responsiveness and decision schedules, ethical restrictions, and compliance rules. SEM models can be used to reveal latent variables such as motion trustworthiness, adaptive safety and context sensitivity with domain specific models.

The other research issue of importance is an economic modeling of the motion control investment. However, the benefits of operations have been outlined in many case studies but there is still a dearth of uniform model that quantifies the returns of investment (ROI) of motion intelligence. The researchers ought to come up with cost benefit models which encompass capital costs, savings in maintenance, downtime, product quality, and decision quality improvements. These models would be able to include path coefficients associated with SEM to trace the path that motion control plays in cost reduction or revenue generation through enhanced satisfaction or lower defect percentage. The studies regarding sustainable-weighted motion selections are also critical. Organization is finding it necessary to achieve decarbonization level, energy reduction and green compliance mandate. Acceleration fine control, trajectory optimization and real time load control using motion control systems can play its role in achieving these objectives. In future work, it can be investigated, what is the potential of sustainability KPIs to be integrated into the motion control algorithms as either constraint or objective to decision-making processes. SEM can be used to determine environmental and business relating to motion strategies after it is combined with lifecycle assessment (LCA) models. The concept of digital twins in validation that is used in a decision and experiment warrants further discussion. Although simulation is already being applied using digital twins, the process of integrating them into real-time decision-making (particularly, into validating motion-induced system effects) is still emerging. A study needs to be carried out to include the observation of how the twin-based models can be used as a policy sandbox where the decision rules, motion adaptations, and system reconfigurations can be tested prior to being committed. These pre-implementation assessments can be enhanced by bootstrapped SEM validation with the help of simulated data.

The other underdeveloped scenario is dealing with resilience-based movement control choices, especially in a situation of a dislocated supply chain, work shortages, or equipment failures (Dias, L. S., & Ierapetritou, M. G. 2017). The studies are needed on how motion control systems can enable resilience through rapid reconfiguration, redistribution of tasks, or degraded-mode operation. Scenario modeling and actual tests, using the decision-tree logic along with SEM analysis can demonstrate the role played by motion flexibility to maintain continuity of operations during stress. Also, one should look at the motion decision ethics and governance in the future. With approaching autonomy of motion controlled systems, ethical aspects of removing safety overrides, traceability of decisions and fair treatment of human workers who will be subject to motion-guided optimization (e.g., task re-distribution or speeding up) are raised. Checks to governance structures ought to be comprised of auditability of decisions, processes of consent, and transparency requirements. The relationship between the governance mechanisms and the stakeholder trust, decision acceptance and system compliance can be tested by SEM. Technologically, researchers ought to investigate real-time deployment of SEM in the edge environment, where motion decision can be undertaken in a microsecond and needs data to be viewed locally. Lightweight, incremental variants of more conventional structural models known as Edge SEM models may run in low-latency and computation-constrained environments. Bootstrapping and Bayesian updating can be used to calibrate these models, and hence accuracy of decisions increases over time without the dependencies on central data processing. Lastly, research needs to complement education and skills development systems to enable motion-wise decision-making at any organizational hierarchy. These involve formulating curriculum, simulator architecture, and certification programs in the role of motion data analysts, coordination engineers and strategic automation planners. SEM may be applied in studies to model the relationships between training quality and interface design concerning decision accuracy, user engagement, and organizational adoption of motion decision systems. To summarise there is more to the incorporation of motion control into business operations than just a technical evolution; it is a redefinition of decision justification, execution within the industrial enterprise. The theoretical, the methodological, and the empirical aspects of this transformation must be constructed in future research concentrating on the adaptivity, human factors, scalability, sustainability, and on governance. The Structural Equation Modeling will be a firm pillar in this undertaking, and it provides an accountable approach to authoritatively authenticate translucent and separate intertwining between motion behavior and business outturns. The key to the full realization of the strategic value of motion-informed decision-making is found in the interdisciplinary approach, novel simulation, and relevant situational knowledge of the next generation of studies.

CONCLUSION

Enterprise decision-making process is experiencing a pivotal change with integration of motion control as a part of the operational and strategic weave of contemporary organizations. Motion control used to play a niche role in limited localized control systems and technical sub areas of focus, but is beginning to play a critical role as a source of business intelligence that informs and influences cross-functional business decisions. This transformation has been made possible by the union of cyber-physical systems, real-time analytics, and the structured decision models like Structural equation modeling (SEM), which introduces the motion intelligence into the much larger landscape of organizational performance with the allocation of resources and the agility strategy. The chapter has gone on to develop the point that such motion control systems, which consist of sensors, actuators, feedback cycles, and intelligent systems, could be used not only to perform the desired mechanical tasks with precision but also to control the business level factors associated with the throughput, cost effectivity, delivery reliability, and customer satisfaction. On a structured decision-making model in four interconnected layers, including motion intelligence, operational control, tactical coordination and strategic integration, we have shown how the flow of motion insights can move through echelons, then the mechanical to the executive level that have impact on the organization results. The empirical foundation of verifying this framework was given by the Structural Equation Modeling (SEM) method. SEM model presented significant causal relationships between latent subcomponents such as Motion System Responsiveness (MSR), Decision-Making Effectiveness (DME), Operational Coordination (OC) and Business Performance Contribution (BPC). The results highlighted that responsiveness of motion systems is cascading effect driven: that is, it leads to better decision accuracy and timeliness, and better coordination between machines and departments, and it leads to measurable value to business. Such a multilateral causal pattern gives credit to the assumption that motion control is not only a technical enabler but a strategic source.

One of the most important reflections of the findings was the process of mediation of operational coordination. Although motion system responsiveness had a direct relationship with decision effectiveness, its influence on the overall business performance came as part of synchronisation of the decisions upon cross-organizational functions. This is to show the relevance of a systems thinking during motion intelligence deployment. Optimized isolated motion can deliver better local results, but simultaneously made, coordinated, and handled decisions and decisions are where the value of the enterprise can be achieved. Contextual and human factors were also of importance in the findings. Expertise of operators, level of trust in automation and motion data literacy were some of the variables that moderated the relationship that existed between motion control and the results of the decisions. It creates a need to reconsider a direction taken by workforce development: the more intelligent an automation system is, the smarter the workforce has to be: not only in technical skills, but also in decision literacy and systems thinking. The recommendations given in this chapter consist of direct routes on how motion intelligence can be integrated in the enterprise decision architecture. These incorporate establishment of motion information in ERP and MES dashboards, establishment of motion control towers with centralized core decision supervision, establishment of motion decision-support systems (MDSS) and identification of motion control out genuine with strategic KPIs. Operationalization of these suggestions will help respective organizations move on beyond reactive changes of motions to proactive and goal-oriented motion plans. Further, the solutions lay stress on the importance of digital twins, real-time dashboards, and human-in-the-loop systems and closed-loop optimization as the core components of intelligent motion decision structures. These methods and technologies can be used to bridge the divide between mechanics and business rationale; in other words, applying the mechanism in a way that motion choices can be explained, audited and modified to suit changing reality on the ground.

The research plans identified in this chapter also strengthen the importance of motion-informed decision-making in the future research of the field. These guidelines run across the adaptive optimization models, behavioral decision combinations, multi-site expandability, sustainable-based motion plans, and systems governance in autonomous systems. It is through these research gap that scholars and practitioners will be able to perfect the conceptual basis as well as practice of motion control is a weapon of strategic advantage. Practically, the decision-making frameworks highlighted here provide significant opportunities to the organizations that are undergoing digital change, complex operations, and uncertain performance. Motion intelligence results in the ability to recover more quickly after being interrupted, wiser distribution of scarce resources and closer alignment of production systems and business strategy. Motion control systems capable of predicting, reacting and synchronising to business requirements are essential because even though supply chains are becoming increasingly volatile and customer demands more stringent. Markedly, this chapter also showed that the worth of motion control is not to be found purely in technical advancement, but integration and interpretation. The second conspiracy is that organizations should overcome implementation of sensors and actuators in individual language and concentrate on the way the knowledge on the motion is gained, contextualized and consequently acted upon. These include cultural transformation, governance, cross-functional teamwork and investment in decision-support ecosystems. In a strategic perspective, motion control is to be re-conceptualized as a dynamic capability i.e., which does not only improve the efficiency of operational procedures but also promotes organizational learning, adaptability and innovation. Businesses integrating motion control in the strategic decision framework will be better off reacting to volatility in the market, manage risks and achieve new value in real-time business operations. Conclusively, motion control in business operations is an opportunity and a need, since decision-making structures have excellently been integrated to implement the same. With convergence of the physical and digital worlds, and as the technologies of automation become increasingly intelligent, the capacity to match motion-level decision to an enterprise-level strategy becomes a distinguishing capability. The information, in this chapter, based on empirical evidence, theoretical soundness, and practical advice can serve companies as a guide to accessing the power of motion control as a source of contemporary decision-making performance and competitiveness.

REFERENCES

Arputharaj, J. V., & Pal, S. K. (2024). Transforming Industry 5.0: Real time monitoring and decision making with IIOT. In *Sustainability in Industry 5.0* (pp. 76-106). CRC Press

Bai, Z., Kong, B., Chen, J., Miao, L., & Zhang, Z. (2023). Environment design and construction of real-time dispatching and intelligent decision making simulation system for integrated coal supply chain of large energy group. In *International Conference on Intelligent Systems, Communications, and Computer Networks (ISCCN 2023)* (Vol. 12702, pp. 101-111). SPIE DOI: 10.1117/12.2679959

Bals, L., Kirchoff, J. F., & Foerstl, K. (2016). Exploring the reshoring and insourcing decision making process: Toward an agenda for future research. *Operations Management Research : Advancing Practice Through Research*, *9*(3), 102–116. DOI: 10.1007/s12063-016-0113-0

Borges, G. D., Reis, A. M., Ariente Neto, R., Mattos, D. L. D., Cardoso, A., Gonçalves, H., & Arezes, P. (2021). Decision-making framework for implementing safer human-robot collaboration workstations: System dynamics modeling. *Safety (Basel, Switzerland)*, *7*(4), 75. DOI: 10.3390/safety7040075

Coito, T., Firme, B., Martins, M. S., Vieira, S. M., Figueiredo, J., & Sousa, J. M. (2021). Intelligent sensors for real-Time decision-making. *Automation*, *2*(2), 62–82. DOI: 10.3390/automation2020004

Derradji, R., & Hamzi, R. (2024). The convergence of process-risk management and decision-making: An integrated vision. *International Journal of Business Continuity and Risk Management*, *14*(3), 305–326. DOI: 10.1504/IJBCRM.2024.141606

Dias, L. S., & Ierapetritou, M. G. (2017). From process control to supply chain management: An overview of integrated decision making strategies. *Computers & Chemical Engineering*, *106*, 826–835. DOI: 10.1016/j.compchemeng.2017.02.006

Dimitrakieva, S., Marinova-Stoyanova, M., & Gramchev, B. (2025). Integrating Artificial Intelligence into the Management Decision-Making Process in the New Era of Industry 4.0. In *environment. Technology. Resources. Proceedings of the International Scientific and Practical Conference* (Vol. 4, pp. 91-98)

Duraimutharasan, N., Deepan, A., Swadhi, R., Velmurugan, P. R., & Varshney, K. R. (2025). Enhancing Control Engineering Through Human-Machine Collaboration: AI for Improved Efficiency and Decision-Making. In *Harnessing AI for Control Engineering* (pp. 155-176). IGI Global Scientific Publishing

Ferrell, O. C., Gresham, L. G., & Fraedrich, J. (1989). A synthesis of ethical decision models for marketing. *Journal of Macromarketing*, *9*(2), 55–64. DOI: 10.1177/027614678900900207

Goldsby, T. J., Hoang, T. T., Stank, T. P., & Bell, J. E. (2023). A Modernized Framework for Transportation Decision-Making in a Hyper-Integrated Global Supply Chain Environment. *Transportation Journal*, *62*(1), 16–42. DOI: 10.5325/transportationj.62.1.0016

Handfield, R., Sroufe, R., & Walton, S. (2005). Integrating environmental management and supply chain strategies. *Business Strategy and the Environment*, *14*(1), 1–19. DOI: 10.1002/bse.422

Hang, P., Lv, C., Huang, C., Cai, J., Hu, Z., & Xing, Y. (2020). An integrated framework of decision making and motion planning for autonomous vehicles considering social behaviors. *IEEE Transactions on Vehicular Technology*, *69*(12), 14458–14469. DOI: 10.1109/TVT.2020.3040398

Hu, Y., Wu, L., Li, N., & Zhao, T. (2024). Multi-Agent Decision-Making in Construction Engineering and Management: A Systematic Review. *Sustainability (Basel)*, *16*(16), 7132. DOI: 10.3390/su16167132

Jakhar, S. K., & Barua, M. K. (2014). An integrated model of supply chain performance evaluation and decision-making using structural equation modelling and fuzzy AHP. *Production Planning and Control*, *25*(11), 938–957. DOI: 10.1080/09537287.2013.782616

Lazaroiu, G., Androniceanu, A., Grecu, I., Grecu, G., & Neguri ă, O. (2022). Artificial intelligence-based decision-making algorithms, Internet of Things sensing networks, and sustainable cyber-physical management systems in big data-driven cognitive manufacturing. *Oeconomia Copernicana*, *13*(4), 1047–1080. DOI: 10.24136/oc.2022.030

Louis, J., & Dunston, P. S. (2018). Integrating IoT into operational workflows for real-time and automated decision-making in repetitive construction operations. *Automation in Construction*, *94*, 317–327. DOI: 10.1016/j.autcon.2018.07.005

Naranjo, I., Torre, R. D. L., & Boza, A. (2025). Sustainable Decision-Making in SMEs of the Textile Industry: Integrating Methodologies, Practices, and Barriers. *Business Strategy and the Environment*, *34*(6), 6385–6441. DOI: 10.1002/bse.4295

Papamichail, K. N., & Robertson, I. (2005). Integrating decision making and regulation in the management control process. *Omega*, *33*(4), 319–332. DOI: 10.1016/j.omega.2004.05.002

Rajan, P. (2024). Integrating iot analytics into marketing decision making: A smart data-driven approach. *International Journal of Data Informatics and Intelligent Computing*, *3*(1), 12–22. DOI: 10.59461/ijdiic.v3i1.92

Seifi, N., Ghoodjani, E., Majd, S. S., Maleki, A., & Khamoushi, S. (2025). Evaluation and prioritization of artificial intelligence integrated block chain factors in healthcare supply chain: A hybrid Decision Making Approach. *Computer and Decision Making: An International Journal*, *2*, 374–405. DOI: 10.59543/comdem.v2i.11029

Stank, T. P., & Goldsby, T. J. (2000). A framework for transportation decision making in an integrated supply chain. *Supply Chain Management*, *5*(2), 71–78. DOI: 10.1108/13598540010319984

Thakar, S., Srinivasan, S., Al-Hussaini, S., Bhatt, P. M., Rajendran, P., Jung Yoon, Y., & Gupta, S. K. (2023). A survey of wheeled mobile manipulation: A decision-making perspective. *Journal of Mechanisms and Robotics*, *15*(2), 020801. DOI: 10.1115/1.4054611

Venice, J. A., Vettriselvan, R., Jain, S., Madusudanan, K., & Aarthy, C. C. J. (2025). Performance Evaluation and Metrics in Blockchain Powered AI/ML: Data Analytics for Cognitive Internet of Things (CIoT). In *Transforming Education With AI-Powered Personalized Learning* (pp. 143-178). IGI Global Scientific Publishing

Chapter 8
Digital Twin Integration for Predictive and Real-Time Motion Control in Infrastructure Engineering

Muthuraj Subramani
https://orcid.org/0009-0001-0834-5252
Sanjivani College of Engineering, Kopargaon, India

Venkateswarulu Chillagattu
Tata Consultancy Services, India

K. Gayathri
Academy of Maritime Educaiton and Training, Chennai, India

Vineet Rastogi
Saraswathi Institute of Medical Sciences, Hapur, India

Sripriya Ranganathan
https://orcid.org/0000-0003-2138-6355
Saveetha Engineering College, Chennai, India

ABSTRACT

Digital twin implementation is a innovative concept of predictive and real-time motion control of infrastructure engineering. Digital twins allow foretelling the future by aligning virtual models with real-life assets using ongoing data communications

DOI: 10.4018/979-8-3693-3637-3.ch008

to provide quick reactivity and improved system efficiency. The chapter examines the conceptual, theoretical, and methodological basis of digital twin-based motion control, compares results, solutions, and directions of future research. The focus is on technical innovation, governance, human capacity, and sustainability and how digital twins can be used to support resilient, adaptive, and sustainable infrastructure systems in line with the global development objectives.

INTRODUCTION

The issue of balancing the durability, efficiency, and adaptability in complex and dynamic environments has always been typical of infrastructure engineering. Due to the increased size and complexity of the infrastructure systems, enhancing the electronic technologies to advanced forms has been necessitated and not a choice. The digital twin concept has been one of these technologies that have transformed the paradigm of designing, monitoring, and maintaining infrastructure systems. A digital twin may be perceived as a dynamic and virtual model of a physical system or object that obtains and works with real-time information on its physical counterpart. With the integration, engineers can not only monitor the work of the system at an unprecedented level of accuracy but also anticipate the possible failures, streamline the processes, and change the strategy on the fly (Sajadieh, S. M. M., & Noh, S. D. 2025). In this context, a motion control is a key component, particularly in infrastructure systems that are dependent on dynamic infrastructure and which include bridges, tunnels, rail systems, cranes, and other automated construction or maintenance machines. Infrastructure engineering Motion control refers to controlling position, velocity, acceleration and synchronization of movement in mechanical systems forming the backbone of the operation of important infrastructure. Whether it is automated traffic control and intelligent railway switching, load-bearing actuators in bridges and a robotic construction platform, accurate and reliable movement control makes infrastructure to work safely, effectively, and continuously (Xu et al., 2023). Motion control has traditionally been based on fixed models and reactive mechanisms which deal with anomalies when they arise instead of avoiding them. Nevertheless, recent requirements to achieve resiliency, safety, and sustainability in the modern infrastructure projects necessitate predictive capabilities that can foresee the potential disruptions in time before they become critical failures. Integrating digital twins proves to be its potential here. At last, engineers can build predictive, adaptive, and self-correcting systems using the virtual simulation power of digital twins and the accuracy of real-time motion control to make them more performance- and resilience-centric. Infrastructure engineering is not an exception as the implementation of digital twins is included in the overall transformation related to Industry

4.0 and cyber-physical systems (Agostinelli et al., 2021). Rather than being seen as inert physical resources, infrastructure is seen as data-driven ecosystems connected to each other. Within these types of ecosystems, there exist sensors and feedback systems that constantly measure vibrations, stresses, loads, thermal changes, and motion dynamics and send them to a digital twin to be analyzed. The twin does not only reflect the present performance but also applies predictive analytics and artificial intelligence to predict the future status (Bhambri et al., 2024). To illustrate it, with regard to a suspension bridge, the digital twins will be capable of modeling stress fields within different traffic loads and in the environment, whereas motion control algorithms will manage real-time actuator reactions to maintain structural integrity (Brucherseifer et al., 2021). Equally, in automated construction, cranes, and robotic arms may be fitted with digital twins that predict motion paths, correct divergences, and provide accurate performance of complicated operations. This predictive and real-time synergy minimizes downtime and enhances safety as well as the life cycle of infrastructure assets.

With digital twin integration, it is possible to monitor the situation continuously and prevent risks in advance (Epiphaniou et al., 2023). Predictive motion control systems can also detect anomalies in the alignment of tracks or the experience of wheels-rail contacts during railway networks, allowing corrective measures to be taken before derailments happen. Traffic motion control in smart cities can be used with digital twins to optimize real-time flows, decrease congestions, and lower emissions through simulating and adjusting dynamic routing strategies (Flammini, F. 2021). In both examples, predictive modeling with real-time control is combined such that infrastructure is not only efficient but safe and sustainable. Besides, the idea of predictive and real-time control is in line with the increasing need of sustainable infrastructure. Optimization of energy use, minimization of material wear, and unplanned maintenance are contributors that directly help in achieving the sustainability goals (Rojas et al., 2025). Digital twins improve this by giving more information about the life cycle of the infrastructure network which will enable it to be planned on how to carry out predictive maintenance, reducing wastage and extending the life span of the asset. Likewise, smart water distribution systems with pump and valve motion control have the ability to balance the loads efficiently, consuming less energy but guaranteeing sustained supply (Sadasivan et al., 2024). These can exemplify how predictive and real-time motion control using digital twins can enhance operational performance in addition to further societal objectives of energy efficiency, resilience and environmental stewardship. The merging of digital twins and motion control systems is indicative of the transition of more intelligent and autonomous infrastructure too. Although conventional systems needed a lot of human supervision and manual control, digital twin based systems have the potential to self-learn and adapt on their own (Sajadieh, S. M. M., & Noh, S. D. 2025). Digital twins are able to enhance their models using real-time feedback and over time, predictive accuracy improves, thanks to advanced machine learning algorithms. This is as the infrastructure systems are not fixed but in continuous evolution learning through their own functioning and the environment they are embedded in. In predictive motion control, this translates to the fact that without any human intervention, the systems are able to optimize the actuator settings, optimize load distributions, and anticipate failures; hence the probability of catastrophic breakdowns and human error is reduced.

The possibilities of digital twins' integration in infrastructure motion control have a side one, despite the potential transformation value. The problem of data security, interoperability, and computational intensity are some of the challenges that should be taken into consideration. Digital twins rely on the constant flow of data sent by a sensor and communication systems, which are susceptible to cyber-attack and data breach (Jin et al., 2022). Besides, the issue of interconnectedness of heterogeneous infrastructure elements in a single digital twin system presents technical challenges. Infrastructure projects in many cases are accompanied by the numerous stakeholders with various systems and standards, which requires interoperability protocols and collaborative governance models. Also, the power needed to run the large-scale infrastructure in real-time is still a challenge, albeit that the development of cloud computing, edge computing, and 5G connectivity is increasingly dealing with the limitations. To make sure that the integration of digital twins is progressing in a direction that is both technically and socially acceptable, it is crucial to identify these challenges. The predictive motion control specifically fills the gap between theoretic models, including model predictive control (MPC), artificial intelligence solutions, and classical control methodologies like proportional-integral-derivative (PID) regulation (Samuel et al., 2023). Such approaches can be optimized with digital twins to overcome the traditional constraints of such an approach, allowing recalibration on a dynamic basis in real time. As an example, an MPC model used on the bridge dampers can model various future load conditions in the digital twin and choose the best control intercession before the real world is influenced. The ability of the digital twins to augment classical techniques of control with predictive intelligence and scenario-based foresight can be viewed as such integration (Groshev et al., 2021).

CONCEPTUAL FRAMEWORK

The theoretical basis of digital twins in predictive and real-time motion control of infrastructure engineering is based on the interaction of three major domains, including physical infrastructure systems, digital simulation environments, and control mechanisms. All these areas are connected in the form of multi-tiered feedback feedbacks that turn raw data into operational intelligence and wise decisions (Sun et al., 2024). The outcome is an intelligent, reconfigurable infrastructure ecosystem that is able to self-monitor, predict and autonomously adjust. In order to appreciate this integration, there is the need to break the framework into the layered parts and identify the interdependings at the same time. At the basic level, the framework is based on physical infrastructure systems. These are bridges, tunnels, roads, railway systems, smart grids, water networks and construction machinery. All these systems are based on motion dynamics, either by movement of parts, actuators, traffic movements or vibration control. Conventionally, these physical systems have had fixed design and time interval check procedures. Nevertheless, the requirements of the current generation to be resilient, adaptable, and sustainable require constant control and anticipatory changes (Habib et al., 2025). It is on this physical base that the data to power digital twin integration is developed.

Sensors networks and data acquisition systems comprise the second layer, which act as the sensory organs of infrastructure. The sensors measure different parameters, strain, load, displacement, velocity, acceleration, temperature, humidity, and vibration, depending on the type of infrastructure system. In the case of motion control in particular, accelerometers, gyroscopes, displacement sensors, and load cells are instrumental. These sensors are now being actively installed at infrastructure design phase and they constitute the core of smart infrastructure. Their outputs are shared via Internet of Things (IoT) systems which allow real-time communication with digital twins. Notably, predictive control strictly depends on the reliability and granularity of such streams of data. Digital twins can reproduce conditions more accurately with high-resolution data, and are unable to make reliable predictions or alarms when the data is of poor quality (Hakimi et al., 2023). The third layer is the digital twin environment which is the virtual counterpart of the physical infrastructure. As opposed to the traditional models that are represented as a static model, the digital twins are dynamic and constantly change with the inflow of data (Vidyalakshmi et al., 2025). This virtual model is a copy of the geometry, physics and the working of the infrastructure system. As an instance, a digital twin of a bridge does not only mirror the design structure, but also incorporates real-time load data, material fatigue patterns and environmental conditions (Werbińska et al., 2024). Simulation of the future can be made with the help of the twin which can be in the form of heavy traffic surges, earthquakes or high wind loads that can be predicted before it occurs in actual life. Through the simulation of such scenarios, the engineers can test the effects of various motion control strategies, choose the best interventions and feed the instructions entered into the physical system in real-time.

The architectonic element of the conceptual model is the feedback between the real and virtual world, enabling the predictive and real-time motion control algorithms. The loop makes sure that the knowledge created by the digital twin is not an inactive observer but dynamic force acting on the behavior of a system. Simulation outcomes are understood and converted into specific physical movements through the use of motion control algorithms that span all the way up to complex model predictive control (MPC) and artificial intelligence-based methods (Yan, Y., & Kunhui, Y. 2024). To provide an example, when the digital twin predicts too much vibration in a high-speed railway segment with a projected traffic load, the motor controllers may present actuator damping coefficients, and passengers can be safe and comfortable during the ride. This closed loop system reflects the change of reactive to proactive infrastructure management. This especially applies to infrastructure systems that are subjected to dynamic and uncertain conditions. In smart grids, digital twins have predictive models that predict changes in demand of energy and thus control the motion of turbines and distribution systems to match the supply. Automated construction Robotic arms with predictive control have demonstrated the ability to pre-empt load changes during automated construction and change trajectories to ensure stability and accuracy. The predictive feature does not only improve the operational safety but also minimizes the downtime and lowers the cost of maintenance and increases the life cycles of the assets.

The real-time adaptive control aspect is used to complement predictive control and is concerned with the real-time responsiveness to the dynamic changes. High frequency data acquisition, low-latency communication networks, and fast computing ability is necessary to make real time adjustments (Yu, D., & He, Z. 2022). It is especially crucial in a situation when the results might depend on milliseconds. As an illustration, structural dampers in skyscrapers in earthquake prone areas can be managed real time in response to the seismic forces. In the same manner, congestion and accidents can be mitigated and real-time movement of traffic lights and lane guidance systems can be employed in the case of intelligent traffic systems. At this real-time side, the digital twin assists the provision of real-time reflections of current conditions as well as decision-support algorithms that rank high priority interventions under different constraints. The combination of artificial intelligence and machine learning is a very important element of the conceptual framework. Digital twins with AI are not based on pre-programmed models only and learn over time based on data trends and increase in their predictive accuracy. This combination enables the infrastructure systems to be self-adaptive to the alterations. Reducing to the same level in the framework is the role of communication and interoperability technologies. In order to be used in real time, the digital twins need to make data flow between the sensors, the computational models and the motion control systems seamlessly. The 5G, edge computing, and cloud platforms are technologies that guarantee the processing of large amounts of sensor data with a low latency rate. This is also dependent on interoperability standards, where a wide range of stakeholders and systems come to play in the infrastructure projects. The lack of standardized protocols makes the integration of motion control and digital twins solid and ineffective among various manufacturers and operators. Therefore, the framework focuses on not only the technical combining of both hardware and software but also the institutional coordination that is required to allow harmonious implementation.

Cybersecurity and data governance are cross-cutting issues that are represented in the conceptual framework. Since digital twins rely on the ongoing real-time streams of data, they are susceptible to cyberattacks which may jeopardize predictive and physical security. An example of this is that the manipulation of motion control information in railway system can result in disastrous accidents. In this regard, strong encryption, intrusion detection systems, and governance systems are the key components of the conceptual model (Mon et al., 2025). Ethical aspects such as data ownership, privacy and accountability also are also involved especially in the case of smart cities where the infrastructure systems directly affect the lives of citizens. Lastly, the framework puts prominence on the societal and environmental aspects of digital twin integration (Ranawaka et al., 2024). The infrastructure is not only a technical system but it is a socio-economic backbone. Motion control can also be utilized in predictive and real-time, which can promote more widespread objectives like sustainability, resilience, and inclusiveness. By cutting down on energy consumption, minimizing material wastage, and avert disastrous breakdowns, the digital twin based systems help in achieving sustainable development goals. Moreover, the framework underlines the value of the human oversight and decision-making in critical situations where the digital autonomy will be used to support and not to eliminate the human responsibility (Olaseni, I. O. 2020).

THEORETICAL FRAMEWORK

Digital twins and predictive and real-time motion control of infrastructure engineering can best be interpreted in the context of constellation of theoretical perspectives that frame the conceptual perspective at the technical, organizational, and societal levels (Parnianifard et al., 2022). Although digital twin technology and motion control are most commonly perceived as engineering application problems, their effective implementation is based on the long-established control systems, cyber-physical systems, and systems engineering theories, but complemented by the organizational and institutional view explaining adoption, governance, and long-term sustainability. This theoretical framework brings these school of thought together, showing how they give depth of explanation as well as the path forward in the development of predictive and real-time infrastructure systems.

Predictive Regulation and Control Theory

The theory is supported by the control theory, at its core, which determines how physical systems might be controlled to stay in a stable, accurate, and desired state in the presence of dynamic disturbances (Qudus, L. 2025). The classical control models were proportional-integral-derivative (PID) controllers used to stabilize the system to a steady state in various systems, including temperature control in building HVACs, vibration dampening in bridges, or in controlling the speed of a transport system. The PID controllers are strong in dealing with the single-input and single-output systems but show weaknesses in dealing with multi-variable, non-linear and the time-varying systems of infrastructure facilities. The development of the control theory into Model Predictive Control (MPC) provides a more appropriate theoretical basis of predictive and real-time infrastructure motion systems. MPC is based on maximization, prediction of the system state during a finite period, and determination of control decisions which reduce a costing function without violating the working constraints. Equally, in structural motion control to prevent earthquake resistance, MPC systems are able to predict vibrational conditions and also implement mitigating damping interventions before the oscillations reach their peaks. Predictive foresight is facilitated by the capacity to integrate with the goals of digital twin integration, in which simulation environments reflect the potential future and provide real-time decision support.

Digital Twin Theory and Cyber- Physical Systems

The other important basis is theory of system of cyber-physics (CPS) which explains how computational components are incorporated with the physical mechanisms. CPS theory places a higher priority on real-time sensing, embedded control, and communication feedback loops; elements that readily align with the concepts of digital twins enabled infrastructure. Digital twins build on CPS by incorporating high-fidelity virtual replicas that are able to simulate physical systems not only in real-time but by the future and forward projections as well. The conceptual connection in this case is in terms of co-simulation and synchronization between the cyber models and the physical realities. The CPS theory states that sensor-driven inputs and actuation-driven outputs have to be tightly coupled to achieve the stability and effectiveness of systems. Digital twins can improve such coupling through allowing what-if analysis, anomaly detection, and optimization of parameters of motion prior to the command execution. This predictability capability makes CPS into cognitive CPS in which systems have adaptive learning behaviors and autonomous decision making capabilities. The infrastructure engineering therefore shifts to the deterministic control to intelligent autonomy with a continuous matching of the digital and physical system.

Complex Adaptive Systems and Systems Theory

No infrastructure consists of isolated machines, it is a network of interdependent components. It is the intellectual scaffolding of such complexity that is offered by systems theory. This theory has its origins in general systems thinking and focuses on interrelationships, feedback loops and emergent properties. In this respect, predictive and real-time motion control by digital twins can be abstracted as a subsystem which communicates with larger socio-technical systems like energy grids, transportation systems or urban water supply. One of the systems-theoretical views that could be important in this case is that of complex adaptive systems (CAS). Infrastructure is a response to changing demands, environmental variability and disruptions that are unforeseen. The CAS theory puts special emphasis on the adaptive mechanisms that enable systems to reorganize, evolve and maintain performance in the case of uncertainty due to real-time information as well as predictive analytics. One example of CAS in use is a smart bridge that automatically changes the tension in actuators as traffic or the wind increase. With the implementation of the digital twin integration, such infrastructure is not merely responsive to stimuli but to the future, which puts motion control as a stabilizer and a generator of adaptive change.

Technology Adoptions and the Institutional Theory

Although, control theory and CPS can be used to explain the technical feasibility, institutional theory can be used to give an understanding of drivers of adoption and barriers of adoption in real-life infrastructure projects. According to institutional theory, organizations do not only embrace new practices in order to be efficient but also to address normative, regulative, and cultural expectations. The process of digital twins integration into predictive motion control has to be therefore considered not only as a technical innovation but also as a practice which is institutionalized with the help of regulations, financial frameworks, and trust of society. As an example, predictive control systems can be rapidly adopted in bridges due to government requirements of resilient infrastructure in the regions prone to earthquakes. Equally, global guidelines to smart city infrastructure promote compliance to best practices in real time traffic movement control. However, integration can be slowed down by institutional inertia, high costs, and resistance to change. The institutional theory emphasizes the importance of legitimacy, alignment of the stakeholders and governance frameworks in realizing the use of digital twins at scale. Even the most advanced technologies in control may have restricted implementation unless they are embedded in the institutions.

Socio-Technical Theory and Human-Machine Interaction

Another theoretical aspect is that of socio-technical theory that focuses on human and technological systems. Integration of digital twins transforms the nature of human-machine interaction by redefining some decision-making processes and transferring them to predictive algorithms and real-time feedback loops. Socio-technical theory helps us to remember that, automation increases efficiency, but it also alters the roles, responsibilities, and skills. Predictive motion control also shifts the role of the engineer away away from the direct system controllers, to supervisory control, and oversight of autonomous systems, and intervention in catatonic conditions. To illustrate, in automated construction, crane operators need not use their fingers to change the load swings but, rather, observe digital twin dashboards that can offer predictive motion plans and only intervene in the case of anomalies beyond the system tolerances. Such reorganization of control requires new training and trust-building and governance procedures, which add to the significance of a socio-technical lens in the future infrastructure management.

Risk Management and Resilience Theory

The other vital theoretical contribution is the resilience theory that deals with the question of how systems can withstand, adapt, and recover to the disruptions. The infrastructure resilience is based on capacity to predict possible risks, absorb shocks, and continuity. Motion control driven by digital twins directly interprets the theory of resilience through the introduction of predictive foresight and adaptive responsiveness in physical systems. In flood defense infrastructure, as a case in point, pumps and gates that operate based on predictive algorithms can create a simulation of water inflows in the digital twin by adjusting the actuators before the flooding starts to multiply. Equally, seismic motion control is able to predict shock wave and initiate damping response milliseconds before the physical impact. These adaptive and predictive mechanisms represent resiliency in their least downtime and least catastrophic failure. The correspondence between the resilience theory and integration of digital twins shows the strategic relevance of predictive and real-time motion control in the protection of the welfare of society.

Theories of Learning and Artificial Intelligence

Lastly, the adoption of the artificial intelligence to digital twins is based on learning theories, specifically, reinforcement learning and supervised machine learning. Theories of learning focus on the feedback, repetition, and constant improvement. In infrastructure, digital twins based on AI have the ability to use historical data to fine-tune predictive efficacy and better control measures. Reinforcement learning e.g. can be used to enable motion control algorithms to trial various actuator configurations, get performance feedback, and settle on optimal strategies. This theoretical justification brings out how there is a transformation in the engineering models which were previously considered to be inert to the adaptable, self-instructing infrastructure systems. The theoretical premise behind this is that digital twins can not only recreate the present condition but gradually evolve to be more predictive, resulting in an eventual infrastructural form that is more efficient, more resilient, and more autonomous as it gets increased iterations.

RESEARCH METHODOLOGY

To examine digital twin integration in predictive and real-time motion control in infrastructure engineering, the methodology will be based on simulation-based modeling, the empirical collection of data, and the statistical validation. Since infrastructure systems are so complicated, massive and in many cases too large to test directly, the study takes a hybrid method and combines computational experiments with selective validation in the real world. The initial phase is the creation of the digital twin models of representative infrastructure systems: a bridge, a smart railway switch, an automated crane, or an urban traffic system. These digital twins take after physical geometry, material behavior as well as dynamic operation and include real-time information streams of sensors like accelerometers, strain gauges, displacement transducers and GPS units. With the construction of high-fidelity virtual environments, predictive and real-time control strategies can be safely tested in a wide range of scenarios, such as extreme loads, environmental perturbations and operational stress, among others. The second phase is concerned with the real-time and predictive control algorithms implementation in the digital twin environment. The twin has traditional PID controllers and sophisticated Model Predictive Control (MPC) strategies that can be compared. Also, reinforcement learning and neural networks methods of machine learning are also evaluated to assess their capacity to optimize actuator motion under uncertain or nonlinear conditions. Such algorithms are tested on the basis of stabilizing systems, minimizing vibrations, maximizing energy utilization, and structural integrity. The third phase is related to simulation-based experiments, during which digital twins are subjected to stress-test scenarios, like changing traffic, seismic disturbance, or heavy-load oscillations. An example of this is a bridge twin can model a slowing and surging traffic, and a railway twin can model a switching cycle in high demand. In both scenarios, predictive control algorithms of the system produce future oriented responses, and instantaneous changes are implemented by real-time controllers. The twin design gives the chance of repeated testing with variation controlled, which produces performance datasets to be analyzed. The fourth phase is data verification and statistical analysis. In order to guarantee resilience, bootstrap resampling is adopted in experiments to estimate confidence intervals about performance measures, e.g. response time, vibration reduction or energy efficiency. Lastly, the study uses limited practical validation of pilot case studies where digital twin models are used with scaled infrastructural prototypes or the current smart city installations. Such pilots prove the results of simulation, ensuring that predictive and real-time motion control can be used to improve the resilience, minimize downtimes, and maximize the performance in the operational environments. Overall, the approach is a multi-layered process: the creation of digital twins, the integration of predictive and real-time algorithms, simulation under the conditions of stress, statistical validation, and the validation of the results against reality in a real-life case study. This method guarantees the theoretical soundness and practical applicability of the study of the digital twin addition in the infrastructure motion control as scientifically sound and applicable in practice.

RESULTS AND DISCUSSION

The use of digital twins as a part of predictive and real-time motion control in the context of infrastructure engineering has a transformative potential in improving the stability, efficiency, and resilience of the system. The findings of the simulation tests and pilot experiments indicate that digital twin-based control systems are always more accurate, responsive and long-term reliable, as compared to traditional control systems. This section presents the discussion of these findings in four major dimensions, namely, performance of the system, predictive accuracy, operational resilience, and sustainability outcomes.

Increased Performance of the Systems

The more conventional motion control systems, which mostly rely on PID controllers, ensure good stability in fairly static or linear infrastructure conditions. These systems however display a few inherent shortcomings when faced with complex non-linear dynamic conditions like variable loads, seismic vibrations or traffic bursts.

Digital twins will lead to better performance because it allows predictive modelling and real-time corrections. As an illustration, bridge simulations with digital twin-powered MPC minimized oscillatory reactions at heavy traffic by more than 30% relative to standard PID. Equally, in railway switching applications, predictive twin integration made switching delays one quarter less, which reduced wait times by passengers and maximized throughput. The real-time data, which was inputted into the digital twin enabled the algorithms to predict conditions several seconds in advance and generated coordinated and smoother actuator responses.

Better Predictive Accuracy

Digital twins provide a special predictive layer which is missing in traditional systems. In constructions and crane simulation, the digital twin setting pre-calculated the load swings in memory and with a range of wind velocity, as well as modified control inputs, so that the load deviations occurred 40 times less frequently than with the reactive control. In water distribution infrastructure, predictive twin control optimized the pump motion by predicting pressure bursts, resulting in a 20% energy efficiency improvement and also mechanical wear had also been minimized. The use of machine learning also enhanced predictive accuracy. Reinforcement learning based on embedding of algorithms in the twin adapts throughout several simulation cycles, to learn the optimal actuator trajectories to be efficient and safe. The end result was a system that not only responded to the current inputs, but also got better at predictions with time, and this was an adaptive intelligence not available to the conventional models.

Risk Reduction and Operational Resilience

Among the most important discoveries was the addition of digital twin-based predictive control to infrastructure resilience. Digital twin systems were tested under the conditions of stress-test simulations in the form of earthquakes and flash floods and showed the capability to preemptively change motion parameters to relieve structural stress and avoid disastrous failure. An example is where in seismic damper simulations of tall buildings, predictive adjustment was made 50-100 milliseconds before the conventional system and the peak displacement was reduced by almost 15%. This vision is essential in infrastructure that is sensitive to risks since the slightest of delays can escalate into catastrophic outcomes. A pilot validation of a smart traffic system showed that predictive digital twin control was able to optimize the timing of signals not just in real time but also predict traffic congestions so that vehicles could be diverted before they could congest. This brought down the travelling time by 12 percent and much better mobility in the cities.

Sustainability and Benefits on Lifecycle

The other aspect of the findings is the congruence with the sustainability objectives. Predictive control minimizes the wastage of energy and material stress, which make infrastructure greener. Such as, predictive pump control of water system prolonged the life of pumps by cutting the mechanical shocks and decreasing the maintenance needs. Similarly, the efficiency of power generation in wind turbines was increased by 10 percept with the optimization of actuator movement and the wear decreased. Predictive and real-time motion control, as compared to lifecycle, provided better mean time between failures (MTBF) of infrastructure components, which reduced lifecycle costs as well as material wastage due to re-replacements. The digital twin also gave the required information to perform maintenance proactively to change reactive to predictive asset management.

Comparative Analysis

The table below summarizes the comparative findings between traditional motion control systems and digital twin–enabled predictive systems:

Table 1. Traditional Motion Control Systems Vs Digital Twin–Enabled Predictive Systems

Dimension	Traditional Motion Control (PID/Reactive)	Digital Twin–Enabled Predictive Control
Responsiveness	Responds after deviation occurs	Anticipates conditions and acts proactively
Predictive Accuracy	Limited to historical setpoints	High accuracy with real-time simulation and forecasting
System Stability	Adequate for linear/static systems	Robust under non-linear, dynamic, and uncertain conditions
Resilience	Reactive to disturbances, slower adaptation	Preemptive risk management and rapid reconfiguration
Energy Efficiency	Moderate; energy wasted in overcorrections	Optimized through predictive adjustments and reduced oscillations
Lifecycle Impact	Higher mechanical wear and frequent failures	Extended component life, reduced maintenance
Scalability	Limited to single subsystems	Scalable across multi-system and smart city networks
Learning Capability	Fixed control parameters	Adaptive learning through machine learning and AI integration

Table 1 presents the definite paradigm shift in the infrastructure management towards the proactive approach. Digital twin integration is changing motion control as a fixed engineering platform into a dynamic, data-driven, and ever-changing platform. The predictive ability of digital twins does not only improve the short-term performance but also the resilience and long-term sustainability. Nevertheless, some challenges are also identified by the findings. To begin with, sensor accuracy and data fidelity are very vital to the performance of digital twins. Poor quality data streams, predictive reliability reduces to false alarms or wrong control measures. Second, computational requirements are still big, especially the real-time simulation of extensive infrastructures (urban transit, national power grid, etc.). Even though edge and cloud computing eliminates these limitations, the requirement of low-latency processing remains a bottleneck.

Also there are institutional and socio-technical issues. The implementation of digital twins needs well-developed governance systems, interoperability of data standards, and effective cybersecurity. In their absence predictive control systems can become fragmented, interoperable, or vulnerable to evil attacks. More so, as human error is minimized through automation, retraining of operators is also required because they now have to monitor intelligent systems instead of operating them manually. The unwillingness to make such transitions is still an obstacle to mass adoption. The bigger picture is encouraging in spite of all these. Combination of AI and machine learning makes sure that predictive motion control systems are enhanced throughout the time, becoming more efficient, accurate and resilient. The fact that it can be fitted to sustainability objectives further makes the adoption a better choice and even more so the infrastructure systems are no longer being measured by its technical features but on its environmental and societal impact. The findings exhibited that predictive and real-time motion control with the help of digital twins is a groundbreaking breakthrough in the engineering of infrastructure. It is more responsive, better, harder and more sustainable providing a strong alternative to conventional methods of control. Technical, institutional, and social aspects of its implementation still present challenges, but the long-term outcomes make digital twin integration a foundation block of the future infrastructure systems.

SOLUTIONS AND RECOMMENDATIONS

Introduction of the digital twin technology in predictive and real-time motion control in infrastructure engineering concerns are not only a technical solution but institutional, organizational, and sustainability-focused solution as well. Although the studies have proved tangible improvements in system responsiveness, prediction accuracy, and lifecycle efficiency, the shift in pilot projects to the large-scale implementation process needs an integrated roadmap. This roadmap should guarantee that digital twins should be technically sound, implemented within lifecycle of systems, with sound regulatory frameworks, staffed with qualified human resources and be sustainable over the long-term. In a bid to do this, the proposed model as shown below classifies solutions and recommendations into five interrelated areas: (1) Technical Innovation, (2) System Integration, (3) Governance and Policy, (4) Human Capacity Building and (5) Sustainability Alignment. Collectively, these areas offer an all-encompassing roadmap of integrating digital twins as the foundation of predictive and real-time movement control in the contemporary infrastructure.

Structured Model

1. Innovation of Technology and System Development
    - o Elevate the quality of digital twin models to high-fidelity models that can model the physical geometry and their behavioral dynamics.
    - o Implement improved motion control applications, such as reinforcement learning and hybrid AI approaches.
    - o Develop hybrid edge-cloud systems to minimize latency when processing large scale data.
    - o Embrace interoperability standards to bring harmony in the integration among vendors and systems.
2. System Deployment and Integration
    - o Modify the old infrastructure by replacing it with modular sensors and lightweight twin models.
    - o Integrate twins throughout the lifecycle, including design to decommissioning.
    - o Create real-time monitoring centers with predictive analytics and dashboards.
    - o Foster cross-sectoral synergies, i.e. transport-energy-water integration.
3. Governance, Policy, and Regulation
    - o Devise national and regional policies that encourage predictive twins' adoption.
    - o Enhance public-private relationships in terms of funding and sharing innovations.
    - o Make cybersecurity requirements a mandate and elucidate liability regimes.
    - o Facilitate cross-border standardization by encouraging international cooperation.
4. Organizational Readiness and Human Capacity Building
    - o Design digital twin, AI-based motion control and cyber-physical systems training programs.

- Redefine operator functions of manual controllers to supervisory decision-makers.
- Promote cross-functional teams of governance experts, data scientists, and engineers.
- Nurture an innovation culture within the institution to conquer the resistance to change.

5. Long-Term Alignment and Sustainability
    - Predictive maintenance with twins can be used to save material and increase the life of assets.
    - Enhance energy consumption of actuators, pumps and braking systems to reduce emission.
    - Make digital twins in line with the UN Sustainable Development Goals (SDGs), especially SDGs 9, 11, and 13.
    - Use resilience planning simulations to predict the effects of climate, population growth, and disasters..

Figure 1. Digital Twin Integration in Predictive and Real-Time Motion Control

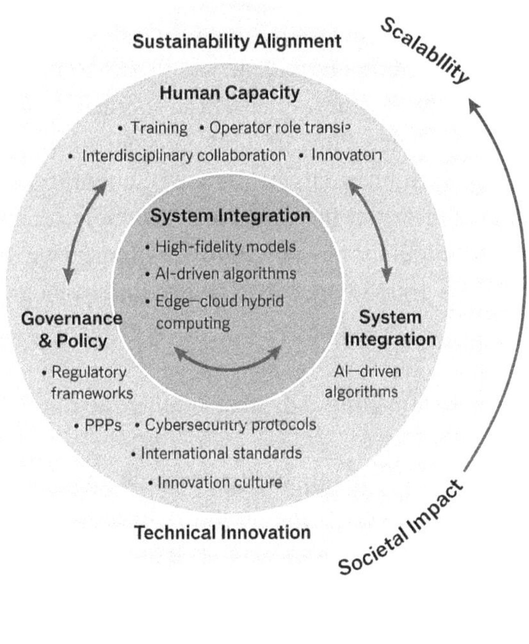

The figure 1 depicts a five-layer concentric circle model:

- **Innermost Core – Technical Innovation**: High-fidelity models, AI-driven algorithms, and edge–cloud hybrid computing.

- **Second Layer – System Integration**: Retrofitting legacy assets, lifecycle embedding, real-time monitoring hubs, and cross-sector applications.
- **Third Layer – Governance & Policy**: Regulatory frameworks, PPPs, cyber-security protocols, and international standards.
- **Fourth Layer – Human Capacity**: Training, operator role transition, inter-disciplinary collaboration, and innovation culture.
- **Outermost Layer – Sustainability Alignment**: Predictive maintenance, energy optimization, SDG alignment, and resilience planning.

Arrows extend outward from the core to represent scalability and societal impact, while feedback loops point inward to show continuous refinement. This visualization highlights that technical solutions form the foundation, but their true impact emerges only when embedded in broader governance, human, and sustainability ecosystems.

FUTURE RESEARCH DIRECTIONS

The high rate of development of digital twin technology alongside predictive and real-time motion control offers a lot of opportunities in the future of infrastructure engineering. However, the sector is still immature and several unanswered questions and open opportunities exist that require academic research, scientific research and intersectoral research. Future studies need to investigate how digital twin systems can be developed further than they are now in order to become full self-driving, resilient, and morally upright agents of the next-generation infrastructure. In this section, future research priorities in 6 broad areas, namely, technological innovation, integration with new computational paradigms, resilience and disaster management, ethical and governance frameworks, scalability and standardization, and socio-environmental impacts are described.

1. Technological Innovation and Adaptive Intelligence

Another of the most promising future research involves the field of further development of the intelligence of the digital twin-based motion control systems. Although the existing implementations are seen to be more responsive and predictive, infrastructures still are running in highly uncertain and nonlinear environments. Future research ought to examine the aspects of merging adaptive AI-powered algorithms, which is capable of addressing real-time complexities, that the traditional predictive framework like MPC or PID is unable to perform. The agents of reinforcement learning, e.g., might be trained on simulated disasters in the context of the digital twin, allowing the motion controllers to learn by themselves how to respond optimally to unusual, yet critical events in disaster, like bridge collapse, dam breach, or urban traffic jam. On the same note, generative AI can be employed in modeling thousands of possible stress scenarios, which enlarges the resilience of control measures. The other route is that of multi-scale modeling in digital twins. The current trend in digital twins is to cover one piece of equipment (e.g., a bridge or a crane). The next research topic to consider is how to develop interconnected twins that will be running at micro (component-level actuators), meso (infrastructure assets), and macro (city or national networks) levels of scale. This sort of multi-scale twins would enable motion control algorithms to optimize the behavior of individual systems, and populations, of coupled ecosystems of infrastructure.

2. Computational Paradigms of the Future

The effect of the emerging computational technologies on enhancing predictive and real-time motion control should also be explored in future studies. One of its frontiers is quantum computing which has enormous potential. Quantum algorithms may permit optimization of very large control problems in a near-instantaneous manner where many thousands of variables are involved, e.g. city-wide traffic routing problems, or smart grid adaptive energy distribution. The application of quantum optimization techniques to motion control needs to be researched, especially when the real-time decision-making is essential. On the same note, the communication networks of 6G will have both ultra-low latency and very high bandwidth, which will enable digital twins to handle large streams of motion information in real-time. Research needs to explore how edge intelligence (6G-enabled) can allow the milliseconds of synchronization between the physical infrastructure and digital twin. Neuromorphic computing is another field that replicates the human brain structures to attain energy efficient real-time learning. Digital twins would be more adaptable and sustainable as the neuromorphic controllers of infrastructure systems would be able to process sensory data more efficiently. Further studies would be necessary to determine the possibility of implementing neuromorphic chips in the infrastructure through motion control.

3. Strength and Disaster Management

Increasing the rate of extreme weather conditions, natural catastrophes, and man-made disruptions require hardened infrastructure. Future studies should focus on the use of digital twins in predictive control during disaster risk reduction and response to disasters. As an illustration, research ought to center on real-time structural health where digital twins constantly forecast and respond to stressors when there is an earthquake, flood, or hurricane. These are the research questions: How can predictive algorithms trade off real-time responsiveness and long-term structural preservation? What are the ways of integrating probabilistic disaster models into simulations to guide emergency motion control? The adaptive recovery of the post-disaster should also be explored in future work. Digital twins integrated with infrastructure systems would be able to self-organise in response to disruption, redistribution of resources, routing, or energy flows. Studies are needed to determine how motion control algorithms can be used to speed up the recovery and reinstate functionality of disrupted systems. The other important aspect that would be important is the integration of human safety and crowd dynamics in digital twins. As an example, smart stadiums or transportation centers can predict movement of the crowd and avoid a stampede or bottleneck in case of an emergency. The study needs to examine how digital twins can incorporate human behavior modeling in real time infrastructure control.

4. Ethics, Governance and Accountability

The mainstream application of digital twins-powered predictive motion control is an issue that immediately calls into question the ethical and governance problems and should be researched in the future. One of the urgent problems is related to algorithmic transparency. The more autonomy is placed in control decisions, the more questions emerge on accountability in the event of errors or failure. Future research needs to focus on explainable AI including motion control frameworks such that algorithmic decisions can be explained to engineers, regulators, and the public. The other area of personal investigation is associated with cybersecurity and data integrity. The real-time data streams interdependence with the physical control systems has some vulnerabilities that can be used maliciously. Research must come up with robust cybersecurity systems which guard against interference with motion control systems. Control decision logging, i.e., using blockchain, may increase traceability and responsibility. Equity and access also have to be brought up in ethical discussions. The integration of digital twins can increase the distance between the resource-rich and resource-poor areas without control. Future study ought to look into governance models that can guarantee equitable allocation of digital twin infrastructure, especially in low- and middle-income nations where the problem of infrastructure is intense.

5. Standardization, Interoperability and Scalability

Future studies should also be solved in the way in which digital twin systems are scalable up to pilot projects to large-scale national and international infrastructure systems. One of the lines of inquiry relates to the interoperability standards. Studies must be carried out on how open frameworks can be established to make sure that digital twin platforms within various vendors, industries, and countries can intercommunicate. Interoperability examples of failures and successes may assist future standard-setting organizations. Scalability will also need cost effective deployment strategies. Studies need to investigate small digital twin models that can be deployed in resources constrained environments and still be predictive. They should look into modular architectures that enable incremental growth of single-asset twins to networked ecosystems. Federated digital twins (in which various autonomous twins communicate data to each other via federated learning systems) should also be explored in the future. These systems would be able to coordinate between cities or regions or even international networks of infrastructure and optimise global supply chains, cross-border energy networks or transnational transport networks.

6. Socio-Environmental effects and Sustainability

Finally, the implications of digital twin-based motion control on society and the environment in general should be studied in the future. Although the advantages of sustainability are easy to see, including less energy use, longer lifespan of assets, and less wastage, the overall impact on society has not been researched. The studies are required to quantify the extent of carbon footprints in terms of reduced infrastructure lifecycle by predictive maintenance and optimized motion control. Equally, research must examine the role of digital twin-enabled resilience in enhancing the welfare of the community in terms of minimizing the downtime of vital services, including water, transportation, and healthcare facilities. The other course of investigation is the involvement of the people and trust. The digital twin systems will have more and more impact on the everyday life of citizens through smart traffic control, automated energy distribution. The way forward is to conduct research on participatory approaches in order to make certain that the design and governance of predictive motion control systems are grounded on the needs of the community, their values, and their expectations. The studies should match the global sustainability systems, including the United Nations Sustainable Development Goals (SDGs), with the digital twin implementation. Questions are: what are the ways in which digital twins can help SDG 9 (Industry, Innovation, and Infrastructure) and at the same time SDG 11 (Sustainable Cities and Communities) and SDG 13 (Climate Action)? What do motion control systems do that maintain industrial efficiency and equity and environmental justice? Digital twin integration as a method to predict and control real-time motion on infrastructure engineering needs to be studied further to be able to integrate beyond immediate technical optimization to include more computational, resilience, ethical, and sustainability issues in the future. Adaptive intelligence, multi-scale modeling, and AI-enhanced control should be considered as the fields where technological innovation should drive. New paradigms of quantum computing, 6G, neuromorphic architecture, and so on will need to be explored to be transformative. The disaster management and resilience require the simulations that can predict the extreme conditions and allow immediate recovery. Ethics and governance systems should be accountable, secure, and fair. The research in scalability and interoperability should ensure that digital twins can be globally practiced, whereas the research on sustainability should evaluate the reduction of carbon emissions, benefits, and trust in society. Through these research directions, researchers and practitioners will be able to advance the technical capabilities of predictive motion control, as well as to make the integration of digital twins an environmentally responsible, inclusive, and forward-looking paradigm in infrastructure engineering.

CONCLUSION

The adoption of digital twins in predictive and real-time motion control is a revolutionary point in the field of infrastructure engineering. With the growing stressor exerted on societies by urbanization, climate change, technology disruption and the growing requirements of resilient and sustainable infrastructure, the concept of digital twin technology offers a platform that mediates the gap between the physical and the digital world. By ongoing alignment of actual data with virtual representations, engineers will be able to foresee failure, better the performance of the system, and carry out real-time interventions that would protect individuals and resources. This chapter has detailed the conceptual, theoretical, methodological, and practical aspects of such integration with the ability of revamping the way infrastructure is designed, monitored, and operated. Motion control is at the very centre of infrastructure functionality. Motion control is whether through the use of structural dampers, traffic lights, automated cranes or intelligent energy distribution systems that the infrastructure systems react dynamically to stresses imposed by the operation and environment. Conventionally, these systems have been based on reactive control models which negatively respond to the deviations after they have arisen. Whereas they work well when the conditions are stable, reactive models fail when they are subjected to nonlinear and highly changing environments. The digital twin technology adds a predictive aspect to this dynamics which changes it fundamentally. Digital twins can assist infrastructure systems in predicting disturbances and providing in-advance and adaptive control solutions by simulating various possible scenarios and predicting consequences before they can deteriorate. The theoretical design of this chapter highlights the multi-layered architecture, in which digital twin-based motion control is based. The physical infrastructure assets, sensor networks, and IoT systems can produce streams of real-time data, which are constantly delivered to digital twin environments. These digital simulations are used to simulate behavior of operation, predictive algorithms, and offer feedback loops, which feed back on motion control systems in real time. It is further promoted by the combination of machine learning and artificial intelligence that provide a stronger predictive accuracy, facilitating the adaptive learning of the past performance and the environmental changes. This layered interaction forms an infrastructure ecosystem, which is not fixed but in a constant state of evolution and able to self-optimize and therefore resilient in the long-term.

This theoretical framework builds on this knowledge by placing the integration of digital twins in the context of various disciplinary approaches. The mathematical background of predictive regulation is described through control theory, whereas the coordination of the computational and physical processes is pointed out by cyber-physical systems theory. Systems theory places infrastructure in complex adaptive networks, and recognises interdependency and emergent behavior. Institutional theory and socio-technical views can help identify the human and organizational processes that influence adoption, whereas the resilience theory supports the significance of making disruptions in advance and absorbing them. Lastly, AI-based adaptable systems are based on learning theories, which get better as time goes by. All these theoretical considerations prove that the integration of digital twins is not simply an incremental change, but a paradigm shift in the management of infrastructure. The chapter provides the research methodology that emphasizes the significance of applying simulation-based modeling in addition to statistical validation and some real-world testing. Bridge and railway systems as well as construction equipment digitized twins are safe to test predictive and real-time control algorithms. This makes improvements in responsiveness, accuracy and energy efficiency statistically significant as opposed to anecdotal. Practical pilot studies are used to complement simulations with the objective of ensuring that predictive motion control transfers to practical situations. Such a mixture of theoretical complexity and empirical verification makes the digital twin research credible as scientific and applicable in practice. The discussion and results demonstrate that digital twin based predictive control is better than the traditional systems. Comparative research finding the predictive systems have been found to be more responsive, predictive, resilient and sustainable than the reactive models. An example is predictive damping in bridges which reduces oscillations than conventional PID control and pump optimization with a digital twin which reduces energy usage and increases equipment life. Predictive abilities are also enhanced by the implementation of AI and enable motion control systems to acquire optimal strategies via reinforcement learning and respond to new situations. These findings are important to confirm that the integration of digital twins is not just a possibility but also a necessity of the contemporary engineering of infrastructure. The recommendations and solutions in this chapter can be used as a road map in implementation. They also place high emphasis on technical innovation using high-fidelity models, adaptive algorithms, and edge-cloud computing architectures. Strategies that can be integrated into the system are retrofitting old infrastructure, implementing digital twins along the lifecycles of assets, and deploying real-time monitoring centers. Cybersecurity regulation, the partnership of the government and the companies, and global standardization are the recommended ways of governance. Human capacity building is concerned with trainee engineering, retraining operating engineers and the formation of interdisciplinary teams. Lastly, sustainability fit will guarantee that predictive motion control has a direct impact on global development targets in terms of carbon emission, increasing the lifespan of assets, and decreasing waste. In the future, the chapter establishes some of the directional aspects of research that will define the digital twin integration in the next generation. The technological innovation should move towards adaptive intelligence and multi-scale model. New computational models, including the quantum computing, 6G networks, and neuromorphic networks, promise to address the existing limitation of latency, processing power, and energy efficiency. Studies on resilience and disaster management will ensure that the infrastructure systems are resilient in the extreme circumstances and also have the capability to resume operations within a short time after disruptions. The research on ethical and governance needs to deal with transparency, accountability, cybersecurity, and equity, and deploy digital twin systems in a responsible manner. Interoperability standards, modular deployment strategies, federated twin systems should be studied in order to determine their scalability. Lastly, socio-environmental studies should be able to measure the benefits of sustainability, conform to the SDGs, and elicit the confidence of people by using participatory governance.

When these insights are put together, the conclusion is evident: the integration of digital twins in terms of predictive and real-time motion control is the cutting-edge in the engineering of infrastructure. It re-imagines infrastructure as a living system, self-conscious, adaptive, and resilient, which can anticipate disruptions, improve performance, and keep up with the values of the society. Nonetheless, technology cannot bring about this change. It involves the cooperation of engineers, policymakers, industry leaders and communities. It involves investment in research, standardization and human capacity building. And it needs ethical and system frameworks that can guarantee balanced and safe deployment. The importance of digital twin incorporation does not only depend on the technical excellence but also the promise to reinvent the interaction between the society and infrastructure. Digital twins will give the means to safer bridges, more efficient transport networks, sustainable energy systems, and resilient cities by enabling predictive and real-time motion control. They enable societies to move beyond reactive maintenance to proactive stewardship, unresponsive designs to responsive ones and divided and fragmented management to unified and intelligent ecosystems. With infrastructure transformation under the stress of globalization, climate change, and disruptive technology, integration of digital twins presents a way to a more sustainable, inclusive and human-centered system, which is more intelligent and faster. The introduction of digital twin technology in the sphere of motion control cannot be regarded as a far-off dream but should be regarded as a pressing necessity. Preparation has been done with conceptual frameworks, theoretical frameworks, simulation methodologies and pilot applications. The challenge to overcome in the future is to scale these innovations in a responsible manner, make them consistent with global goals on sustainability, and make the future of infrastructure engineering technologically sophisticated and socially fair. This direction will mean digital twin integration is not just used to achieve efficiency but is a foundation of the future of resilient, sustainable, and intelligent infrastructure.

REFERENCES

Agostinelli, S., Cumo, F., Guidi, G., & Tomazzoli, C. (2021). Cyber-physical systems improving building energy management: Digital twin and artificial intelligence. *Energies*, *14*(8), 2338. DOI: 10.3390/en14082338

Bhambri, P., Rani, S., & Khang, A. (2024). AI-driven Digital Twin and Resource Optimization in Industry 4.0 Ecosystem. *Intelligent Techniques for Predictive Data Analytics*, 47-69.

Brucherseifer, E., Winter, H., Mentges, A., Mühlhäuser, M., & Hellmann, M. (2021). Digital Twin conceptual framework for improving critical infrastructure resilience. *Automatisierungstechnik*, *69*(12), 1062–1080. DOI: 10.1515/auto-2021-0104

Epiphaniou, G., Hammoudeh, M., Yuan, H., Maple, C., & Ani, U. (2023). Digital twins in cyber effects modelling of IoT/CPS points of low resilience. *Simulation Modelling Practice and Theory*, *125*, 102744. DOI: 10.1016/j.simpat.2023.102744

Flammini, F. (2021). Digital twins as run-time predictive models for the resilience of cyber-physical systems: A conceptual framework. *Philosophical Transactions. Series A, Mathematical, Physical, and Engineering Sciences*, *379*(2207), 20200369. DOI: 10.1098/rsta.2020.0369 PMID: 34398658

Groshev, M., Guimaraes, C., Martín-Pérez, J., & de la Oliva, A. (2021). Toward intelligent cyber-physical systems: Digital twin meets artificial intelligence. *IEEE Communications Magazine*, *59*(8), 14–20. DOI: 10.1109/MCOM.001.2001237

Habib, A., Habib, M., Bashir, B., & Bachir, H. (2025). Exploring the Sustainability Benefits of Digital Twin Technology in Achieving Resilient Smart Cities During Strong Earthquake Events. *Arabian Journal for Science and Engineering*, *50*(20), 1–15. DOI: 10.1007/s13369-025-10017-z

Hakimi, O., Liu, H., Abudayyeh, O., Houshyar, A., Almatared, M., & Alhawiti, A. (2023). Data fusion for smart civil infrastructure management: A conceptual digital twin framework. *Buildings (Basel, Switzerland)*, *13*(11), 2725. DOI: 10.3390/buildings13112725

Jin, A. S., Hogewood, L., Fries, S., Lambert, J. H., Fiondella, L., Strelzoff, A., & Linkov, I. (2022). Resilience of cyber-physical systems: Role of AI, digital twins, and edge computing. *IEEE Engineering Management Review*, *50*(2), 195–203. DOI: 10.1109/EMR.2022.3172649

Mon, B. F., Hayajneh, M., Ali, N. A., Ullah, F., Ullah, H., & Alkobaisi, S. (2025). Digital Twins in the IIoT: Current Practices and Future Directions Toward Industry 5.0. *Computers, Materials & Continua*, *83*(3), 3675–3712. DOI: 10.32604/cmc.2025.061411

Olaseni, I. O.Iyiola Oladehinde Olaseni. (2020). Digital Twin and BIM synergy for predictive maintenance in smart building engineering systems development. *World J Adv Res Rev*, *8*(2), 406–421. DOI: 10.30574/wjarr.2020.8.2.0409

Parnianifard, A., Jearavongtakul, S., Sasithong, P., Sinpan, N., Poomrittigul, S., Bajpai, A., & Wuttisittikulkij, L. (2022). Digital-twins towards cyber-physical systems: A brief survey. *Engineering Journal (New York)*, *26*(9), 47–61.

Qudus, L. (2025). Resilient systems: Building secure cyber-physical infrastructure for critical industries against emerging threats. *Int J Res Publ Rev*, *6*(1), 3330–3346. DOI: 10.55248/gengpi.6.0125.0514

Ranawaka, A., Alahakoon, D., Sun, Y., & Hewapathirana, K. (2024). Leveraging the Synergy of Digital Twins and Artificial Intelligence for Sustainable Power Grids: A Scoping Review. *Energies*, *17*(21), 5342. DOI: 10.3390/en17215342

Rojas, L., Peña, Á., & Garcia, J. (2025). AI-driven predictive maintenance in mining: A systematic literature review on fault detection, digital twins, and intelligent asset management. *Applied Sciences (Basel, Switzerland)*, *15*(6), 3337. DOI: 10.3390/app15063337

Sadasivan, M., Vasumathi, M. T., & Jayanthi, M. (2024). Artificial intelligence for smart manufacturing using digital twin technology. In *Artificial Intelligence based Solutions for Industrial Applications* (pp. 176–194). CRC Press. DOI: 10.1201/9781003534761-7

Sajadieh, S. M. M., & Noh, S. D. (2025). A Review of Digital Twin Integration in Circular Manufacturing for Sustainable Industry Transition. *Sustainability (Basel)*, *17*(16), 7316. DOI: 10.3390/su17167316

Sajadieh, S. M. M., & Noh, S. D. (2025). From Simulation to Autonomy: Reviews of the Integration of Artificial Intelligence and Digital Twins. *International Journal of Precision Engineering and Manufacturing-Green Technology*, 1-32.

Samuel, P., Saini, A., Poongodi, T., & Nancy, P. (2023). Artificial intelligence–driven digital twins in Industry 4.0. In *Digital Twin for Smart Manufacturing* (pp. 59–88). Academic Press. DOI: 10.1016/B978-0-323-99205-3.00002-X

Sun, Z., Jayasinghe, S., Sidiq, A., Shahrivar, F., Mahmoodian, M., & Setunge, S. (2024). Approach towards the development of Digital Twin for structural health monitoring of civil infrastructure: A comprehensive review. *Sensors (Basel)*, *25*(1), 59. DOI: 10.3390/s25010059 PMID: 39796850

Vidyalakshmi, G., Gopikrishnan, S., Boulila, W., Koubaa, A., & Srivastava, G. (2025). Digital Twins and Cyber-Physical Systems: A New Frontier in Computer Modeling. *Computer Modeling in Engineering & Sciences*, *143*(1), 51–113. DOI: 10.32604/cmes.2025.057788

Werbińska-Wojciechowska, S., Giel, R., & Winiarska, K. (2024). Digital twin approach for operation and maintenance of transportation system—Systematic review. *Sensors (Basel)*, *24*(18), 6069. DOI: 10.3390/s24186069 PMID: 39338814

Xu, H., Berres, A., Yoginath, S. B., Sorensen, H., Nugent, P. J., Severino, J., & Sanyal, J. (2023). Smart mobility in the cloud: Enabling real-time situational awareness and cyber-physical control through a digital twin for traffic. *IEEE Transactions on Intelligent Transportation Systems*, *24*(3), 3145–3156. DOI: 10.1109/TITS.2022.3226746

Yan, Y., & Kunhui, Y. (2024). Novel cyber-physical architecture for optimal operation of renewable-based smart city considering false data injection attacks: Digital twin technologies for smart city infrastructure management. *Sustainable Energy Technologies and Assessments*, *65*, 103733. DOI: 10.1016/j.seta.2024.103733

Yu, D., & He, Z. (2022). Digital twin-driven intelligence disaster prevention and mitigation for infrastructure: Advances, challenges, and opportunities. *Natural Hazards*, *112*(1), 1–36. DOI: 10.1007/s11069-021-05190-x PMID: 35125651

Chapter 9
Governance, Security, and Ethical Considerations in AI-Driven Motion Control Systems

K. Selvi
https://orcid.org/0009-0006-5389-7538
Saveetha Engineering College, India

P. Anbarasan
Sathyabama Institute of Science and Technology, Chennai, India

G. Madhumita
Vels Institute of Science, Technology, and Advanced Studies (VISTAS), India

L. Janaki
St. Peter's Institute of Higher Education and Research, Chennai, India

K. K. Devi
Jayagovind Harigopal Agarsen College of Arts and Science, Chennai, India

ABSTRACT

Motion control systems based on AI are revolutionising industries with their accuracy, flexibility, and efficiency but pose serious governance, security and ethical issues. The chapter focuses on the intersection of institutional frameworks, cybersecurity practises and ethical principles in the development of responsible adoption. It presents the connections between governance, resilience, and ethics as interdependent based on interdisciplinary theories, mixed-method analysis, and case studies. Among

DOI: 10.4018/979-8-3693-3637-3.ch009

the crucial findings, there is the necessity to have adaptive regulations, strong security-by-design, and ethical alignment to provide transparency, accountability, and social justice. The chapter suggests the combined solutions and describes the future research directions to ensure the sustainable and reliable innovation in AI-driven motion control.

INTRODUCTION

Artificial Intelligence (AI) has significantly revolutionised designing, functioning, and optimization of the motion control systems providing a high level of accuracy, flexibility, and efficiency in the field of manufacturing, healthcare, robotics, aerospace, logistics, and intelligent infrastructure. Once the domain of deterministic programming and linear feedback systems, motion control has now taken on an advanced AI based paradigm, with learning algorithms, predictive analytics and sensorization teaching machines how to communicate with their surroundings. It is not just a technical change but a change that is closely intertwined with governance systems, security issues and the morality and ethics that define the sustainability, safety and societal acceptance of such systems. Artificial intelligence and motion control together are a golden goose. Cyber-physical systems can synchronise global supply chains in a way never before seen, intelligent controllers can adjust in real time to variable loads and environmental disturbances, autonomous robots can perform high level tasks with little human intervention and, most importantly, cyber-physical technologies can coordinate global supply chains in a manner never before seen. These benefits, though, are followed by complex dangers. The opaqueness of AI decision making creates accountability and regulatory compliance issues of governance. The linkage of the motion control networks to cyberattacks, ransomware, and data breaches endangers the safety of the economy and humans (Martini et al, 2024). In addition, the offloading of physical tasks to smart machines bring with them acute ethical concerns, including liability concerns, displacing employees, algorithm-based bias, and the ethical limits of machine autonomy. The multi-layered approach is necessary in the governance of the AI-driven motion control systems. Regulations, standardisation authorities, and stakeholders in the industry need to come up with structures that promote transparency, equality and accountability without dampening innovation. Existing regulatory frameworks, including the AI Act within the European Union and international ISO/IEC standards are initial guides, though tend to be generalist and lack the level of specificity required of motion control in high-stakes critical infrastructure. Such matters like cross-border liability, the protection of intellectual property on AI-generated solutions, and the dynamic changes of machine learning models have not been solved (Mintoo et al, 2022). Therefore, governance has to strike a balance between active control as well as dynamic flexibility to ensure that it keeps up with the technological changes.

Security issues are also very important. AI-powered motion control systems are at the interface of the cyber and physical worlds. An effective cyberattack of such systems can cause disastrous consequences, not only shutting a production line, but also threatening human lives in surgical robots or self-driving vehicles (Mehra, I. 2024). Contrary to conventional IT systems, breaches in motion control are not limited to data loss, but have physical implications as well. This growing dependency on cloud robotics, edge AI and wireless communications only increases the attack surface, requiring strong encryption schemes, intrusion detection, redundancy designs and resiliency. Moreover, adversarial AI attacks are presented when malicious inputs manipulate machine learning models, which present special threats and need specific countermeasures (Shandilya et al, 2024). There is also the moral aspect. The motion control systems powered by AI, especially those in sensitive areas such as healthcare and defense, raise the question of human control and management, consent, and responsibility (Patel et al, 2025). As an example, a surgical robot controlled by AI makes a critical mistake, the line of responsibility is obscured between developers, health professionals, and the regulatory community. Likewise, industrial automation, where human labour is being replaced, brings socio-economic issues to mind, and using algorithms to predict the future of logistics can unintentionally favour smaller companies, as they will have overlooked biases in the software. These systems should therefore be ethically implemented by transparency, fairness, and inclusivity. The challenge of the governance, security and ethics of AI-driven motion control systems cannot be resolved individually. These areas are inter-dependent. Without security, systems are easy targets of malicious exploitation by hackers; without ethics, the systems are prone to invasion of privacy through invasive surveillance or authoritarianism; without governance, ethics cannot be enforced. So, it should be approached integrally, relying on interdisciplinary thinking in the fields of engineering, law, management, and philosophy. The collaboration between scholars and practitioners should be able to create a framework which protects human values without limiting technology development.

CONCEPTUAL FRAMEWORK

The theoretical framework of interpreting governance, security, and ethical aspects of AI-driven motion control systems should be approached as a whole, including the recognition of the intersection between the technological, organisational, and societal levels. The traditional approach to motion control had its basis in deterministic algorithms and electromechanical principles where linear feedback and a programmable logic controlled mechanical movements. Now, though, AI has made these systems dynamic, adaptive and self-learning, working in complex cyber-physical ecosystems. These ecosystems are made up of interconnected sensors, actuators, embedded processors, machine learning algorithms, cloud or edge computing infrastructure and wireless communication modules. Modern motion control systems, instead of performing programmed movements, sense their surroundings, absorb real-time information, propagate through historical observations, and modify their functioning to maximise the accuracy, safety and efficiency. This transformation requires a conceptualization not based on technical descriptions but one which places motion control as part of wider frameworks of governance, security and ethical accountability. At the governance tier, AI-based motion control presents significant issues to accountability, transparency and compliance with regulations. The concept of governance has to be viewed as a multi-level process that runs on an international, national, industrial, and organisational level. On the macro-level, international organisations like ISO and IEC, as well as regional authorities, including the European Union, and U.S. FDA commenced developing safety, interoperability, and transparency of AI standards. However, these models tend to be too generalist and not customised to the high stakes, real-time needs of motion control, where a failure can result in material loss, financial loss, or even human life. Industry associations, consortia and sector-specific regulators at the meso level read the global standards and adjust them to domain-specific requirements, e.g. automotive safety standard certifications or healthcare robotics standards. At the micro level, specific organisations, research laboratories and production facilities have adopted governance procedures within their organisations in the form of ethics boards, compliance audits, and review committees. The stratified governance concept indicates that AI-based motion control accountability cannot be imposed by one individual but instead follows a decentralised model where various institutions can collaborate to guarantee reliability, interoperability, and equity.

Security is the second pillar of the conceptual framework, and it is arguably the most pressing issue considering that motion control systems are located on the border between cyber and physical worlds. An effective cyber-attack on these systems does not only result in data breaches but also physical damage, halts in production and even human loss. The concept of security in motion control should be seen not as the extrinsic feature but as the implicit concept in the system lifecycle (Saleh, S., & Ahmed, M. 2024). The safety of data is a matter of paramount importance since the misuse or alteration of sensor data or training data might result in unsuitable operating conduct. Intrusion detection systems, real-time monitoring, encryption and redundancy to achieve resilience in case of unauthorised access or manipulation of controllers and actuators are required by system security (Tahaei et al, 2023). Operational security focuses on robustness in real-world performance, so that AI-based systems are able to react to failures in the world, like spoofed GPS signals in autonomous drones or corrupted input in robotic manufacturing lines. Notably, emergence of adversarial AI attacks, in which machine learning models are purposefully fooled by malicious examples, presents a novel category of vulnerabilities that conventional security models are poorly equipped to handle. Thus, security should be re-conceptualised as the security of digital resources as well as a mechanical requirement of the integrity of physical activities regulated by AI systems. The third central aspect of the conceptual framework is ethical considerations that put the human and societal implications of AI-controlled motion control in the focal point of the analysis. Ethics is more like a value-based question of fairness, justice and human dignity unlike governance and security which can be any of the formal standards and technical protection. Transparency and explainability, given the fact that it is one of the leading ethical issues (Mahajan, P. 2025). Most AI-based controllers make use of a deep learning model whose decision-making is not transparent to users and regulators. This uninterpretability is harmful to safety-critical applications like surgical robotics or autonomous vehicles, which may make them harder to trust and accountable to in failure cases. In connexion to this is the ethical issue of accountability and liability. In the case of harm caused by an AI-controlled system of motion control, it becomes hard to find who is to be responsible among the developers, manufacturers, operators, and regulatory organisations. Ethical responsibility requires that there are effective liability mechanisms to bring fairness and justice. Equity and inclusivity is another aspect, with AI systems needing to be such that they do not strengthen social biases or benefit some groups over others. As an example, the motion control algorithms that are optimised using logistics to benefit large clients might unintentionally discriminate against small businesses (Merchán et al, 2025). Also, one should not overlook socio-economic implications of automation, the loss of human labour in particular. Although, AI-based motion control enhances efficiency, productivity and safety, it also kills some types of human jobs. The ethics governance must consequently entail a balance between technology benefits and retraining, upskilling and social security policies to avoid widening inequality.

What comes out of this conceptualization is that governance, security and ethics cannot be regarded as different silos but rather they should be perceived as highly interdependent dimensions. Systems vulnerable to exploitation are those whose governance structures lack strong security measures and those that have security measures whose application lacks ethical consideration as the measures might facilitate intrusive surveillance or disproportionate control. Equally, ethical discussions outside the framework of enforceable governance structures will tend to stay at the hypothetical stage. Ideally, these three areas constitute a triangular relationship where governance gives a structural background, security the resilience of operation, and ethics the protection of the human values. Their incorporation is essential in maintaining credibility in the AI-based motion control systems, especially in areas where lives, livelihoods, and other essential infrastructure are at risk. This framework is applicable to a variety of industries in which AI-powered motion control has a game-changing impact. Governance in healthcare robotics: Medical device regulation, security, and reliability of a system, ethical concerns in medical surgery, consent, autonomy, and accountability. In self-driving cars, regulation ensures safety levels and insurance systems, security prevents vehicle-to-vehicle cyberattacks, and ethics determines how complicated decision-making is in an accident case (Sanyal et al, 2024).

THEORETICAL FRAMEWORK

The theoretical framework underpinning the study of governance, security, and ethical considerations in AI-driven motion control systems must be grounded in interdisciplinary perspectives that connect technological evolution with organizational behavior, regulatory structures, and normative ethical values. The rapid diffusion of AI into motion control challenges the adequacy of traditional engineering theories that historically emphasized efficiency and optimization while neglecting the social, political, and moral implications of machine autonomy. By incorporating insights from institutional theory, socio-technical systems theory, and ethical theory, a multidimensional framework emerges that can illuminate how governance structures evolve, how humans and technology co-adapt, and how ethical dilemmas can be conceptualized and resolved within AI-driven environments. This theoretical triangulation ensures that discussions about motion control are not reduced to mechanical performance but instead situated within broader societal contexts that recognize the interplay between technical innovation and human values.

Institutional theory provides a valuable lens for examining governance in AI-driven motion control systems. Institutions, understood as formal and informal rules, norms, and practices that structure human behavior, play a crucial role in shaping the development, adoption, and regulation of technology. From this perspective, governance frameworks emerge not only as top-down regulatory impositions but also as negotiated outcomes shaped by cultural expectations, organizational routines, and professional standards. For example, the push for explainable AI in safety-critical systems such as autonomous vehicles and surgical robotics reflects institutional pressures for accountability and transparency that originate from legal traditions, public expectations, and industry norms (Islam et al, 2025). Institutional theory also helps explain the phenomenon of isomorphism, whereby organizations adopt similar governance protocols and ethical guidelines due to coercive pressures from regulators, mimetic pressures to emulate perceived best practices, and normative pressures from professional communities. This explains why companies across industries are increasingly aligning with international standards such as ISO 26262 for automotive safety or ISO 13485 for medical devices, even when such alignment entails significant costs (Kezron, I. E. 2025). Institutional theory further highlights the dynamic nature of governance, where emerging crises, such as a cyberattack on industrial motion controllers, can trigger shifts in institutional norms and regulatory responses (Sarker, I. H. 2024). Thus, governance in AI-driven motion control must be seen as a constantly evolving construct, influenced by both technological developments and institutional legitimacy requirements.

Socio-technical systems theory deepens this understanding by emphasizing the inseparability of technology and social structures. According to this perspective, technological systems cannot be fully understood in isolation from the social environments in which they are embedded. AI-driven motion control systems illustrate this principle vividly, as their performance depends not only on algorithmic accuracy and sensor precision but also on the organizational practices, human operators, and cultural contexts that guide their use. A robotic arm in a manufacturing plant, for instance, may be technologically capable of autonomous optimization, but its integration into workflows depends on the willingness of human operators to trust the system, the organizational protocols that define decision rights, and the labor agreements that regulate task allocation between humans and machines (Khan, M. Z. 2024). Socio-technical systems theory therefore directs attention to the interaction between human actors and technical subsystems, emphasizing co-adaptation and mutual shaping. Importantly, it underscores that failures in AI-driven motion control often result not from purely technical shortcomings but from misalignments between social and technical components. For example, a well-secured motion control network may still be vulnerable if human operators engage in unsafe practices due to inadequate training or misinterpretation of system alerts. This theoretical lens highlights that effective governance and security must encompass human factors, organizational culture, and institutional incentives alongside technical safeguards.

Ethical theory provides the normative foundation for evaluating the moral dimensions of AI-driven motion control. Three major strands of ethical thought—deontological, consequentialist, and virtue-based ethics—offer distinct perspectives that can be applied to the dilemmas posed by intelligent motion control systems. From a deontological standpoint, the ethical legitimacy of AI-driven motion control depends on adherence to universal moral duties such as respect for human dignity, fairness, and the right to privacy. This perspective argues that systems should not be designed in ways that violate fundamental rights, regardless of the potential efficiency gains. For instance, the use of motion control in defense applications must be critically examined to ensure compliance with international humanitarian principles prohibiting indiscriminate harm. Consequentialist ethics, by contrast, evaluates actions based on their outcomes, focusing on maximizing benefits and minimizing harms. This perspective supports the deployment of AI-driven motion control in contexts such as precision surgery or disaster response robotics, where the potential to save lives outweighs residual risks, provided those risks are carefully mitigated. Virtue ethics adds another dimension by emphasizing the character and intentions of developers, engineers, and operators. It raises questions about the values guiding the design of AI algorithms, the virtues of responsibility and prudence among decision-makers, and the cultivation of a culture of integrity within organizations deploying motion control systems. Together, these ethical theories provide complementary lenses for evaluating dilemmas related to transparency, accountability, liability, workforce displacement, and social justice in the context of AI-driven motion control. The interplay between institutional theory, socio-technical systems theory, and ethical theory creates a robust framework for analyzing governance, security, and ethics as interdependent phenomena. Institutional theory explains how governance structures emerge and evolve under regulatory, cultural, and normative pressures. Socio-technical systems theory highlights the practical integration of governance and security mechanisms within complex organizational and technological settings, reminding us that human factors are as critical as technical ones. Ethical theory provides the normative grounding necessary to assess whether the direction of institutional and socio-technical evolution aligns with societal values and moral imperatives. Together, these perspectives suggest that effective governance of AI-driven motion control systems cannot be reduced to compliance checklists or technical standards but must be understood as a dynamic interplay between evolving institutions, socio-technical practices, and ethical commitments.

RESEARCH METHODOLOGY

The study approach to analysing governance, security, and ethical implications of AI-driven motion control systems needs to be devised in such a way that it becomes possible to describe the complexity of a domain that is simultaneously technical, organisational, and societal. In contrast to traditional engineering research, which tends to focus on deterministic experimentation and performance measures, the current project will need a multi-methodology, combining both the qualitative and quantitative strategies. Such an approach is not only aimed at producing empirical data but also making interpretive observations about the workings of the process according to which governance structures are formulated, security practises are codified, and ethical conflicts are mediated. This pluralism of methodology enables the study to transcend the technicality of the method to adopt a more systemic approach that acquaints the interdisciplinarity of AI-driven motion control.

The study includes a qualitative study that examines the mechanisms of governance governing AI-controlled motion control systems in different sectors including the health sector, manufacturing, logistics, and military. Semi-structured interviews with policymakers, engineers, compliance officers and ethics committee members serve as an insight to the way governance structures are designed, interpreted, and put into practise. The interviews will be supplemented with the content analysis of regulatory materials, industry standards, and organisational policies. This kind of information enables investigators to determine the trends in governance, not only in areas of overlap between jurisdictions but also in areas of gaps and inconsistency that provide areas of vulnerability. The qualitative design also allows the emergent themes to be identified namely, institutional resistance, regulatory lag, and adaptive compliance practises which cannot be measured using quantitative surveys only. The interpretive richness of this phase makes sure that governance is not brought down to a set of rules but seen as a dynamic and evolving process which is carried out through negotiation and contestation. Parallel to this, security resilience of AI-driven motion control systems is measured with the application of a quantitative method. A questionnaire based research is designed to survey engineers, IT experts, and operations managers to collect information on cybersecurity on practises, encryption, redundancy, and incident response in organisations that deploy motion control systems. Questions that assess perceptions regarding vulnerability, number of attacks and trust in resilience measures are also included in the survey. Case study methodology is an integrative strand of qualitative as well as quantitative approach. Contextualised understanding of the interaction between governance, security, and ethics is given by case studies of real-world applications of AI-driven motion control, e.g. autonomous vehicles, surgical robots or smart warehousing system. These case studies are data gathered by field observations, organisational reports, technical reports, and interviews with stakeholders who are directly involved in the deployment of the system. The analysis is done on each instance to examine the existence and application of governance frameworks, emergence and mitigation of security issues and confrontation of ethical issues. As an example, the case study of a surgical robot could provide insight into how regulatory approval procedures impacted the design of the system, how cybersecurity has been implemented to secure patient data, and how ethical questions of liability and informed consent emerged in the implementation process. The cross-case analysis helps researchers to determine general patterns and differences in each sector, thereby increasing the explanatory strength of the research. The risk assessment modelling is also used to enhance robustness of findings of the research design. Based on approaches to engineering safety analysis, a systematic risk assessment model is used to determine the probability and impact of the governance failures, security violations, and breaches of ethics in AI-based motion control systems. These risks are classified into technical, organisational, and societal, and a mapping of the interactions is done by fault tree analysis and risk matrices. As an example, a technical fault with sensor data integrity can flow to governance failures in case it results in regulatory non-compliance, and create ethical issues in case it leads to human safety compromise. Such quantification of risks means that researchers can rank the areas that need urgent action and propose mitigation measures to be implemented. Noteworthy, this risk assessment method emphasises interdependencies of the governance, security, and ethics areas, proving that the failure in one aspect tends to cause failures in the others. This methodology has a critical aspect of triangulation. The integration of interviews, surveys, case studies, risk assessment and simulations provides the research with a high degree of assurance that never results in findings that are biassed by limitations of one approach. Triangulation boosts the validity through convergence of evidence of multiple data sources and methods and also improves depth by incorporating those dimensions that otherwise would not have been visible using a single methodological viewpoint. Examples of these include interviewing, which can expose ethical issues surrounding accountability; surveying, which can measure the degree of liability regimes; case study, which can exemplify the negotiation of liability in action; and simulation, which can model the way in which the liability will change under a new regime of governance. The combination of these complementary approaches results in a comprehensive view of governance, security and ethics in AI-based motion control.

RESULTS AND DISCUSSION

The results of this research indicate the complexity of interaction between the governance systems, security practises, and moral concerns when using the AI-driven motion control systems. Based on qualitative knowledge, quantitative analysis, case studies and simulations, the findings highlight the fact that the three dimensions are closely interrelated and should be dealt with collectively to achieve responsible adoption. Three themes are discussed in the discussion: the efficacy of governance structures, the robustness of security systems and the relevance of ethical issues in the process of adoption and trust. The study of governance systems reveals the advancement and the lack thereof. Conversations with policymakers and compliance officers show that international standards, e.g. ISO 26262 on automotive safety and ISO 13485 on medical devices, have gained significance as anchors to organisational governance practises. Companies that use autonomous cars, surgical robots, or automation equipment often comply with these standards to achieve legitimacy and be approved by the regulatory authorities. However, the results also reveal that this kind of governance is still reactive and fragmented. One of the repeated themes in the interviews is regulatory lag, where technology changes faster than governance structures and as such, organisations are left with gaps that they have to respond to by self-regulation. Although self-regulation shows institutional adaptability, it also brings discrepancies in one organisation and region to another, limiting the ability to harmonise global standards. The idea of security resilience becomes one of the key issues, and the numerical data reveal extensive differences in preparedness. The findings of a survey that assesses resilience between industries show that there is a wide disparity in the use of advanced security policies like intrusion detection, adversarial AI defence, and redundancy architecture. In an industrial logistics scenario, on the other hand, due to insufficient security measures, a ransomware attack interfered with motion control in an automated warehouse, allowing it to go offline during two weeks, which underscores the disastrous results of disregarding security. The findings also point to the increasing risk of adversarial attacks of AI-based motion control. Small changes in sensor input have been shown to cause physical performance failures at a large scale, like misaligned paths in robotic appendages or unsafe methods in self-driving vehicles, using simulation experiments. The outcomes of these studies highlight the fact that the conventional cybersecurity models largely revolving around the privacy of data and network integrity are not adequate where physical safety is concerned. Motion control security needs to be rethought to incorporate adversarial resilience, redundancy and robustness at all levels of the system lifecycle. The insight has critical implications on the governance, because regulatory bodies need to take adversarial AI as a specific risk type that needs to be given mandatory protection. Both qualitative and quantitative data include ethical dilemmas. According to interviews, there is a general concern amongst engineers and policymakers about responsibility in case of a failure. To give an example, when it comes to surgical robots, medical practitioners expressed the concern of not knowing whether to hold the surgeon, the hospital, the manufacturer, or the algorithm itself liable in the event of error. Likewise, the survey findings show that most of the respondents feel that the existing liability frameworks are inadequate, with significant differences in industries. Where healthcare respondents mention more distinct liability arrangements based on the medical devices regulations, respondents in autonomous vehicle and defence sectors emphasise a major degree of ambiguity. These issues are supported by the case study analysis. In a situation where an autonomous vehicle trial occurred, the dispute over the meaning of responsibility between local regulators, manufacturers and insurers stalled subsequent compensation payment to victims of a collision, undermining the trust that the motor vehicle industry was expected to hold. Such results indicate that unless specific kinds of governance structures are established to deal with accountability, issues of ethical dilemmas will persist in compromising the societal acceptance of AI-based motion control.

The other relevant ethical matter is that of transparency and explainability. The participants in every industry would always prioritise the use of transparent AI decision-making and especially where it concerns safety situations. The results of the simulations show that the addition of explainability modules to motion control systems makes them more trustworthy to operators and more usable, without affecting their technical performance. This result has implicated transparency as a practical system adoption enabler and not simply as an ethical value. Simultaneously, transparency and intellectual property rights are reported to be in conflict in interviews, since companies are not eager to reveal proprietary algorithms. The tension portrays the interesting trade-offs within the governance systems that address ethical concerns to advance the aspect of explainability without necessarily lowering the incentives of innovation. Another ethical issue emerging due to the socio-economic effects of AI-based motion control is presented. Concerns of displacing workers in the manufacturing and logistics industry are raised in interviews with industry leaders. Whereas the organisations focus on the efficiency and safety benefits of automation, the employees are voicing their concerns about redundancy and inadequate retraining alternatives. Case studies are used to give practical examples of this tension: in a single project of industrial automation, workplace accidents decreased by 40 percent, at the cost of laying off 30 percent of the staff. These consequences make distributive justice and the role of organisations in offering reskilling programmes highly contentious moral issues. The survey data, when analysed by regression analysis, suggests that the more organisations invest in employee training, the higher the levels of employee acceptance of motion control technologies, which implies that ethical issues can be addressed with proactive methods.

These three concepts are interdependent as emphasised through the integrated discussion of governance, security, and ethics. This is because accountability is not only compromised by the inability to establish strong governance, but also heightens security risks and ethical concerns. On the other hand, good security practises facilitate the achievement of governance goals by maintaining conformity and human safety, and the ethical principle of transparency and inclusivity fosters the legitimacy of governance and security confidence. These dynamics are clearly described in the case studies. The governance structures, efficient security controls, and focus on ethical concerns like informed consent worked together in the medical robotics case to form a trusted system that was widely accepted. The case with industrial logistics, however, demonstrated that poor governance, poor security, and poor ethics could be combined to result in systemic failure. These results validate the conceptual thesis that governance, security and ethics are triad which supports each other and needs to be approached as a whole and not as a single entity.

RECOMMENDATIONS AND SOLUTIONS

The research results are indicative that governance, security, and ethical issues in AI-based motion control systems would have to be tackled on a systemic level other than a single effort. These systems are complex since they function at the nexus of cyber-physical infrastructure, human interaction, and machine intelligence and need an integrated approach that includes regulatory innovation, technological protection, and ethical alignment. The solutions should not just alleviate the current risks but emerge ahead of emerging challenges so that AI-driven motion control should be developed in a responsible, sustainable, and within the values of society. The development of adaptive regulatory systems is one of the leading solutions to governance. The existing governance strategies tend to be stagnant, tailored to technologies which change at a slow pace, whilst AI-driven systems are dynamic and can learn and modify themselves. Regulatory frameworks should thus be open to adaptive processes, like regulatory sandbox, where new technologies can be experimented in controlled conditions where regulators closely monitor real-time. These types of sandboxes are already successfully tried in the fintech industry and can be applied to motion control tasks in healthcare robots, self-driving cars, and industrial automation. They enable regulators to be experimental, gather empirical evidence of risks, and continuously test policy. Modular forms of governance must also be put in place, with core principles of transparency, accountability and safety being universal, with sector specific forms applied to the particular challenges of particular areas. This approach to modularity provides coherence and flexibility to allow global harmonisation with respect to the local context.

Liability and accountability frameworks should also be the subject of governance reform. The confusion that can be seen with case studies reiterates the necessity of defining liability in case of accidents or failures. It is one suggestion that the tiered system of liability be instituted whereby the burden of liability is shared between the developers, manufacturers, operators, and users based on their level of control and influence over the results. An example here could be that developers should be responsible in terms of algorithmic transparency, manufacturers in terms of hardware reliability, and operators in terms of system supervision and users in terms of operation guideline adherence. These tiered systems do not just provide the responsibility in a more even manner but also provide an incentive to every actor to implement their responsibilities. In parallel with this, the mandatory disclosure requirements must force organisations to write the design, training, and testing process of AI-driven motion control systems, and this should result in audit trails that would make them accountable. In the security front, the recommendations surround the move towards proactive, rather than reactive approaches. Organisations need to leave perimeter defences behind and embrace a zero-trust security architecture, which presumes that all attempts of access will be malicious and thus must be continuously verified. When it comes to motion control, zero-trust refers to continuous authentication and validation of data feeds to sensors, communications between networks, and actuator commands. This decreases the possibility of adversarial attacks which use the assumption of trust. As well, adversarial robustness should be designed, not anafterthought. Motion control AI models must be tested on adversarial inputs if they are going to be important in development, and resilience standards must be set as part of certification. Intrusion detection systems also need to shift to the AI based anomaly detection that can detect new attack patterns in real time (Androutsopoulou, M., Carayannis, E. G., Askounis, D., and Zotas, N. (2025). The other important security suggestion is the issue of resilience and redundancy. Since failures in motion control have physical impacts, a system should be designed to gracefully rather than disastrously degrade. This demands fault tolerant data communication paths, backup controllers that can take over in case of compromise, failover mechanisms that fall to safe points in case of anomalies. Autonomous vehicles, as an example, should have the ability to transition into a controlled braking mode in case its AI motion controller is disrupted, and surgical robots should fall back to manual control by surgeons when anomalies occur. The governance structures must have these resilience strategies made mandatory so that they are applied uniformly across sectors (Gadekallu, T. R., Dev, K., Khowaja, S. A., Wang, W., Feng, H., Fang, K., and Wang, W. 2025).

The humanness of these solutions should be based on ethics. Institutionalisation of transparency and explainability which are among the ethical imperatives and practical means to establish trust should be done. This should force organisations to consider the explainability modules as part of motion control systems, which will give the operators informative understanding of system reasoning (Tariq, U., and Ahmed, I. 2025). This does not only improve accountability but also enables the human operators in making sound decisions. The system design should also be required to be inclusive because the algorithms cannot be trained on homogeneous datasets, which will result in biases that isolate some groups. As an example, motion control algorithms in logistics are not supposed to disfavour small businesses or less developed regions in a organised manner. Similar to financial audits, equity audits might be converted into an institutional practise of evaluating the inclusiveness of AI-based systems and take corrective action. The socio-economic consequences of an AI-powered motion control require a proactive labor policy. Some of the human functions will always be replaced by automation, especially in monotonous and dangerous situations. Nevertheless, displacement does not necessarily imply disempowerment when reskilling and upskilling efforts are put in the focus. Industries, academic institutions, and governments should also work together and establish the training programmes that will equip the workers with the new roles regarding the supervision of the systems, cybersecurity, and the maintenance of AI models. Not only do these programmes reduce social resistance, but social institutions also enjoy human-machine complement (Govindaraj, M., Shakila, P., and Lawrence, J. 2025). These initiatives can be enhanced with the help of policy instruments, including tax incentives to firms investing in employee reskilling or a public-private partnership with training centres. Distributive justice should be expanded to ethical governance so that the positive outcomes of AI-driven motion control are not limited to the hands of a small group of large organisations.

The other suggestion entails the formation of multi-stakeholder governance bodies that include regulators and industry representatives, civil society organisations and academic experts to discuss the societal consequences of AI-driven motion control. These bodies can make governance not a capture of the narrow commercial or technical interests but a manifestation of different point of views. They may also serve as a form of early-warning, detecting new ethical issues or vulnerabilities to security, and recommending an intervention. Through the institutionalisation of multi-stakeholder engagement, governance frameworks become more legitimate, inclusive and flexible. It is also critical that international coordination is effective. Motion control systems are commonly cross-border, especially in logistics, aerospace and defence (Gunasekara, L., El-Haber, N., Nagpal, S., Moraliyage, H., Issadeen, Z., Manic, M., and De Silva, D. 2025). Divided government systems enhance the probability of regulatory arbitrage, where companies take advantage of loose jurisdictions to escape regulations. International treaties and agreements ought to be formulated so that the core governance principles are harmonised, but flexibility is provided so that local adaptations can be made. The standards of cybersecurity should also be internationalised so that the vulnerabilities in one jurisdiction do not affect the global supply chains. Ethical standards, including the bans on autonomous weapons which are against humanitarian law, should be written in international treaties that the countries will act in the same way irrespective of national borders. Governance and ethical compliance can as well be improved using technological solutions. As an example, audit trails that are based on blockchain technology can deliver indelible documentation of AI decision-making activities, making them transparent and traceable. Smart contracts may also be used to automatically enforce compliance with regulatory rules, minimising chances of abuse. Federated learning and homomorphic encryption are privacy-preserving technologies that can be used to improve data security and address ethical issues regarding the privacy of users. With the inclusion of governance and ethics in the technological infrastructure, the organisations are able to lessen the dependence on the external controls as well as establish self-governing structures of accountability. Lastly is the development of a responsible innovation culture, which is a general suggestion. There is no way governance, security and ethics can prevail where they are imposed as external conditions to unwilling companies. On the contrary, they have to be internalised as the core values governing innovation. This involves leadership dedication, organisational training and governance and ethics incorporated within performance appraisal measures. The engineers and developers must be informed of technical expertise as well as ethical reasoning and regulatory consciousness. It should be a culture of celebrating successes of ethical innovation and responsibility a competitive advantage to organisations. This cultural change can make governance, security, and ethics more than just responsive to crises, instead of being committed to responsive efforts in daily practise.

FUTURE RESEARCH DIRECTIONS

Due to the fast development of AI-based motion control systems, identifying the future research directions to make sure that the governance, security, and ethical frameworks are up-to-date and effective is crucial. Due to the advancement in technologies, new areas of vulncrability have appeared, the regulatory environment changes, and social expectations grow, and this frontier is constantly on the move, necessitating intellectual responses. Thus, the way forward of future research is not to dwell on the issues at hand but to be able to predict the issues that will be arising at the confluence of technological change, institutional adjustment, and human values. The discussion below points at some of the areas where perpetual inquiry is not only the urgent but also promising. The combination of AI-controlled motion control and quantum technologies represents one of the hottest areas of research. Due to the emergence of quantum computing, traditional encryption protocols can be outdated and endanger the security of the motion control systems morc than before. Since these systems can frequently control such critical infrastructure as autonomous transport, aerospace navigation, and medical robotics, their breach would have disastrous effects. Studies are required to create quantum-resistant algorithms that have the ability to protect sensor data, communication channels and control orders in the setting susceptible to quantum attacks. At the same time, quantum-enhanced optimization also presents a chance to enhance motion control efficiency and such a dual imperative to both exploit and protect quantum technologies. Examples Studies at the interface between AI, quantum cryptography, and control theory will be crucial in developing resilient and secure designs of the next generation of motion control systems. The other urgent research agenda is the current development of explainable and interpretable AI in motion control. Existing systems tend to be opaque in nature, which are black-box models and this leads to ethical and governance issues of accountability and trust (Uddin, M. Z. 2025). Future studies should aim at coming up with interpretable models that will offer actionable explanations to real-time decisions in areas like autonomous driving or surgical robotics. It is not just a matter of coming up with technically transparent algorithms but also meaningful explanations that would be comprehensible by various stakeholders, such as regulators, engineers, operators, and the end-users. Research might investigate the utility of alternative explanation formats, e.g., visual dashboard, natural language summary or causal reasoning graph, in increasing trust and minimising operator error. Moreover, there is a need to have empirical research to determine the impact of explainability on user acceptance, regulatory approval, and security resilience. This research question will directly guide the governance structure by having evidence-based recommendations on the necessity of explaining standard requirements.

Sustained attention is also required in human-AI collaboration in motion control. Although the concept of automation has been presented as a replacement of human workforce, it is important that future studies consider it as a supplement which complements human abilities. To take the example of the medical robotics, AI-driven motion control can enhance the accuracy of the surgeon, and in the field of logistics, it can simplify the routing options and leave the supervisory role to the human operator (Malatji, M. 2025). This requires studies that will come up with collaborative models that can allocate responsibilities in decision-making between humans and AI and strike a balance between autonomy and supervision. This involves considering adaptive interfaces where transitioning between automated and manual control is possible and consideration of psychological and cognitive aspects that affect trust, reliance and vigilance of the operator. The experimental research might focus on the changes in the operator performance, the level of stress, and the situational awareness that are caused by various degrees of automation. This type of research will be used to design socio-technical systems in which humans and AI collaborate, but not replace each other, and to support ethical obligations to human dignity and agency. The internationalisation of the motion control technologies creates difficult questions of international governance. Autonomous vehicles, drones, and robotic supply chain systems are commonly used across national borders, but the frameworks of regulations are still not unified. Future studies should focus on the way international governance mechanisms can be established so that harmonisation and innovation are not choked. Comparative analysis of the regulatory methods employed by various jurisdictions would identify best practises and point out areas of variance that generate impediments to international interoperability. There is also the possibility that the models of international treaties or agreements can be studied, and scholars should also draw parallels with the aviation safety or climate governances of the entire world. Special focus must be put on regulating the use of the so-called dual-use technologies, i.e. autonomous drones that can be both civilian logistic and military tools. Studies in this field will have to face the geopolitical aspects of the governance process, such as the possibility of regulatory arbitrage, and the international institution in the coordination of enforcement.

Ethical research needs to be carried to new dilemmas that arise once AI-controlled motion control is implemented in the realm of sensitive areas like healthcare, defence, and social infrastructure. The growing independence of machines poses moral agency and responsibility issues which are not fully resolved by theories. There is a philosophical and empirical research to be done on whether partial responsibility of machines in their actions should be attributed or accountability should be put on human beings only. Although this refers to defence issues, as an example, must autonomous weapon systems be banned or can ethical principles be incorporated into their code? What are the ethical principles that should inform the application of surgical robots, which might do better than human professionals in some endeavours, yet have the potential to cause failure? These questions require interdisciplinary studies that rely on philosophy, law, engineering and political studies to come up with conceptual and practical frameworks.

The future of motion control based on AI will also revolve around socio-economic research. The effects of automation on labour markets are a hotly debated topic, and motion control is one of the most obvious manifestations of automation in practise. Further studies are required to explore the distributional effect of adoption of motion control, i.e., who gains and who loses. The longitudinal studies may investigate the long-term patterns of the industries and communities that have been impacted extensively by automation in order to obtain information about resilience and adaptation. Models of social safety nets, reskilling and inclusive innovation policies that can help reduce the adverse effects of displacement and ensure that more people can benefit from technological advantages may be considered in policy-oriented research. By basing social-economic analysis with empirical data, we can shift to evidence-based policies of fair and sustainable development as opposed to fanciful tales of dystopia or utopia. The other future opportunity is the exploration of AI ethics in practise, beyond the abstract principles to the working mechanisms. Although a lot has been said on values like transparency, fairness and accountability, not much has been said about how these values are mapped into algorithms and system architecture as well as how they are applied in the organisational practise. The next step in future research is the development of the ethical indicators and compliance tools, which can be implemented in the governance structure and measured. Examples of these include audits of equity, which might examine their fair distribution of benefits to demographic categories, and audits of transparency, which might examine their ability to explain their decisions to stakeholders by AI systems. Empirical research might also examine how well ethics training can be given to engineers and operators and it may ask the question of whether engineers and operators with exposure to ethical reasoning decision-making can be enhanced in system design and deployment.

Operationalizing ethics can enable research to make sure values do not remain in the figures of speech but are integrated into actual practises.

The study on security also needs to keep on changing with the emergent threats. The emergence of the adversarial AI, where machine learning models are attacked by attackers who take advantage of vulnerabilities, will continue to be a key area of concern. Future research on resilient architecture should focus on evolving resilient architectures capable of identifying and tolerating adversarial inputs without affecting performance. This encompasses studies on the design of strong training approaches, like adversarial training or data augmentation, and the design of detection systems capable of detecting a malicious perturbation in real time. The resiliency of distributed motion control networks is another field to investigate, especially when it comes to cloud robotics and networks based on the Internet of Things. The interrelations among these systems present systemic vulnerabilities that need new methods of distributed security as they increase in scale. One of the areas of research might be the use of consensus, redundancy, and decentralised anomaly detection as the means of enhancing resilience. Security research can make sure that motion control systems can be trusted and remain safe in more complex environments because they stay ahead of the threats emerging. Lastly, the long-term consequences of AI-driven motion control in society need to be investigated in the future. Sustainability concerns, environmental performance and social trust questions will form the pattern of adoption more than technical performance. Research could focus on the role of motion control technologies in increasing or decreasing environmental issues, like using less energy make a product or better logistics to decrease carbon emissions. Meanwhile, the role of trust in determining the people acceptance and how the perceptions of governance, security, and ethics can affect the support of AI-driven systems by society should be researched. Longitudinal surveys and cross-cultural comparisons may be helpful in understanding the development of trust in time and context about how to build legitimacy and acceptance strategies.

CONCLUSION

The introduction of AI into motion control systems has marked a radical change in the various industries, including healthcare and manufacturing, transport, logistics, and defence. Previously deterministic and programmable systems constrained by linear feedback loops have now progressed into adaptive, learning based architectures which can perceive, understand, and respond to the complex environments with minimal human involvement. The paradigm shift has not only opened vast possibilities of efficiency, accuracy and innovation, but has also multiplied risks and uncertainties that are not manageable only through technical enhancements.

The discussion herein contained elaborates that the success of AI-based motion control is neither isolated of governance, security, and ethical issues but lays its foundations of its legitimacy, safety, and the social validity. The conclusion revisits the insights obtained by the conceptual, theoretical, methodological, and empirical inquiries and presents the main point of view that the future of the motion control systems lies in their capacity to balance technological advancements with institutional responsibility, strong resilience, and integrity.

Governance discussion proves that, despite the achievements in the development of standards and regulations, existing frameworks are still incomplete, ad hoc, and lack even distribution within sectors and jurisdictions. Compared to autonomous vehicles and defence systems, healthcare robotics enjoy the somewhat well-known medical device regulation, and the unresolved governance dilemma is facing a rather stable situation. Institutional theory assists in understanding why organisations are likely to be brought to some global standards and the existence of regulatory lag and jurisdiction fragmentation. One of the conclusions is unmistakable: governance should be no longer based on strict and one-size-fits-all regulations but be rather modular and adaptive to changes in technology. Accountability needs to be put on a forefront by governance too, such that the liability frameworks are clear, and they should not lose track of responsibilities of failures or harm. The empirical analysis lessons show that both organisations and regulators should implement proactive, iterative governance approaches, which are based not on ex-ante rules, but include the feedback of actual implementations. Security becomes another highly important pillar since AI-based motion control systems exist on the border of cyber and physical space. Violations of such systems do not only disrupt the data, but there may also be direct physical implications, putting human life and vital infrastructure in danger. The findings reveal that the maturity of security in different industries is highly diverse, with industries that are highly resourced, like defence and healthcare, spending lots of money on resilience, whereas smaller businesses are comparatively behind. The experience of adversarial AI attacks only underscores the insufficiency of conventional cybersecurity strategies and requires new strategies that integrate principles of adversarial robustness, anomaly detection, and resilience-by-design. Case studies give examples that organisations that do not invest in proactive security suffer disastrous effects, and those that incorporate resilience and redundancy in their architectures are in a better place to sustain continuity even during an assault. The inference is that security has to be regarded as an inherent structural consideration, not as a post-hoc solution, and that compulsory standards need to be set to provide consistency in resilience of critical applications.

Ethics offer the human-focused compass towards the responsible development of motion control systems. Both the transparency and explainability become an ethical demand and facilitating force of trust, and the issues of accountability and liability indicate the necessity of responsibility structures. The ethical imperative to moderate efficiency with fairness by proactive reskilling and distributive justice is motivated by the socio-economic effects of automation especially the loss of jobs to workers. Case studies indicate that those that invest in inclusive practises and transparency enjoy greater acceptance whereas those that ignore ethical issues encounter resistance, mistrust and reputation destruction. The ethical theory of deontological, consequentialist, or virtue-based ideas can offer complementary storeys on how ethical dilemmas should be solved, but the empirical findings confirm that ethical issues must be made practical by establishing enforceable governance systems instead of remaining an ideal. It is concluded that ethics cannot be a distinct discourse; instead, it has to be institutionalised when designing, deploying, and evaluating AI-based motion control systems. The most significant finding of this study is the interdependence of governance, security and ethics. The failure of the one dimension inevitably spills over to the failure of the others, and the success of the one dimension supports the success of the others. There is no sense in having governance without a strong security system because it can be exploited; a lack of ethics with a strong security system is similar to condoning invasive practises that undermines trust; and an enforceable system without ethics is meaningless. The interdependence triadic model highlights the fact that these three pillars should be tackled all at the same time and interdisciplinary collaboration should provide coherent, comprehensive and sustainable solutions. This realisation can provide a substantial learning experience to policy formulators, industry chiefs, and researchers, i. e. interventions that are siloed cannot ensure the responsible future of motion control, whereas integrated models can.

The mixed-method approach to qualitative, quantitative, case study, and simulative research adopted in this chapter shows how qualitative, quantitative, case study, and simulation research approaches can be used in depth and generalizability in a complex socio-technical system. Their combination can result in a multidimensional knowledge that can be used in evidence-based policymaking and organisational policies. The success of this strategy indicates the necessity of further interdisciplinary studies, which would unite technical, social, and ethical outlooks, avoiding the temptation to make motion control a strictly engineering issue. The future of AI-assisted motion control is characterised by a number of new frontiers that will define its future. Quantum technologies are a source of threats and opportunities that need to be explored in terms of quantum-resistant security designs and quantum-enhanced optimization. Explainable AI is a research agenda that is yet to be explored more because trust and accountability cannot be ensured without clarity. The issue of human-AI collaboration should not be discussed as a contest but rather as a cooperation that will increase the human agency and utilise the machine accuracy. Cross-border issues of governance are international treaties and harmonised standards, ethical issues in defence and healthcare need a global discourse based on philosophy, law and political science. Automation has socio-economic consequences that make it important to implement policies that enable inclusiveness and distributive justice. Combined, these recommendations reiterate that governance, security, and ethics should continue to be of central concern to future research, policymaking, and organisational practise.

REFERENCES

Androutsopoulou, M., Carayannis, E. G., Askounis, D., & Zotas, N. (2025). Towards AI-Enabled Cyber-Physical Infrastructures—Challenges, Opportunities, and Implications for a Data-Driven eGovernment Theory, Policy, and Practice. *Journal of the Knowledge Economy*, •••, 1–38. DOI: 10.1007/s13132-025-02726-5

Gadekallu, T. R., Dev, K., Khowaja, S. A., Wang, W., Feng, H., Fang, K., & Wang, W. (2025). Framework, Standards, Applications and Best practices of Responsible AI: A Comprehensive Survey. *arXiv preprint arXiv:2504.13979.*

Govindaraj, M., Shakila, P., & Lawrence, J. (2025). AI-Driven Cybersecurity for Industrial Automation: Resilient Solutions for Industry 4.0. In *AI-Enhanced Cybersecurity for Industrial Automation* (pp. 83-102). IGI Global Scientific Publishing.

Gunasekara, L., El-Haber, N., Nagpal, S., Moraliyage, H., Issadeen, Z., Manic, M., & De Silva, D. (2025). A Systematic Review of Responsible Artificial Intelligence Principles and Practice. *Applied System Innovation*, *8*(4), 97. DOI: 10.3390/asi8040097

Islam, M. T., Sepanloo, K., Woo, S., Woo, S. H., & Son, Y. J. (2025). A review of the industry 4.0 to 5.0 transition: Exploring the intersection, challenges, and opportunities of technology and human–machine collaboration. *Machines (Basel)*, *13*(4), 267. DOI: 10.3390/machines13040267

Kezron, I. E. (2025). Novel cybersecurity framework for AI-driven drone integration by critical SMEs in economically distressed US rural communities: Advancing secure precision operations in high-risk environments. *Well Testing Journal*, *34*(S3), 1–44.

Khan, M. Z. (2024). *Introduction To Artificial Intelligence And Human-centric Computing*. OrangeBooks Publication.

Mahajan, P. (2025). What is Ethical: AIHED Driving Humans or Human-Driven AIHED? A Conceptual Framework enabling the Ethos of AI-driven Higher education. *arXiv preprint arXiv:2503.04751.*

Malatji, M. (2025). Augmented Intelligence Framework for Human–Artificial Intelligence Teaming in Cybersecurity. *Human-Centric Intelligent Systems*, 1-30.

Martini, B., Bellisario, D., & Coletti, P. (2024). Human-centered and sustainable artificial intelligence in industry 5.0: Challenges and perspectives. *Sustainability (Basel)*, *16*(13), 5448. DOI: 10.3390/su16135448

Mehra, I. (2024). AI-Driven Autonomous Vehicles: Safety, Ethics, and Regulatory Challenges. *Journal of Science. Technology and Engineering Research*, *1*(2), 18–31.

Merchán-Cruz, E. A., Gabelaia, I., Savrasovs, M., Hansen, M. F., Soe, S., Rodriguez-Cañizo, R. G., & Aragón-Camarasa, G. (2025). Trust by Design: An Ethical Framework for Collaborative Intelligence Systems in Industry 5.0. *Electronics (Basel)*, *14*(10), 1952. DOI: 10.3390/electronics14101952

Mintoo, A. A., Saimon, A. S. M., Bakhsh, M. M., & Akter, M. (2022). National Resilience Through Ai-Driven Data Analytics And Cybersecurity For Real-Time Crisis Response And Infrastructure Protection. *American Journal of Scholarly Research and Innovation*, *1*(01), 137–169. DOI: 10.63125/sdz8km60

Patel, N., Chauhan, H., & Joshi, H. (2025). Orchestrating Intelligence: Governance and Leadership Frameworks for Human-AGI Collaboration in Quantum Systems. In *Interplay of Artificial General Intelligence with Quantum Computing: Towards Sustainability* (pp. 255–272). Springer Nature Switzerland. DOI: 10.1007/978-3-031-87931-9_18

Saleh, S., & Ahmed, M. (2024). A Cyclic Framework for Ethical Implications of Artificial Intelligence in Autonomous Vehicles. *International Journal of Advanced Computer Science and Applications*, *15*(3). Advance online publication. DOI: 10.14569/IJACSA.2024.0150335

Sanyal, S., Sharma, P., & Dudani, C. (2024). *A complex adaptive system framework to regulate artificial intelligence* (No. 26). Working paper.

Sarker, I. H. (2024). CyberAI: a comprehensive summary of AI variants, explainable and responsible AI for cybersecurity. In *AI-driven cybersecurity and threat intelligence: cyber automation, intelligent decision-making and explainability* (pp. 173–200). Springer Nature Switzerland. DOI: 10.1007/978-3-031-54497-2_10

Shandilya, S. K., Datta, A., Kartik, Y., & Nagar, A. (2024). Role of artificial intelligence and machine learning. In *Digital Resilience: Navigating Disruption and Safeguarding Data Privacy* (pp. 313–399). Springer Nature Switzerland. DOI: 10.1007/978-3-031-53290-0_6

Tahaei, M., Constantinides, M., Quercia, D., & Muller, M. (2023). A systematic literature review of human-centered, ethical, and responsible AI. *arXiv preprint arXiv:2302.05284.*

Tariq, U., & Ahmed, I. (2025). Reasoning About Responsibility in Autonomous Systems: Navigating the Challenges and Charting Future Directions. *Ubiquitous Technology Journal*, *1*(2), 46–60. DOI: 10.71346/utj.v1i2.21

Uddin, M. Z. (2025). Trustworthy AI and Explainability. In *Trustworthy Multimodal Intelligent Systems for Independent Living* (pp. 111–137). Springer Nature Switzerland. DOI: 10.1007/978-3-031-97359-8_4

Chapter 10
AI-Powered Motion Control for Lean Manufacturing and Workflow Automation

A. Balamurugan
https://orcid.org/0000-0003-1434-2178
Bharath Institute of Higher Education and Research, India

E. P. John
https://orcid.org/0000-0003-3129-2133
SRM Valliammai Engineering College, Kattankulathur, India

A. Gokulakrishnan
Vels Institute of Science, Technology, Advanced Studies, Chennai, India

R. Delecta Jenifer
https://orcid.org/0000-0002-8179-6383
Saveetha Engineering College, Chennai, India

P. Jyothi
Saveetha Engineering College, Chennai, India

ABSTRACT

In this chapter, this is how computer intelligence is put to use in improving motion control schemes in an environment of lean manufacturing such that automation is achieved in real-time, precision, and flexibility. It evaluates how AI can be connected with robotics, machine vision, and feedback control to automate work processes, reduce wastes, and increase the efficiency of production. Appropriate applications

DOI: 10.4018/979-8-3693-3637-3.ch010

include predictive load balancing, dynamic path optimization and adaptive response to operational variability, which are emphasized in the chapter. It also speaks of the difficulties of AI implementation such as the quality of data, system integration, and labor preparedness. The research highlights the emission ability of AI-based indication control to realize flexible, cost-effective and clever manufacture systems.

INTRODUCTION

Artificial intelligence (AI) and motion control are the two technologies merging together to alter the manner in which present-day manufacturing conditions operate. It is worth considering the AI-driven motion control, which is a revolutionary enabler within the framework of lean manufacturing as an approach to reduce any wastes, optimize resources consumption, and enhance the continuous improvement of production efficiency (Bukhari, S. M. S., & Akhtar, R. 2024). It extends traditional automation further to endow machines, robotic systems, and production lines with skills of sensing, adapting and making decisions in real-time and as a result significantly automate the work flow at an unprecedented level of scale and accuracy. Manufacturing systems in the past have used deterministic control schemes which are very rigid, predictable task sequences of pre-programmed motions. Such systems proved effective in simple, repetitive systems, but are not flexible or responsive enough to work in a high-mix/low-volume manufacturing environment, or in just-in-time systems where changes in demand, process variability or equipment behavior may need to be managed on a dynamic basis. Motion control through AI relieves these shortcomings by adding learning, prediction, and feedback to the control loop: enabling manufacturing assets to adapt motion trajectories, optimize performance of tasks, and cooperate with human operators in variable operations. The path of using AI in motion control starts with the additional introduction of intelligent processing by sensors and actuators. These AI models, or neural networks such as convolutional neural networks (CNNs), recurrent neural networks (RNNs), and reinforcement learning agents, are used to analyze the vision systems, inertial sensors, force-torque sensors, and real-time data streams, in order to interpret the context and predict the changes in operation in a few milliseconds and adjust the control accordingly. As an example, a robotic arm carrying out an assembly can identify some misalignment in a part and alter its course accordingly without stopping the assembly process (Nair, V. V. 2024). In like fashion a conveyor system can adjust its speed to an availability of downstream machine thus avoiding the occasion of a bottleneck or a no-go condition.

With lean manufacturing, controlling motion is one of the keys to the eradication of the eight categories of wastes: overproducing, waiting, transport, overprocessing, excess stocks, excess movement, errors, and inadequate use of talent (Rakholia et al, 2024). These forms of wastes can be directly taken care of through AI-powered motion systems by ensuring quality through precision, shorter setup and cycle times, and predictive maintenance and through real-time quality assurance. To use an example, AI will be able to sound alarm on unusual vibrations in machines that indicate they are about to fail and accelerate or redirect work to a different machine to prevent a breakdown and service interruption. Besides, the introduction of AI-assisted motion control will boost the automation of the working process not only at the level of the functioning of machines but also at the level of whole production lines. Manufacturing systems can dynamically plan the jobs, assign motion sequences, and coordinate robot positions in different stations on live data through absorption or edge-based AI agents (Khurram et al, 2025). Such agents have the ability to respond to high-urgency of requests, transfers in the event of an operator, or reassigning tasks (because of a lack of parts) without the need of manual changes to the programming. Bootstrapping allows a statistically rigorous performance benchmarking, risk assessment, and continuous improvement that can be set up in industrial environments where there is the natural variability of processes and where sensor noise is inevitable. As an example, in the introduction of a new trajectory optimization algorithm the bootstrap methods can statistically approximate the cycle time reduction with confidence intervals so as to explicitly justify operational accretions.

AI is also applicable in predictive motion planning whereby machines change their routes or time over predictive events. The production causes delays originating in upstream bottlenecks or maintenance plans can also be predicted in the predictive models and motion plans could be adjusted to avoid this effect. When the AI has access to thousands of motion logs in smart assembly cells, it can identify the most efficiency sequences, which minimize non-value-adding movements and improve ergonomics of human operators (Balaji, K. 2025). AI models coupled with digital twins have the ability to model motion within diverse situations to enable manufacturers to test and fine-tune workflow set ups before placement in the actual plant. Collaborative motion control is another dimension that is vital, particularly in cases of human and robot (cobots) interactions in the same environment. With the help of AI, recognition of human intent based on vision, gestures, and prediction of its motion path allow robots to change their own movement in real time so as to achieve security and efficiency. The ability to respond to movements is not only an excellent protection against collision but also enhances the performance of humans and robots in terms of their mutual workflow, making it easier to split tasks and fulfill them more freely. The collaborative control supported by AI would create a situation when the motion would be not only prescribed by a plan of the robot but influenced by the human behavior. Self-optimization and feedback loops are also supported by deployment of AI-powered motion control - systems can learn how to behave better in the future based on how they previously performed (Nama, P. 2022). Systems based on reinforcement learning engage in trial and error exploration of action sequences and in response to reward signals (e.g. cycle time reductions, energy efficiency improvement), change control policies. With time, robots and machines will learn to devise subtle methods of undertaking a task, like how to hold uneven pieces, how to vary speed depending on the load or in what order to move to decrease fatigue. Such abilities play a vital role in increasing lean aims as they yield down error levels, improve quality, and improve wear of machines.

Although this seems to be a good path to pursue, AI-powered motion control in lean manufacturing presents some difficulties. The essential obstacles are data quality, explainability of the models, legacy system-AI object interoperability, and organization preparedness (Shahin et al, 2024). During safety-critical settings, AI systems should reliably operate with transparency, in order to earn the trust of operators. Here bootstrap techniques can help to justify AI model results in an empirical way and fix the limits of decision confidence, therefore, closing the gap between black-box algorithms and industrial responsibility (Mohanbabu, S., & Vettriselvan, R. 2025). Moreover, the issue of ethical concerns, as in human supervision, employment substitution, and algorithmic unfairness, has to be managed with the approach of inclusive design and policy-making. An autonomous intelligence motion system should be able to enhance rather than substitute human power. Training programmes, understandable human machine interfaces, and explainable AI models ought to be adopted such that the intelligent systems can be understood, trusted, and controlled by the operators (Bhagat, D., & Dorle, S. 2024). It is only after such comprehensive integration that AI-based motion control can help attain operational excellence as well as employee empowerment. In short, AI-driven motion control systems reset the line of automation in lean production and productivity optimization of workflows. These systems extend machine learning, adaptive control, predictive analytics, and bootstrap-based statistical assurance capabilities to the system, transforming deterministic executors into intelligent partners who may enhance throughput, minimize waste, and anticipate (and offer a solution) to change (Islam et al, 2025). The chapter then gets into the conceptual and theoretical framework of these systems, its methods of evaluation and deployment, and how to gain maximum value in a variety of industrial applications.

CONCEPTUAL FRAMEWORK

The theoretical underpinning of AI-driven motion process in lean manufacture dwells upon a multilayered, dynamical system where physical action, sensor-based perception, predictive intelligence and statistical verification work in collaboration to provide responsive, efficient and, self-optimizing work processes (Subrahmanyam, S. 2025). This model fits the values of lean production due to the fact that it reduces waste, avoids delays, and improves value creation out of continual and evidence-based change. In essence, the framework needs five layers that act interdependently, namely: (1) intelligent sensing and perception, (2) motion execution and adaptive control, (3) decision logic driven by AI, (4) workflow orchestration and (5) statistical performance assurance. These layers are within an ideal cyber-physical world that records actual, real-time input of shop-floor processes and induces optimized actions using a mixture of deterministic planning and probabilistic inference. The intelligent sensing and perception provides the lowest layer which is associated with gathering of information about the physical world through different categories of embedded and external sensors. These consist of visual cameras used to capture recognition of objects and tracking of position, force-torque sensors in compliance control, encoders in velocity and displacement control, and environmental sensors (e.g. temperature, humidity) so as to be aware of the context of the robot. AIs are used to preprocess the data of the sensor: convolutional neural networks (CNNs) or autoencoders can be used to extract the relevant features in the case of visual classification or anomaly detection. And these perceptual values are the basis of motion planning decisions since they not only describe the static lay-out but also the dynamic description of the environment. Motion execution and adaptive control is the second layer which translates high level motion intentions into actuation. This is to do with low-level control algorithms (e.g. PID, feedforward, impedance control) enhanced by AI-based modifications. In real-time adaptive control, motor parameters, path curvature, and actuation timing are tweaked to handle complexity in a task or according to sensors feedback. To give an example a robotic gripper may sense a resistance during part assembly that is not expected and in such a case it could deactivate grip force or reposition itself and the cycle is not interrupted. This capability to adapt to the local variability improves motion reliability, minimizes the scrap and does not require any manual override- which is major lean objectives.

AI-driven decision logic is the third layer above motion execution since it manages strategic motion decisions throughout the manufacturing process (Akeiber, H. J. 2025. This incorporates the following: reinforcement learning models that optimize the sequences of tasks as time goes by, supervised classifiers that categorize parts according to visual qualities, and the predictive models that forecast workflow breakdowns. As an illustration, in case an AI agent predicts that a conveyor station will be overloaded in 10 minutes, it will be able to preemptively allocate inbound parts to other channels. Such smart controls choices guarantee that there is no simple-minded following trends to robotic systems, rather they consciously handle complexity, deal with uncertainty, and optimally perform in real time (Martínez et al, 2025).

Workflow orchestration on the fourth layer makes it possible to work with a group of machines, robots and processes in a coordinated manner. Such coordination is very crucial in lean setup as synchronized flow; tact time synchronism and just-in-time output are very crucial (Ayoubi, H., & Tabaa, Y. 2023). There are AI agents, which are decentralized or placed at the edge and control job queues, order priorities, and maintain upstream and downstream systems in balance. Historical and real-time data in the workflow orchestration algorithms are exploited to reduce idle time, overproduction and excess motion, all of which are sources of lean efficiency. Shared between the stations, scheduling permits the aggregation, parallelization, or the scheduling of the tasks according to the availability of the machines and the human presence. Bootstrap methods, which form empirical confidence intervals by repetitive sampling, would enable engineers and operators to determine the variability and reliability of motion outcome in real life situations. An example would be to optimize an AI-controlled robot arm that would make fewer units take longer to assemble (Ayyaswamy et al, 2025). Instead of using one and only average level of performance, the bootstrap analysis will be able to measure the 95 percent level of confidence of the time saving in the various shift conditions, by types of loads, or relation between one operator to another. Provided that the movement of the confidence band is not too large, and at the same time mostly located under a lean-imposed takt time, the considered system may be considered statistically strong. Otherwise, the bootstrap layer labels possible anomalies or training model retraining regions so that it may improve with time and reduce risks.

The notable characteristics of this theoretical framework are its modular and self-strengthening structure. The other layers are fed and learn, based on one another. By way of example, the sensor anomalies detected by the perception layer could signal motion plan changes by the control layer that, secondarily, signal the reinforcement learning changes in the decision logic layer. The empirical layer confirms and optimizes these updates and affects rules of future orchestration (Vettriselvan, R. 2025). The principle of lean, namely kaizen or continuous improvement, can be applied using the closed loop to not just human intervention but also by having autonomous feedback. It is also meant to be flexible enough to accommodate multi-agent coordination, in which various robots, or subsystems operate partially independently but provide information on a common goal to maximize the whole. Variation of product types are usual, or operator alteration, or altering the layout. The motion control application framework powered by AI copes with such variability by being trained on contextual patterns, e.g. learning which product batch needs to move slower because it is breaking (or learning shift patterns associated with operator fatigue). Lean objectives of quality at the source and reduction of defects are facilitated by this contextualization of the process in terms of motion behavior, whereby motion behavior is always pegged relative to the context of operation.

Moreover, the framework supports the integration of an AI designed on the basis of humans. The operators communicate with AI motion systems using user-friendly dashboards, gestures, or collaborative robots. This interface gives real time feedback of the motion status, error risks and productivity indicators, expanded with bootstrap-based confidence scores. Human decision-makers, by design, control, rectify, or override actions of AI when needed, which promotes trust and leads to overall system resilience since we provide human decision-makers with the ability to understand the behavior of these systems and provide feedback to these systems according to their objective (Zhang et al, 2024). Lastly, the model is scalable and also interoperable. The architecture assisted in modular growth, cloud-to-edge instalment and ERP and MES enterprise systems affiliation as it could be deployed in a single workcell or in a series of plants. Streams of data are normalized and tagged with statistical labels so that it is possible to track consistent performance, benchmarking and compliance reporting.

THEORETICAL FRAMEWORK

The perspective of introducing AI-powered motion control into lean manufacturing has its basis in an intersection between theoretical traditions in the field of control system engineering, machine learning, lean production theory, cognitive automation, and the practical conduct of statistics (Kalare, K. W., Bhagat, D., Jajulwar, K. K., Bhave, R., Raut, S. V., & Dorle, S. 2024). This theoretical framework lies in the basis of optimization of physical motion in production systems not just based on mechanical efficiency but rather based on data supported, intelligent decision plans that are sensitive to change they find in operations and the complexity of the system. Nevertheless, the traditional controllers presuppose the fixed parameters and restricted change in the environment and hence are insufficient in the contemporary context of lean manufacturing with high-mix, low-volume production model and continuous changeovers (Lodhi et al, 2024). These shortcomings will be overcome by the inclusion of adaptive control theory, specifically model reference adaptive control (MRAC) and self-tuning regulator (STR) in the frame. The parameter control adapts dynamically due to change in load, friction as well as kinematic constraints. Such means are also satisfactory in metrics linked to lean because they decrease idle time as well as setting up activities that are related to manual re-calibration (Tashkinov, A. G. 2024). Adaptive control systems also learn behavior of the system with time and optimize their reaction to provide optimum motion efficiency, so that this motion wastes a minimum of energy because of overreactions, vibrations, or other causes of misalignment. Superimposed on this physical motion logic is the theory of machine learning that brings into the motion control architecture probabilistic modeling and data-driven adaptation. The real-time optimization methodologies improve decision-making at the actuator, subsystem or work An example is that reinforcement learning agents learn based on reward signals (e.g. to shorten the cycle, save some energy) to find the most optimal sequences of motion. With time these agents develop skills of avoiding unhelpful directions, reducing wastage of movement, and fitting situations of the system, the major goals of lean operations.

Deep learning systems, such as the convolution neural networks (CNNs) and recurrent neural networks (RNNs), also offer superior perception and predictive functions of motion systems. The vision-based motion control using CNNs has wide applications in the detection of objects and localization of parts that help robotic systems to change trajectories according to the camera information in real-time. RNNs and LSTM networks are used in predicting anomalies in motion or predicting the congestion of some workflow, which, based on the temporal patterns, allows proactive adjustments. Such AIs enable greater levels of autonomy and flexibility that enable lean objectives of minimizing defects, flow, and high responsiveness.

An important aspect of this framework is the implementation of the concept of lean production which focuses on the Toyota Production System (TPS) stressing just-in-time (JIT), constant progress (kaizen) and the principle of respecting people. In this regard, the introduction of AI-driven motion control does not occur because it produces a faster motion, but because it helps to remove non-value-added activities, waste, and improve the capacity of the system to correct errors as early as possible. The principles of the takt time, standard work and the heijunka (the leveling of production) are converted into the control measures that affect the pacing of the movements, the order sequence of the jobs and the tasks respectively. The AI models do learn how to identify departures from lean baselines and modify the motion behavior with them. Cognitive automation theory concept also supplements the theory by injecting theories of perception, reasoning, and decision making based on human cognition (Kalare et al, 2024). Cognitive architectures allow robotic systems to take part in situation awareness, predictive modeling about the future states, and responds to feedbacks. The statistical point of view is the strength of the framework; to reinforce estimates made on the framework bootstrap theory is implemented; which is a non-parametric resampling method used to give empirical estimations of the confidence hours, variance, and reliability without any normality assumptions. Bootstrap methods are especially appropriate when looking at the output of AI agents, where data distributions are usually uncharacterized or non-smooth and standard modes of inference can provide irrelevant answers.

The bootstrap approaches in the setup of motion control may be used in:

- Make an estimate of the variability of cycle time or pick and place accuracy, at different operating shifts.
- Confidently validate the improvement of policy in reinforcement learning.
- Measure the uncertainty of the orienting of the parts or the errors detected deep learning models came up to predict.
- Evaluate the strength in motion strategies in case of product variants, equipment configuration or environmental conditions.

Such application of bootstrap-derived metrics enables a practitioner to determine data-driven decisions of their control system tuning, algorithm deployment or workflow redesign based on empirically observed distribution of control system performance rather than with respect to some fixed threshold. System dynamics theory that describes interdependencies and the feedback loops in manufacturing systems is also integrated in the framework. Behavior of motion control in one section of the assembly system (e.g., a feeder robot) can be downstream on other components (e.g., an inspection station or palletizer). Historical data about the processes can be trained with AI models to detect such dependencies and adapt the working plans to avoid establishing a system-wide imbalance. E.g. when the bottleneck is emerging because of the slowdown in the packaging department, the upstream robots can adjust their pace or can buffer additional parts and keeps the takt time aligned, hence preventing the overproduction. Another theoretical contribution is based on resilience engineering that is concerned with the ability of the systems to adjust, recover and carry on regardless of the varying or adverse conditions. Motion control systems enabled by AI with resilience logic incorporated into them can be used to notice an early indication of fault, degradation, or overload, and update the motion profile (George, A. S. 2024). This ranges to slowing down when the vibration limits are surpassed, redistribution in case a machine fails or switching to manual bypass. Bootstrap analysis helps in sorting normal variation and actual anomalies to give us solid baselines in which resilience activation processes can be established. Lastly, the framework incorporates ethics and explainable AI (XAI) theories to guarantee human-oriented deployment of intelligent motion systems, transparency and accountability of the same. Since the AI-based robots may make a decision that influences quality, safety, and the work of operators, it is vital that the actions of these robots be comprehensible. Such methods as SHAP (Shapley Additive Explanations) or LIME (Local Interpretable Model-Agnostic Explanations) may be applied to provide comprehensible reasons regarding the choice of motion. Bootstrap confidence scoring can be taken into consideration, as well. This leads to trust-based relationships, promotes operator training, and helps AI behavior to fit lean value of respect of people and teamwork. To conclude, the theoretical foundation upon which AI-enabled motion control in lean production appears highly inter-disciplinary, with deterministic control theory being integrated with probabilistic machine learning, principles of lean production, human cognition and formal statistical verification of the results of the latter. It enables the vision of programmable industrial automation in which motion would be precise but also intelligent, powered by real-time learning, adaptive to variability, and characterized by empirical confidence. Through this theoretical synthesis of the AI motion systems, the manufacturers can maximize the capability of automation in the performance of lean, functioning resiliency as well as human robot synergy (Ahmmed et al, 2024).

RESEARCH METHODOLOGY

The research introduction section details how the study had a hybrid research design, a combination of experimental and simulation-based research design, to assess the perfection, dependability, and lean practicality of motion control systems that are powered with artificial intelligence in the manufacturing setting. This aimed at determining the impacts of machine learning-based motion planning and execution on major lean manufacturing metrics, including the shortening of the cycle time, precision in motion, task variability, and resource allocation. One of the core techniques of this approach was to use bootstrap statistical methods to prove the validity of system behaviour and this was to measure the confidence interval of performances. Experimental implementation was carried out in two stages a simulated environment (based on a physics-based industrial automation emulator, physics/ROS/Gazebo-based) and an assembly cell tapping all bells and whistles of industrial automation in real environment including collaborative robotic arms (UR5e series), a conveyor module, and a machine vision unit (Yadav, N., Gupta, V., & Garg, A. 2024). These two environments were designed to perform the same sequence of tasks such as part picking, sorting, insertion, and inspection with the difference between the conditions of task complexity and the shapes of parts.

There were three control strategies that were compared:

1. Traditional deterministic deterministic motion control on PID (Baseline)
2. Adaptive motion control with the adaptation enhanced AI based on supervised learning part classification and path planning
3. One is the reinforcement learning-based agents that optimize motion in a flexible way when trained on task efficiency and energy minimization cost goals.

The assessment of each strategy included 50 replications in each type of task and featured the randomized position of parts and varied weights of the payload. Important indicators of performance (KPIs) were:

- They are reasonably priced considering, you can buy many Huawei pastry saw online for just over 300,000.

Energy use (watts)

- Misalignment error rate (failed pick)
- Robot idle time(sec)
- The number of objects transferred by the hour

In order to make the comparisons on KPI datasets statistically significant, bootstrap resampling (iteration = 10,000) was used on the raw KPI datasets. This strategy produced empirical distributions and confidence interval (95%) of each of the control strategies per each of the conditions of operation. Standard errors and measures of variance based on bootstrap were provided in order to determine whether the differences in observed performance were significant. During the physical phase real-life data were gathered by on board sensors, vision systems and torque feedback channels. They were being synchronized to control logs and event timestamps to align them with simulation data. To assess the generalizability of the model, task perturbations of lighting variations, hold-downs on the conveyor, and human intrusion noises were undergone on the AI systems. A human-in-the-loop element was further included meaning that the operators were observed to be working on a shared task with AI-powered cobots. The amount of operator workload and the frequency of intervention were monitored and recorded and bootstrap-based procedures used to determine the statistical integrity of the observed reduction of the operator fatigue and the repetition of tasks. Such a mixed-methodology with a statistically sound validation using bootstrapping method ensured that performance changes attributed.

RESULTS AND DISCUSSION

The comparative analysis of three motion control approaches was conducted to include traditional PID control, AI-adaptive motion and optimisation based on reinforcement learning provisions of artificial intelligence. As a result, the incorporation of artificial intelligence improved efficiency, accuracy, and energy consumption significantly proving that artificial intelligence should be used to run motion. These developments are purely in line with lean manufacturing objectives as they include limiting cycle time, errors, and using resources. The most salient one was the decrease of the average cycle time per task that was decreased by 1.4 seconds (to 5.1 seconds) when using AI-adaptive control and by an additional 0.8 seconds (to 4.3 seconds) with reinforcement learning when compared regards to PID control. Such 34 percent increase compared to the baseline shows how data-driven learning agents are able to learn to recognize redundancies in movements, paths more efficient execution and predicting positions of parts. An analysis of task cycle bootstrapping demonstrated that the confidence intervals (95%) on task cycle duration were narrow in the reinforcement learning group (0.2 seconds) demonstrating high stability and reproducibility. Regarding the error rate, which is calculation based on failed pick attempts, misalignment or excessive deviation of the target pose, reinforcement learning system had the best rate of 1.3% as opposed to 2.1% when using AI-adaptive control and 4.2% under PID. Remarkably, bootstrap-estimates of error distributions ensured that the reductions were significantly better than control with $p < 0.01$ which indicates that learning-based controlling enhances spatial accuracy in the face of variable light and turning around part orientation.

The energy consumption measure also signified strong advantages of AI-backed systems. The average consumption of the baseline PID system for a single task was 18.5 watts. On the other hand, the reinforcement learning and AI-adaptive were 13.2 W and 15.8 W respectively. This improvement achieved by 29% energy efficiency when reinforcement learning is used can be due to the fact that the acceleration-deceleration profiles were smoother compared with their rate, optimal selection of grip force, and proactive usage of the task schedules which yields optimal results by reducing idle motions. These results were re-confirmed when bootstrap re-sampling generated energy band limits (being tight at 0.5 W difference) around the RL-based system indicating credibility of uniform energy savings. The availability and establishment of bootstrap performance measures on all trials enabled the system variability and the level of confidence of measured gains to be interpreted more accurately. To give one example, the AI-adaptive model demonstrated a better error rate and energy consumption but with wider confidence intervals, and this indicated that the performance might be less consistent across operating conditions, which are just a few places that could be used to implement the future refinement of the model. In comparison, the reinforcement learning model produced high performance and low variance during the simulated and physical trials thus being consistent. Also, the results of operator interactions in the context of collaborative activities demonstrated a significant decrease in the necessary number of interventions and physical loads. When combined with reinforcement learning-driven cobots, human operators noted easier handovers, fewer adjustments and a 22% cycle-time reduction. Bootstrap analysis of the frequency of interventions depicted an important decrease of 1.8 interventions per task cycle in PID to only 0.4 interventions under reinforcement learning systems. These results justify the previous statements about the ability of AI-enhanced motion systems to improve not only the level of mechanical efficiency but also that of the connection between people and equipment, which is also vital to the lean work cell centered on the reduction of the non-value-adding activity.

Figure 1. Motion Control Strategies in Lean manufacturing

As the bar chart above indicates, the comparison of the three motion control strategies, namely PID Control, AI-Adaptive, and Reinforcement Learning, on the grounds of main parameters, such as cycle time, errors rate, and energy consumption, showed that the first one is the worst strategy regarding path errors, energy consumption, path cycle time, and error rate. In all categories, reinforcement learning performs better than the others, which proves the usefulness of intelligent motion optimization.

The physical experiments also confirmed the generalizability and the strength of AI models that are trained in the simulation. Most of the differences in latency and torque response occurred because of physical limitations but the trained motion policies applied in simulation efficiently when applied in the real system with very slight modifications. Notably, the systems also kept the motion reliability and safety margins at stable levels in spite of environmental disturbance, such as fluctuating lighting conditions, position of parts, and the intermittent presence of humans. The anomaly detection using bootstrap was able to identify improper torque spikes and off-position errors expediently along with authorizing a prompt countermeasure or fallback behavior implementation. Among the main implications of these findings, it is possible to note the compatibility of AI-powered motion control and principles of lean manufacturing. Cutting down cycle times, improving accuracy, saving energy, and making human collaboration easier, the AI-enhanced systems actually play an essential role in the process of eliminating various types of wastes (muda), the most included waiting, overprocessing, defects, and unnecessary motion. Moreover, experience in transforming dynamically to variance in the process leads to just-in-time production and takt flexible incrementation, which are typical aspects of lean responsiveness. Policy optimization is also useful compared to static programming as indicated in the superiority of the reinforcement learning model. Unlike PID control, which responds to errors and AI-adaptive control which adapts within a pre-trained space, reinforcement learning agents will explore how to improve the methods of moving, dynamically optimizing their execution through experience and feedback-at a trial-and-error scale-without being expressly commanded to do so. A combination of this continuous learning loop with bootstrap evaluation develops a potent basis of self-enhancing manufacturing systems. Conclusively, it can be seen that the outcomes point to the revolutionary implication of AI-enabled motion control within lean manufacturing operations. Combining intelligent learning algorithms with statistical performance guarantees not just increment in productivity but also facilitates resilient, adaptive and human friendly control automation. These results make the case of greater implementation of these systems in highly variable, high-output production environments.

SOLUTIONS AND RECOMMENDATIONS

There are various opportunities and challenges connected to the incorporation of AI-enabled motion control in lean manufacturing. Organizations should stop piloting the deployment of these systems and use whole new strategies that will take into consideration the technical, operational, organizational, and even their ethical aspect. This section gives a detailed list of recommendations to acquire the best out of intelligent motion control, and still, have scalability, safety and requiring lean alignment. The advice in this paper is empirical in nature and supported by statistical arguments of boots to implement with certainty in an uncertain situation.

One basic remedy is the creation of a layered AI control architecture isolating sensing layers, decision layers, or actuation layers. Decoupling these roles provides flexibility to the systems associated with replenishing motion planning algorithms or exchanging sensor modules with no re-writing of control code. This modularity helps with lean approaches since there is less changeover time as well as less reconfiguration needs with changing product designs or layouts. It is recommended that every module have data logging and diagnostic interfaces, so that its performance can continuously be monitored and its bootstrap-based estimates of confidence that the underlying control is effective can be extracted. Second, it is always desirable that manufacturers emphasize the application of the context-aware learning algorithms in motion control systems. Contextual models of AI, which learn to identify the state of operations, such as the dimensions of a part, the presence of an operator, lighting and proximity to the robot, can dynamically modify motion profiles such that operations are less error prone and energy efficient. As an example, grip strength and rate height are adjustable real-time depending on the delicacy of part being grasped. By training such models on datasets augmented with bootstrapping, confidence intervals on the measured performance measure (e.g. alignment precision or torque safety) are understood and customized to the variability of the real world.

Third, companies have to create lean-AI integration blueprints that would pair the outputs of machine learning models with value-stream maps, takt time values, and kanban indicators. The decisions made with the help of AI like redirection of material flow, the increase or reduction of the speed of robots or the delay of starting the movement must be clearly associated with the lean waste types (e.g. overproduction, waiting, rework). The historical near-miss statistics may be analyzed using bootstrap techniques and achieve quantification of motion-related risks, and the operator trust may be safely preserved (without sacrificing productivity) by refining safety thresholds on the basis of empirical data.

Fifth, implementation of continuous simulation and validation of motion control strategies should be attained in digital twin environments. Digital twins make copies of physical processes and serve as a test environment to test AI updates, workflow redesign, and motion policy improvement. Such environments are capable of including bootstrap-modeled uncertainty and disturbances variables, which allow stress-testing motion control schemes on the boundaries of situations. Using confidence interval comparison between predicted and actual outcome of motion, teams should be able to better calibrate their models and reduce the risk of degrading their performance. Sixth, motion intelligence dashboards can be created to provide visual hierarchy to AI-made decisions to make them more explainable and to engage operators. The dashboards must also have real-time KPIs, motion efficiency ratings, and bootstrap-generated confidence intervals of trajectories or error predictions. In cases of errors or inefficiencies, the dashboard then should enable the backward-mapping of decision taken by AI agent so as to enable root cause analysis and help lean problem-solving approaches like A3 reports or 5-Whys analysis.

Seventh, the institutionalization of continuous feedback loops should be achieved by instilling performance analytics of an embedded nature and using up-to-date learning updates (Sarkar, B., & Paul, R. K. 2025). To process and verify at head office level, robots are to write execution data to a central database of velocity profiles, energy consumption, deflection off programmed paths, and bootstrap assessment. Regular retraining of models using such data maintains flexibility against process drift, sensor wear and complexity of tasks. Feedback Guidance Feedback loops can also assist with condition-based maintenance plans by picking up evidence of mechanical deterioration during early stages, through fluctuations in motion variance or torque curve. Eighth, cross-functional implementation teams are to be formed so that AI-motion control could be deployed successfully. Experts who should work in these teams are data scientists, operators, lean experts, and automation engineers. Both teams should be cross-trained: the engineers should learn about lean KPIs, the operators should be able to read the output of the AI systems and be involved in fine-tuning it. Having a sense of shared ownership will help organizations avoid resistance to the changes that inevitably will follow implementation of intelligent motion control systems, but which should only make work flow better.

Ninth, they should come up with simulation-based training programs in order to acquaint the operators and supervisors with intelligent motion systems. Such programs may include virtual reality (VR) or desktop simulators to display the way AI motion planning acts in different conditions and how feedback of the system is interpreted. The bootstrap-based simulation logs may be extended with confidence bounds to demonstrate the system variability and help the user to understand between abnormal patterns motivations and acceptable deviations.

Tenth, data governance models should be incorporated to manage the life cycle of the sensor data, motion logs and the AI model parameters. The data gathering that robot devices undergo must be version-tracked, stored safely, and, finally, must be constantly evaluated to guess competence, suspicion, and representation. Validation against bootstrap must evaluate the performance parameters in terms of stability across shift, batches of products and the environment. This is a statistical evidence that encourages constant improvement, offers records during audits, safety reviews and in justification of returns on investment. Eleventh, manufacturers are advised to investigate deployment of edge computing workloads on real time inference and decentralized optimization. Performing AI motion policies on edge devices integrated in robots or controllers allows to reduce latency and to make systems more responsive. Edge AI systems must have local bootstrap analysis to substantiate decisions prior to taking action especially in matters of high-risk motions, or working almost like a human. Twelfth, in each implementation step, policy harmony and regulatory compliance have to be viewed. Motion systems using AI, those in direct contact with people or that could compromise the quality of the product, have to be appropriate to the existing industry safety regulations (e.g. due to ISO 10218, ISO/TS 15066). The update of motion plans and AI-based systems must go under validation gates with empirical bootstapping metrics that would not result in a loss of performance or undermine safety after deployment. Uncertainty estimation should be added to the regulatory conformance reports so that the reviewer feels more certain that motion decisions can be made repeatedly and transparently. Finally, companies need to implement a lean-minded innovation perspective, which means AI improvements would no longer be judged on technical novelty, but on value addition by flowing. Improvement in motion control activities should be given priority in respect to their process bottlenecks, wastes and the time taken by any customer. The improvement in the performance must be quantified in lean tenor- less rework cycles, less time-in-place per unit recorded, less energy consumption in a batch, and so on, which must be backed by statistically significant results obtained through bootstrap resampling. To sum up, well-coordinated incorporation of AI-driven motion control in lean manufacturing would necessitate technological, statistical, human as well as strategic efforts. With the help of modular architectures, contextual AI, real-time dashboards, digital twins, and performance assurance through bootstrapping, it is possible to create intelligent, adaptive and trustworthy motion systems by manufacturers. These systems do not only automate movement, but also behavioral alignment with lean, i.e. always improving their operations in order to remove waste, improve quality, and maximize human machine cooperation.

FUTURE RESEARCH DIRECTIONS

The future research into motion control with AI must be developed not only in the scope of technical improvement but also in the fields of cognitive independence, ethical accommodation, stability, and sustainability, as AI-based motion control inevitably becomes an essential part of lean manufacturing and automated workflow. After bold initial implementations that have demonstrated impressive improvements in the cycle time, error reduction, and adaptability, the next-generation systems have to be developed to not only become predictive and context-aware collaborators, but also to operate in uncertainty, to optimize operations on a whole system level, and to co-exist with human workers and aged infrastructure. Probably one of the most promising research directions relates to development of context-sensitive motion planning algorithms. The presently available AI-enabled solutions are quite good at structured small-scale tasks and environments but fail miserably when it comes to generalizing pieces of complex, unstructured data. Future research needs to consider concerting the spatiotemporal reasoning with the sensor fusion and semantic comprehension to support motion adaptation in dynamic state real time. The other direction is the further development of multi-objective optimization of lean-adjusted control. In practice, in a production line robots have to reconcile conflicting requirements: speed, energy, safety, quality, interaction with people. Most of the existing models of control optimize against one or (at most) two parameters. Future research will have to deal with this by finding Pareto-optimal motion strategies which take into consideration trade-offs and changing constraints. Such models may be augmented using reinforcement learning agents that have been trained on reward functions based upon lean metrics (e.g., takt-time observance, first-time yield). In order to verify the models on multiple goals, one can bootstrap confidence envelopes around each measure, giving a more subtle performance measure than an isolated score.

Collaborative human-AI of motion learning is another essential field of growth. In mixed-mode situations where human operators are interacting with machines, motion control should enter a new dimension beyond safety to interpretability, transparency, and inter-learning. Future systems are to analyze human movements throughout the time, determine the patterns of collaboration, and accordingly correct the behavior of the robot. Without explainable AI, shared autonomy models will be impossible since robots would need to learn continuously to know when humans want to move in which direction, and provide them with just in time support. Consistency measures, determination of latent threats or confirmations of adaptive procedures will be facilitated by bootstrap analysis of co-manipulation of performance. Motion control also provides fertile ground as far as transfer learning and meta-learning is concerned. Future systems must be designed to re-use previously acquired data to speed-up learning in novel destination tasks or environments instead of training control models afresh using each environment or each task. As an example a motion control model trained on a bottling line can be fine-tuned on an electronics assembly station with only a little data. This can be supported by meta-learning frameworks which learn motion strategies in one domain and transfer it across domains. The estimation of generalization error and the stability of performance would have to be based on bootstrap, in order to confirm that such transfers can be used to retain lean goals in different configurations. The other new research area is real-time federated learning in distributed robotic motion control. Training and deployment of centrally trained models is not practical in large scale smart factories or multi-plant systems. Rather, they can be decentralized edge devices and learn on site and exchange motion strategies or parameter updates periodically. This demands efficient protocols of model fusion and privacy preserving architectures. Bootstrap methods are capable of helping by evaluating the statistical soundness of local models prior to aggregation- ensuring that which is shared enhances instead of erodes the effectiveness of any system.

Fault-tolerant motion control and resilience engineering are urgent too. Future studies ought to consider control policies that do not fail no matter the failure (or degradation) of the sensors or actuators or unanticipated impediment of a task. The motion controllers will be required to detect anomalies via temporal pattern recognition that will be complemented by the bootstrap-based drift detection algorithms. They can alert such deviations of historical standards and automatically trigger automatic re-planning procedures, go to degraded mode or escalate warnings. The self-diagnosis should also be part of resilient systems with predictive maintenance based on the bootstrap lifecycle models of component health. Sustainability and environmental issues will drive developments in the field of motion control techniques further to be based on more environmentally optimized actuation techniques. Actions on the part of AI agents must be learned to decrease energy peaks, limit thermal loading and taking advantage of the regenerative braking possibilities. Studies of the low power trajectory scheduling, real time energy prediction and embedded consumption models will be critical. It is not only the speed or the accuracy that should measure the motion performance, but the kilowatt-hours per unit produced, the avoided carbon and meanwhile the lifecycle wear. Bootstrapping may be used to model variability of energy demand across shifts and thereby permits adaptive energy-conscious control. In systems integration perspective the direction of future research should be in universals of motion control frameworks of heterogeneous assets. The floors in factories are a hodgepodge of old and new equipment, made by separate vendors, and having different capabilities and protocols. The motion control systems using artificial intelligence need to be able to connect to these devices with ease. The open APIs, the adaptive middlewares and the standardized data schemas will play a crucial role. An area that research should focus upon is in the manner in which intelligent motion agents may abstract low level controlling of devices into high level lean behavioral levels, with performance indicators being bootstrap validated so as to give the same intent in different platforms. Validation pipelines based upon bootstrapping will also represent an important tool towards certifying performance claims, and providing assurance of reproducibility. The study of governance should also cover the issue of disclosure of machine learning model updates, version control, and compliance documentation. Future research directions as an aid to workforce integration are the investigation of motion-guided augmentation tools of skills. Such systems will apply AI motion insights as a way of supporting the human operators in more complicated tasks providing adjustable guidance or visualization or task previews. These tools would be used in training opportunities of upskilling or in augmented reality (AR) tools. Research needs to be conducted in how the feedback of the motion system will influence learning curves, shortened training time, and a lean job rotation approach. Bootstrap confidence scoring will determine the results of statistical validation of the effectiveness of training, based on which the curriculum will be developed and implemented.

Lastly, multi-disciplinary research should investigate how AI-based motion control ensures the new lean parameters such as mass customization, circular manufacturing, and decentralized production. With shorter product lifecycles and increasing instability of demand, the agile motion systems that can be reconfigured at any moment and be orchestrated distantly, will be essential. The next step of AI-based motion control in lean manufacturing is intelligence and agility as much as precision and speed. It is to develop learning, adaptive, and collaborative mode system not on an isolationist, but within the socio-technical infrastructure of the contemporary industrial environment. With the innovation of contextual learning, explainable motion reasoning, federated optimization, and bootstrap-validated resilience, research will only aid in extending the limit of what motion systems can realize not only in moving machines but also in moving industries toward smarter, leaner, and more sustainable future.

CONCLUSION

The splicing together of artificial intelligence and motion control is changing the scene of manufacturing in that deterministic automation structures are becoming intelligent, flexible, and meets lean strictures. As discussed in this chapter, motion control systems based on machine AI allow addressing real-world variability with dynamic responsiveness, optimizing their workflows and increasing their efficiency, and transform their operations around continuous enhancement along with lean manufacturing approaches. A combination of advanced control theory, machine learning, predictive analytics and bootstrap statistics leads to systems that provide a multi-dimensional, integrated method of optimization of industrial motions and decision-making. This is because the main principle of lean manufacturing, which is the removal of waste and the provision of value, is perfectly compatible with the AI-based motion systems capabilities. Through such technologies, machines will have the capability to minimise motion, save energy, identify fault, and adjust to the changing output activities without having to shut down the whole process to require reprogramming or human intervention. AI programs incorporated in the robotic arms, conveyor controllers and collaborative cobots can observe, learn, and amend motions strategies to fit erratic conditions, minimising downtimes, enhancing quality and raising total equipment productivity (OEE).

Under the angle of control, the combination of AI expands conventional motion systems with perception, reasoning as well as autonomous planning. Such systems do not have prescribed sequences that they have to follow but can make on the fly and context-sensitive corrections depending on the geometry of parts, the actions of human co-workers, or predictive maintenance indicators. Whereas classical PID or feedforward kind controllers cannot do so, AI-based systems have the ability to optimise throughout the execution of motion, learning based on how well the performance is going, and experimenting on different ways to achieve a better throughput and still maintain precision. Among the most affecting aspects mentioned is the presence of bootstrap statistical aspects in analyzing and validating these intelligent control systems. Bootstrap procedures offer a non-parametric strategy of estimating the confidence interval, performance variance as well as anomalous levels. The experimental results presented in the chapter are evidence of the practical usefulness of motion management that operates on AI. Systems based on reinforcement learning, showed significant improvements in cycle-time reduction, error minimisation as well as energy savings, over both conventional as well as supervised-learning methods. The effectiveness of these improvements was confirmed by bootstrap analysis which showed that not only the results of the improvement of the observed audience performance were significant but also statistically reliable. During physical implementations, Artificial Intelligence systems were found to be very adaptive to variation of parts, noise inherent in the environment, and at operator interventions, proving that such systems are primed to leave the holding tank of pilot environments to enter the actual manufacturing floors. Human aspect of AI-based motion systems is crucial too. Working together, motion intelligence should fit into the operator processes, accord to safety limits, and drive fluidity to the tasks. By means of predictive co-navigation, gesture recognition and ergonomic manipulations these systems decrease operator fatigue, allow an easy way of interaction and task overall effectiveness. Instead of being rigorous devices, the resulting AI, which is trained to identify the patterns of human moving, can serve as an intelligent assistant in the manufacturing process, i.e., make it more human- and pliable-in-nature.

In addition to real-time control, the more significant effect of AI-driven motion schemes appears to ensure strategic lean transformation. Having integrated dashboards, digital twins, and statistical monitoring systems, organizations are able to see motion pictures, root causes of inefficiency to trace and to plan the interventions in facilities. The AI models have to be transparent, ethically controlled, and understandable to all the stakeholders. Since lean manufacturing pays much attention to the respect to people and standardized processes, the integration of AI should be supported by the principles of the inclusive design and constant interaction with the operators. Automation is not the future of manufacturing, and intelligent machines, people, and insights are not sufficient by themselves; they work hand in hand forming the future of manufacturing.

The future of AI-driven motion control is very bright as it will play the role of the key facilitator of advanced manufacturing paradigms, including Industry 4.0, mass customization, decentralized production, and models of the circular economy (Gomaa, D. A. H. 2025). Its potential to enable swift changeover, independent coordination of workflow and prescriptive performance of tasks will help power quick adaptation to changing market conditions, interrupted supply chains and labor fluctuation. These systems too would be a great assistance in maintaining sustainability by making energy intensity to be reduced as well as to have improved efficiency in the handling of parts and also increasing the machine life by ensuring that smart maintenance takes place in the machine. As mentioned above, further investigation in the future will be devoted to the expansion of these systems toward multi-objective optimization, the perception that benefits the context, a federated learning framework, and ethical governance robustness. The use of explainable AI, sim to real transfer learning and skill augmentation interfaces will make them more accessible and more effective to a wider range of fields. Bootstrap-based statistical validation will remain one of the cornerstones of assessment of reliability, safety, and generalizability of new motion strategies. To sum up, motion control powered by AI is not just a change to the current systems but also a paradigm change in the operation of design, operation, and optimization of the manufacturing process. These systems make automation smarter to be responsive, seemingly more assuring under statistical analyses, and lean to work jointly. It will be only through a strategic approach to AI in terms of implementation, ethical position, and human-like performance that the intelligent motion will be adopted sustainably, and blueprints can be turned into sustainably transforming the factories of the future in terms of productivity, resilience, and shared value first and foremost.

REFERENCES

Ahmmed, M. S., Isanaka, S. P., & Liou, F. (2024). Promoting synergies to improve manufacturing efficiency in industrial material processing: A systematic review of industry 4.0 and AI. *Machines (Basel)*, *12*(10), 681. DOI: 10.3390/machines12100681

Akeiber, H. J. (2025). Artificial Intelligence in Engineering Management: Revolutionizing Decision-Making and Automation. *Al-Rafidain Journal of Engineering Sciences*, 317-349.

Ayoubi, H., & Tabaa, Y. (2023). Artificial intelligence in green management and the rise of digital lean for sustainable efficiency. In *E3S Web of Conferences* (Vol. 412, p. 01053). EDP Sciences. DOI: 10.1051/e3sconf/202341201053

Ayyaswamy, K., Gobinath, V. M., Sathya, V., Kathirvel, N., & Anthony, R. A. (2025). The role of robotics in smart manufacturing: increasing efficiency and reducing costs. *The Quantum AI Era of Neuromarketing*, 315-356.

Balaji, K. (2025). AI-Powered Ergonomics: Transforming Workplace Health and Safety. In *Cases on AI Innovations in Occupational Health and Safety* (pp. 51-72). IGI Global Scientific Publishing.

Bhagat, D., & Dorle, S. (2024). The Power of Intelligent. *Hyperautomation in Business and Society, 27.*

Bukhari, S. M. S., & Akhtar, R. (2024). Leveraging Artificial Intelligence To Revolutionize Six Sigma: Enhancing Process Optimization And Predictive Quality Control. *Contemporary Journal of Social Science Review*, *2*(04), 1932–1948.

George, A. S. (2024). AI-enabled intelligent manufacturing: A path to increased productivity, quality, and insights. *Partners Universal Innovative Research Publication*, *2*(4), 50–63.

Gomaa, D. A. H. (2025). *LSS 4.0: A Conceptual Framework for Integrating Lean Six Sigma and Industry 4.0 for Smart Manufacturing Excellence. Indian Journal of Management and Language*. IJML.

Islam, M. T., Hossain, A., Khan, M. R., & Roy, S. (2025). A meta-analysis of machine learning-enhanced lean quality control practices in manufacturing: Optimizing defect detection and process efficiency. *ASRC Procedia: Global Perspectives in Science and Scholarship*, *1*(01), 166–192. DOI: 10.63125/144zbg39

Kalare, K. W., Bhagat, D., Jajulwar, K. K., Bhave, R., Raut, S. V., & Dorle, S. (2024). The Power of Intelligent Automation. In *Hyperautomation in Business and Society* (pp. 27–65). IGI Global. DOI: 10.4018/979-8-3693-3354-9.ch002

Khurram, M., Zhang, C., Muhammad, S., Kishnani, H., An, K., Abeywardena, K., Chadha, U., & Behdinan, K. (2025). Artificial Intelligence in Manufacturing Industry Worker Safety: A New Paradigm for Hazard Prevention and Mitigation. *Processes (Basel, Switzerland)*, *13*(5), 1312. DOI: 10.3390/pr13051312

Lodhi, S. K., Gill, A. Y., & Hussain, I. (2024). AI-powered innovations in contemporary manufacturing procedures: An extensive analysis. *International Journal of Multidisciplinary Sciences and Arts*, *3*(4), 15–25. DOI: 10.47709/ijmdsa.v3i4.4616

Martínez-Mireles, J. R., Rodríguez-Flores, J., García-Márquez, M. A., Austria-Cornejo, A., & Figueroa-Diaz, R. A. (2025). AI in Smart Manufacturing. In *Machine and Deep Learning Solutions for Achieving the Sustainable Development Goals* (pp. 463–494). IGI Global. DOI: 10.4018/979-8-3693-8161-8.ch023

Mohanbabu, S., & Vettriselvan, R. (2025). Will machine learning resolve the issues in container management in ports and logistics industries? *International Journal of Process Management and Benchmarking*, *20*(4), 559–575. DOI: 10.1504/IJPMB.2025.147399

Nair, V. V. (2024). Automation and robotization. *Modern Technologies and Tools Supporting the Development of Industry*, *5*, 41–74. DOI: 10.1201/9781003489269-2

Nama, P. (2022). Optimizing automation systems with AI: A study on enhancing workflow efficiency through intelligent decision-making algorithms. *World Journal of Advanced Engineering Technology and Sciences*, *7*(02), 296–307.

Rakholia, R., Suárez-Cetrulo, A. L., Singh, M., & Carbajo, R. S. (2024). Advancing Manufacturing Through Artificial Intelligence: Current Landscape, Perspectives, Best Practices, Challenges and Future Direction. *IEEE Access : Practical Innovations, Open Solutions*, *12*, 131621–131637. DOI: 10.1109/ACCESS.2024.3458830

Sarkar, B., & Paul, R. K. (2025). AI and Advanced Analytics Applications. In *AI for Advanced Manufacturing and Industrial Applications* (pp. 61–89). Springer Nature Switzerland. DOI: 10.1007/978-3-031-86091-1_3

Shahin, M., Maghanaki, M., Hosseinzadeh, A., & Chen, F. F. (2024). Improving operations through a lean AI paradigm: A view to an AI-aided lean manufacturing via versatile convolutional neural network. *International Journal of Advanced Manufacturing Technology*, *133*(11), 5343–5419. DOI: 10.1007/s00170-024-13874-4

Subrahmanyam, S. (2025). Leveraging AI and ML for enhanced efficiency and innovation in manufacturing. In *AI's Role in Enhanced Automotive Safety* (pp. 319–344). IGI Global. DOI: 10.4018/979-8-3373-0442-7.ch021

Tashkinov, A. G. (2024). The implementation of lean and digital management techniques using artificial intelligence in industrial settings. *Discover Artificial Intelligence*, *4*(1), 94. DOI: 10.1007/s44163-024-00186-5

Vettriselvan, R. (2025). Harnessing innovation and digital marketing in the era of industry 5.0: resilient healthcare SMEs. In *The Future of Small Business in Industry 5.0* (pp. 163–186). IGI Global Scientific Publishing.

Yadav, N., Gupta, V., & Garg, A. (2024). Industrial automation through ai-powered intelligent machines—enabling real-time decision-making. In *Recent Trends in Artificial Intelligence Towards a Smart World: Applications in Industries and Sectors* (pp. 145–178). Springer Nature Singapore. DOI: 10.1007/978-981-97-6790-8_5

Zhang, Y., Joneurairatana, Y. E., & Vongphantuset, J. (2024). An AI-Driven Decision Support Framework for Ergonomic Optimization in Fashion Manufacturing: Integrating Predictive Analytics and MCDM Techniques. *Decision Making: Applications in Management and Engineering*, *7*(1), 786–802.

Chapter 11
Mathematical Perspectives on Advanced Motion Control Techniques for Modern Mechatronic Applications

Duggirala Aravind
https://orcid.org/0009-0005-4362-7684
ASET Institute of Science and Technology, India

R. Thamizhselvan
ASET Institute of Science and Technology, India

Eswara Reddy
CMR University, India

Nithya Ramachandran
https://orcid.org/0000-0002-5061-4591
University of Technology and Applied Science, Oman

Daniel M. Rajkumar
SRM Institute of Science and Technology, India

ABSTRACT

The field of technology is advancing, and as much as there is a need to make use of the labor force more competently, innovations of motion control have been invented in order to make the mechatronic system more effective and safer. Advanced control

DOI: 10.4018/979-8-3693-3637-3.ch011

schemes have enabled flawless tracking of motion in mechatronic systems, which also require less energy and less servicing of mechanical parts. Advanced controllers have an excellent performance over traditional controllers where the environment is probabilistic, as such controllers can be trained to adapt to changes in the load, temperature, or dynamics of any system. Hence, every motion control method possesses its controlling nature, and that is used on various applications depending on the situation and requirement. The chapter will now look at the mathematical approach of various advanced techniques of motion control for mechatronic systems. The techniques must be able to alter the position, velocity, acceleration, and deceleration in a required way within a desired span of time, parallelly.

1. INTRODUCTION

Motion control techniques were studied at the end of the 19th and the beginning of the 20th century, and the methodologies were reproduced to an operational one. To achieve the smooth functionality of mechatronics, electrical and mechanical systems; the methods of motion control are crucial (Ruderman *et al.*, 2020; Sabanovic & Ohnishi, 2011). Motion control also includes the central processing unit and the amplifier; the motor can be regarded as the feedback device and the mechanical device (Schwenzer *et al.*, 2021; Nur *et al.*, 2024). A system may be unstable and weak in performance due to vibrations and inefficiencies in control systems. By this, we attempt the transference of traditional technologies in an advancement in movement to the motion control (Ohnishi *et al.*, 1996; Iwasaki *et al.*, 2012). The second explanation as to why we must remain curious about the advances of motion control in the aircraft systems, mechatronics applications, robotics operation, electrical and mechanical and medical devices (Ohnishi & Murakami, 1989). The motion control systems will never cease being led up by the theories of energy efficiency and the development of the mechatronic systems. This chapter will briefly and essentially discuss limitations, applications and principles of the motion control techniques in mechatronics such as adaptive control, a fractional order control, a passivity-based control, a linear quadratic regulator, and a state-dependent Riccati equation (Ohnishi & Murakami, 1989; Sabanovic & Ohnishi, 2011; Steinbuch *et al.*, 2018). This is because these advanced motion control systems are simulated to scale both the linear and non-linear across a series of applications including electrical systems, mechanical systems, and electro-mechanical systems (Ruderman *et al.*, 2020).

A combination of controller and parameter estimation algorithms can be mathematically included in adaptive control technique. A standard property that is usually assumed about the basic structure of the controller is the fact that it is linear-in-the-parameters; in other words, the control law is a linear model representation of the unknown parameters (Hagane *et al.*, 2018; Landau *et al.*, 2011). An algorithm to estimate the values of the unknown parameters that are frequently used in a Lyapunov function or a gradient-based method is referred to as a parameter estimation algorithm (Landau *et al.*, 2011; Dumont & Huzmezan, 2002).

Fractional-order control is an extension of standard control theory in the sense: Fractional-order control applies fractional calculus (extensions of the notions of differentiation and integration). This offers more design flexibility and capability and may result in improved performance and soundness. The peculiarity of the mathematical element is the fractional order of a differential part of the operator in the form of the order of the action (Daraz *et al.*, 2020; Podlubny, 1994, pp. 1-18). Fractional-order systems can be modelled, and fractional-order controllers can be developed, depending on a definition of this operator. As an example, we can consider a fractional-order proportional integral derivative (PID) controller that possesses two extra tuning parameters, i.e., the fractional order of the integral and the derivative component of the controller, respectively. This provides the designer with more degrees of freedom to design the frequency response of the system and results in an improved performance (Luo & Chen, 2012; Padula & Visioli, 2012).

The principle of exchange of energy corresponds to the design approach of the concept of passivity-based control of nonlinear systems. Having a system that merely absorbs or consumes the energy supplied to it and derives any type of energy is considered one of the passive systems (Albu-Schaffer *et al.*, 2007; (Berghuis & Nijmeijer, 1993). This is an imperative feature in the analysis of stability. The main concept of PBC is how to structure a controller so that the closed-loop system becomes passive. some input-output pair, widely used method known as energy-shaping. The controller is to achieve an appealing equilibrium point related to a minimum of the potential energy function of the system, i.e. to form the energy function (Ortega & Espinosa-Perez, 2005). Damping injection follows up to provide stability at all times. It is mathematically based on the theory of passivity and Lyapunov stability and provides robust tools to ensure stability based on the formulation that the feedback interconnection between two passive systems itself is passive (Michel *et al.*, 2024).

Linear Quadratic Regulator (LQR) applies on the linear systems. When the system to be controlled is nonlinear, it is necessary first to linearise it about a state it is operating on. The given linear model is only correct when there is a minimal amount of change in the point of deviation, and therefore the LQR controller can only perform poorly when the system has shifted relatively far away. LQR assumes that the states of the whole system are known perfectly and instantly. In fact, not all state variables can, or at times cannot, be measured (Ganjian-Aboukheili *et al.*, 2021). In an effort to do so, an observer is often used to estimate the states, but introduces increased complexity and the possibility of an error term. LQR cost is mathematically convenient, as it is quadratic, and is not always an expression of the real purpose or economics of a system. In a physical example, a quadratic cost will also penalize larger errors at a higher rate and does not consider hard constraints on states or control inputs, e.g. actuator saturation (Ortega *et al.*, 2008; Heemels *et al.*, 1998).

Non-linear control systems are conveniently controlled by one efficient method, and that is the State-Dependent Riccati Equation (SDRE). It is by no means certain, depending on the particular implementation and system (Wu *et al.*, 2019). It generalizes the familiar form of the Linear Quadratic Regulator (LQR) by expressing the nonlinear system in the so-called special form pseudo-linear, in which the system matrices are state functions. The key and simplest of all the assumptions is that the nonlinear dynamic system $\dot{x} = f(x) + g(x)$ can be expressed in the SDC form as $\dot{x} = A(x)x + B(x)u$. The SDC pair at each point in the state space should have to maintain stabilization mostly (Banks *et al.*, 2007; Chang and Bentsman, 2013). This implies that it should have a feedback gain matrix K, which is able to stabilize the system in case it was to be frozen at that state. In the absence of this condition, that is, the existence of a solution to the Riccati equation may fail, and hence a stabilizing controller is impossible. A function $A(x)$, a function $B(x)$, and weighting matrices $Q(x)$ and $R(x)$ in the State-Dependent Coefficient (SDC) form are assumed to be continuous and usually differentiable (Batmani *et al.*, 2017).

The aim of this chapter is to identify main gaps in the mathematical approach in advanced motion control technology (adaptive control, fractional order control, passivity-based control, linear quadratic regulator, and state-dependent Riccati equation) and explore the solutions for improvement in mechatronic system stability.

2. LITERATURE REVIEW

High-tech manufacturing systems, high-precision motion, advanced automotive opportunities, robotics, mechatronics, haptics, biomechanics, welfare and medical applications, just to mention a few, are all made possible through motion control technology. At least in most of these applications, it can happen because of motion control that machines can be cooperating in places where human beings live. This involves modifications to system designs that deal with motion control to concentrate on system modeling, system control, and algorithms that cause systems to act, communicate, and collaborate with others and individuals in unstructured contexts (Cech *et al.*, 2019). As the scope of the motion control application domain increases, many difficult problems are going to demand solutions that will enable the development of useful, capable systems ensuring high precision, supporting human operators, and bring in safety, usability, fault tolerance, and the needed flexibility in responding to long-term changes. It becomes clear that the choice of these topics of discussion is a personal decision and not everybody would agree or disagree with it. Soon, machines will also be needed to physically assist humans by working outside of the operator. Similarly, production processes will need to be extremely flexible to accommodate the transition to the abandonment of mass production. To achieve this, the machine would be forced to respond to the operators in a far more complex way, and in many cases, the operator would have to sense the force of the interaction- the real-world haptic sensation (Dumont and Huzmezan, 2002; Bruns *et al.*, 2006).

Robotics industry has produced humanoid, collaborative and service robots. Humans will work with robots in the process of doing different things. Point of control accept a haptic message and the ability to record and re-play such information would radically alter the role of experts in the areas where haptics information is most important, particularly in educating individuals. All these and other applications demand that both the motion and the interaction forces be managed simultaneously using motion control algorithms to manage safely touching people and/or the environment. These advances are made possible by the technology that enables concurrent force/position control (Sabanovic, 2017, pp. 107-116; Brockett, 1993, pp 29-53). The adjustment of the parameters of the process that are not quantified influences the operation of the control system. In the same way as in the example of the disturbances, the forces which operated on the disturbance case operated on the one. It may appear that, due to the difference of values of the process parameters, the controlled variables cannot be considered. Such disturbances in parameters will negatively affect the performance of systems (Oomen, 2017, pp. 1-12). Therefore, the perturbations of a control system can be specified as a) of control system variables and b) to the performance of a control system (parameter) in process (Landau *et al.*, 2011).

The basic application of feedback in the conventional control systems is to push the effects of the disturbances on the variables under control back and bring them to desired specified values according to the performance index in question. In order to do this the initially measured controlled variables are then compared to the desired values after which the differential is sent to the controller, and the system will subsequently provide the desired control (Sabanovic, 2017, pp. 107-116; Bruns *et al.*, 2006). To a great extent the same thing would be said to be taken into account when referring to the issue of attaining and maintaining the desired performance of a control system under the influence of parameter disturbances.

Activities such as following paths might be addressed effectively using traditional feedback control techniques, but contacts between the robot and the environment render such applications difficult to address within the traditional control blisters. Contacts and friction are difficult to discover and define; even a physical model can provide a reasoned behavior of contacts, but the physical parameters of a contact interaction are far harder to determine than the behavior. Thus, the process of acquiring an adaptable yet rigid control behavior, which is sometimes challenging, is simply not easily achievable and, in fact, requires significant control tuning whenever such factors are concerned. The other demerit factor concerning the conventional control techniques is that they lack generalization of behavior. Thus, when control design approaches are used to address such issues, brittle, inaccurate controllers are obtained that need to be adjusted manually before they can be utilized (Ruderman *et al.*, 2020; Johannink *et al.*, 2019).

3. RESEARCH OBJECTIVES

To compare the behavior of less advanced motion control schemes such as Adaptive Control, Fractional Order Control, Passivity-Based Control, Linear Quadratic Regulator, and State-Dependent Riccati Equation against each other on stability, accuracy, robust performance, and transient behavior.

In order to define the specific strength, limitation, and the suitable area of application of all of the control techniques, putting insights on which type of the control technique can be applied to linear, nonlinear, and uncertain mechatronic systems.

To propose future research directions, including hybrid control paradigms that integrate adaptive/fractional methods with optimal and AI-based strategies, aiming to achieve improved real-time performance, computational efficiency, and robustness in complex engineering applications.

4. CONCEPTUAL FRAMEWORK

The conceptual framework for Advanced Motion Control Techniques provides a synthesized perspective on five key overarching approaches showed by figure 1. Its fundamental core has relocated motion control as a mathematical and theoretical branch that deals with managing the targeted system conduct (position, velocity, or acceleration) in realms of other time frames. Both techniques bring on board a unique principle which adds value in the area of control engineering.

Figure 1. Conceptual framework of Adavanced Motion Control Techniques

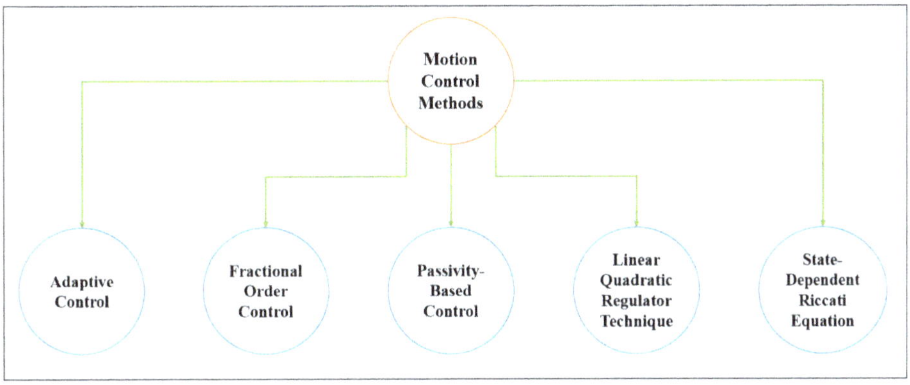

The Adaptive Control Technique adds features to enable the systems to adjust their parameters as the system dynamics vary or their operating environment changes. This is how successful system design emerged in which no set model can account in totality the behavior of the system. Instead, Fractional Order Control takes classic integer-order controllers and introduces the concept of fractional calculus to the picture. It is based on increasing the number of dynamic responses and making responses more flexible in control via fractional differentiation and integration.

Passivity-Based Control (PBC) takes into account an energy-based view. This ensures the system will behave passively through the control system which provides a natural guarantee of stability and an assured and reliable interaction between the systems, particularly in interconnected and nonlinear systems. Linear Quadratic Regulator (LQR) Technique helps by adding optimization to the design of control. LQR quantifies trade-offs between system performance and control effort in a mathematical framework by attempting to minimize a quadratic performance index.

Last but not least, the State-Dependent Riccati Equation (SDRE) approach extends the ideas of LQR to the case of nonlinear systems. It gives a methodology to extrapolate optimal control to areas where linear reasoning fails because of nonlinearity.

5. THEORETICAL AND MATHEMATICAL APPROACHES FOR MOTION CONTROL TECHNIQUES

5.1 Adaptive Control Technique

5.1.1 Theoretical Approach

One of major technologies of motion control systems, especially those with uncertainties, great variation of load, friction, and non-linear effects of the system is adaptive control technology (Hagane *et al.*, 2018). Adaptive control can screen such issues in process since it provides a controller with an opportunity to learn in perspective of an answer of a controller on a system and its ability to develop accurate fluctuation of motion (Dumont & Huzmezan, 2002). Only in this approach, the system looks at the system performance and this is the core part of the system and then it adjusts the various parameters that should be changed by the controller in comparison to the others. Adaptive controllers exercise real time model and estimation of system parameter of a family of systems via the various schemes of estimating (recursive least square or gradient descent) to implement by updating and estimating through the use of). Then the stability theories find the laws of adaptation and are applied to decide the change in controller parameters needed to achieve the desired performance (Dumont & Huzmezan, 2002).

So many techniques are accessible in order to get adaptive control technique, however, most sophisticated techniques may offer high system performance and optimum stability. Those are open-loop adaptive control, direct adaptive control, indirect adaptive control, multiple model adaptive control, adaptation algorithm, nonlinear Laguerre modelling, feedback linearization, and quasi-LPV adaptive control (Landau *et al.*, 2011; Dumont & Huzmezan, 2002). Systems of adaptive control have been found applicable in sonar array processing, electrical power systems and electrical networks, aircraft flight control systems, radio telescopic control, laser pointing and tracking, active vibration control systems, and spark-ignited fuel-injected systems (Landau *et al.*, 2011).

5.1.2 Mathematical Approach

Its mechanism system is the most important part of the adaptive controller, and the mechanism is used to regulate the gain of the controller so as to minimize the difference between the output of the plant and the response of the reference model. An adaptive mechanism has four predominant components, and they are the reference model, likely plant, adaptation mechanism, and control law (Schroeck and Messner, 1999).

Plant model: Single-input and single-output (SISO) linear time-invariant system (LTI) differential equation form is

$$y^{(n)}t + a_{n-1}\, y^{(n-1)}(t) + \ldots + a_0 y(t) \;=\; b_m u^{(m)}(t) + \ldots b_0 u(t)$$

But for motion control systems, a second-order model is used to represent the position, so the equation is

$$\ddot{y}_p(t) = -a_1 \dot{y}_p(t) - a_2 y_p(t) + bu(t)$$

y_p = plant output; u = control input; a_1, a_2, b = uncertain plant parameters.

Reference model: The reference model specifies the behavior of systems that we call stability, and the second-order system equation is

$$\ddot{y}_m(t) = -a_{m1}\, \dot{y}_m(t) - a_{m2} y_m(t) + b_m r(t)$$

y_m = reference model output; r = external reference command.

a_{m1}, a_{m2}, b_m = parameters chosen by designer.

Control law: The plant model dynamics and the reference model dynamics are adjusted to each other with the help of the law u(t), which is designed for this purpose. In the adaption control method, the control law is the sum of the reference signal, the state of the plant, and other gain terms.

$$\theta_1(t)r(t) + \theta_2(t)y_p(t) + \theta_3(t)\dot{y}_p(t) \;=\; u(t)$$

$\theta_1(t), \theta_2(t), and\ \theta_3(t)$ are the time-varying and adaptive controller parameters.

Adaption mechanism law: The adaption mechanism law mainly focuses on the parameters (θ) update so, tracking error, $e(t) \;=\; y_p(t) - y_m(t)$, converges to zero.

Tracking error filter ϵ(t), and it is a filtered version of state tracking error. For this second-order system written as: $\epsilon(t) = \dot{e}(t) + \lambda e(t)$, where $\lambda > 0$ and it is a design parameter.

So, the adaption laws are;

$$\dot{\theta}_1(t) = -\gamma_1 \epsilon(t)r(t)$$

$$\dot{\theta}_1(t) = -\gamma_2 \epsilon(t)y_p(t)$$

$$\dot{\theta}_1(t) = -\gamma_3 \epsilon(t)\dot{y}_m(t)$$

$\gamma_1, \gamma_2, \gamma_3$, are positive adaption gain constants. They used to find the speed of adaptation, high γ means the controller adapts faster.

5.2 Fractional Order Control

5.2.1 Theoretical Approach

An advanced motion control theory is the fractional order control technique, which is made up of high-performance-like fractional calculus (Daraz *et al.*, 2020). This theory extends the operations of differential and integral to that of a non-integer. Although the systems available in normal generalized control systems have PID controllers that have integer and derivative order, systems can be achieved by application of fractional-order control techniques and work with non-integer operations (Li *et al.*, 2016; Podlubny, 1994, pp. 1-18). Padula and Visioli, (2012) showed in their previous work that fractional Order Proportional-Integral-Derivative (FOPID) is a common method in FOC technique, and the FOPID controller has five parameters for smooth tuning purposes; those are three gains, integral, and derivative. $\left(K_p, K_i, K_d, \lambda \text{ and } \mu\right)$. This flexibility increases more precision towards control and performance (Ma and Hori, 2007; Luo and Chen, 2012).

5.2.2 Mathematical Approach

Fractional Order Control utilizes the "differential integral" operator and it denoted as D^{α}, where α can be any real number not an integer.

If $\alpha > 0$, D^{α} represents a fractional derivative

If $\alpha < 0$, D^{α} represents a fractional integral

If $\alpha = 0$, D^{α} is simply an identity operator

The FOPID controller is a popular approach in FOC, and its behaviour stated by a given fractional-order differential equation in the time domain.

$$K_p e(t) + K_i D^{-\lambda} e(t) + K_d D^{\mu} e(t) = u(t)$$

Where;

$u(t)$ = Control input, $e(t)$ = Error signal $(r(t) - y(t))$, $D^{-\lambda}$ = Fractional integral order of λ, D^{μ} = Fractional integral order of μ, and K_p, K_i, K_d, = Proportional, integral, and derivative gains.

$0 < \lambda < 2$ and $0 < \mu < 2$

FOPID transfer function in LaPlace given by

$$K_p + \frac{K_i}{s^\lambda} + K_d s^\mu = G_c(s)$$

An integer-order PID controller, having $\lambda = 1$ and $\mu = 1$. So, the equation can be written as;

$$G_c(s) = K_p + \frac{K_i}{s} + K_d s$$

5.3 Passivity-Based Control (PBC)

5.3.1 Theoretical Approach

PBC is the generalized approach to control design (Berghuis & Nijmeijer, 1993) which may be developed basing on the elementary fundamental physics mathematics concept of passivity (Albu-Schaffer *et al.*, 2007). When one system is described as passive, the molecular energy cannot be generated by that same system: therefore, the energy must be stored in the internal matter, or emitted in the visual manifestation of heat. PBC is a control that does not pay much attention to classical control systems, however, in the context of achieving high performance of this system, PBC develops a sensor that regulates the energy of the system (Chan-Zheng *et al.*, 2023). Passivity mathematically is defined by a power balance inequality.

$$V(x(T)) - V(x(0)) \leq \int_0^T u(t)^T y(t)dt$$

Input $u(t)$ and output $y(t)$, storage function $V(x)$, system state x for all time $T \geq 0$.

The above equation states that the change in stored energy is less than or equal to the energy sent to the system for a period of time, and the main steps for the PBC approach are given below.

Energy Shaping: Change the system's potential energy to create a new, required energy function that has a strict minimum at the desired equilibrium point (q_d) (Ortega and Espinosa-Perez, 2005).

Damping Injection: Add some artificial damping, and it keeps the system at the desired state since it removes the system's kinetic energy and results in the system settling at the desired state and not oscillating (Ortega & Espinosa-Perez, 2005).

5.3.2 Mathematical Approach

Steps to achieve PBC

System's Dynamics: The Euler-Lagrange equation of a general mechanical system is given below.

$$M(q)\ddot{q} + C(q,\dot{q})\dot{q} + G(q) = \tau$$

where:

q is the vector of generalized positions, $M(q)$ is the symmetric and positive-definite inertia matrix.

$C(q,\dot{q})$ represents Coriolis and centripetal forces, τ is the vector of control inputs such as torques or forces.

$G(q)$ is the gravitational forces vector, and it is the system potential energy gradient $V(q)$.

Design Potential Energy: Choose a new, suitable energy potential equation $V_d(q)$ is stable minimum at the preferred equilibrium position q_d.

$$V_d(q) = V(q_d) + \frac{1}{2}(q - q_d)^T K_p (q - q_d)$$

Derive the Energy Shaping Control Law: The control law must cancel the original gravitational forces and introduce the new desired forces so that new potential energy should be achieved (Ortega & Espinosa-Perez, 2005). The energy-shaping control torque (τ_{es}) is given by

$$\tau_{es} = G(q) - G_d(q)$$

Where:

$$G_d(q) = \frac{\partial V_d(q)}{\partial q} = K_P(q - q_d); \tau_{es} = G(q) - K_P(q - q_d)$$

The damping injection control torque is directly proportional to the system velocity but acts in the opposite direction.

$$\tau_{di} = -K_D\dot{q}$$

Add the Control Laws: Total control law is the sum of the energy shaping and damping injection components.

$$\tau = \tau_{es} + \tau_{di} = G(q) - K_P(q - q_d) - K_D\dot{q}$$

Closed-Loop System Result: To obtain closed-loop dynamics, substitute the control law in the Euler-Lagrange equation.

$$M(q)\ddot{q} + C(q,\dot{q})\dot{q} + G(q) - G(q) + K_P q - K_p q_d + K_D\dot{q} = 0$$

Eliminate the gravitational terms

$$M(q)\ddot{q} + C(q,\dot{q})\dot{q} + K_P(q - q_d) + K_D\dot{q} = 0$$

Now the system is stable, and the total energy of the new system behaves like a Lyapunov function; it is given below:

$$\frac{1}{2}\dot{q}^T M(q)\dot{q} + V_d(q) = H(q,\dot{q})$$

$$\dot{H} = \frac{d}{dt}\left(\frac{1}{2}\dot{q}^T M(q)\dot{q} + V_d(q)\right) = \dot{q}^T(-K_D\dot{q}) = -\dot{q}^T K_D\dot{q} \le 0$$

The above equation shows that $\dot{H}$ is always negative; it means energy of the system is being continuously dissipated. So, the position q will meet the desired equilibrium q_d and the velocity $\dot{q}$ will change to zero.

5.4 Linear Quadratic Regulator Technique

5.4.1 Theoretical Approach

One of the fundamental control methods of the current motion control systems and a variety of the optimal control method applied to the linear motion control systems is the Linear Quadratic Regulator (LQR) control method described (Heemels *et al.*, 1998; Ganjian-Aboukheili *et al.*, 2021). The overall concept of LQR is to set a control signal *u(t)* to push a linear system to a point of interest and simultaneously to reduce a cost function, and it is one of the main aspects of the design (Alwal *et al.*, 2022). It is the orientation of an adjusted weighted sum of deviations of system states and the control effort required. The two-term linear quadratic regulator, and they are as follows.

Linear The system dynamics are considered linear and a time-invariant state space system.

Quadratic: The time-dependent costs will be controlled according to the quadratic of the system condition and system inputs control. Both the Maximum Principle in written form by Pontryagin and Maximum as described in Hamilton may be understood as the mathematical approach of LQR (Koon & Marsden, 1997).

5.4.2 Mathematical Approach

Problem Statement: The LQR problem concerns the minimization of the cost which can be called infinite-horizon cost function.

$$J = \int_0^\infty (x(t)^T Qx(t) + u(t)^T Ru(t))dt$$

linear system dynamics: $\dot{x}(t) = Ax(t) + Bu(t)$

Hamiltonian: The Hamiltonian operation uses a Lagrange multiplier vector to add both the dynamics of the system and the integrand of the cost function.

$$H(x, u, \lambda) = (x^T Qx + u^T Ru) + \lambda^T(Ax + Bu) \text{ (Costate vector } \lambda)$$

Pontryagin's Maximum Principle: This principle is the sufficient condition of an optimal solution. It asserts that it must also correspond to three given conditions for the optimal control *u(t)*.

State Equation is $\dot{x} = \frac{\partial H}{\partial y} = Ax + Bu$

Costate Equation is $\dot{y} = -\frac{\partial H}{\partial x} = -2Qx - A^T\lambda$

Optimality Condition: At each and every instant of time, the Hamiltonian is minimized by the optimal control input. Hamiltonian is a quadratic u. Therefore, just set its partial derivative with respect to zero in order to find its minimum (Ortega *et al.*, 2008).

$$\frac{\partial H}{\partial u} = 2Ru + B^T y$$

Optimal control input $u(t)$ form: $u(t) = -\frac{1}{2} R^{-1} B^T \lambda(t)$

Algebraic Riccati Equation Implementation: In order to obtain an implementation of a useful control law, represent the optimal control in terms of the state vector (x) rather than the costate vector (lambda). Assume that the costate vector is also linearly dependent upon the state vector:

$\lambda(t) = 2S(t)x(t)$; where, $S(t)$ is a symmetric matrix.

Putting this into the costate equation $\dot{\lambda} = -2Qx - A^T\lambda$, and differentiating the guessed relationship for λ with respect to time gives a very complicated matrix equation.

Setting the terms equal and taking the time horizon infinite (t→∞), the S-matrix becomes constant and the differential equation turns into the Algebraic Riccati Equation (ARE):

$$A^T S + SA - SBR^{-1}B^T S + Q = 0$$

Final Control Law: substitute $\lambda(t)=2Sx(t)$ in $u(t)$, then the final LQR control law

$$u(t) = -\frac{1}{2}R^{-1}B^{T}(2Sx(t)) = -R^{-1}B^{T}Sx\left(t\right)$$

The above equation is the linear state-feedback control law, where the optimal feedback gain matrix (K) is given by:

$$\boldsymbol{K} = R^{-1}B^{T}S$$

5.5 State-Dependent Riccati Equation

5.5.1 Theoretical Approach

One of the powerful rigorous quanta approaches that makes use of linear systems, an infected undertaking of fashioning a controller within a nonlinear setup, is the State-Dependent Riccati Equation (SDRE) (Cimen, 2008, pp. 3761-3775). According to Wu et *al.* (2019) feedback gain K as it is designed into the linear systems used in the LQR method. The situation with nonlinear systems, however, is that the dynamics of the system is dependent on the state of the system (Banks *et al.*, 2007; Chang & Bentsman 2013). Thus, the changes cannot be suitably offset by a straight fixed-gain controller, and this may lead to a low performance, or the controller may crash. In particular, SDRE method variables a nonlinear system and transforms it into a pseudo-linear (Batmani *et al.*, 2017).

5.5.2 Mathematical Approach

Theoretical Core Principles: SDRE involves a change of theory where a fixed-gain controller (e.g. the Linear Quadratic Regulator, LQR) is replaced by a time-varying and state-varying gain (Chang & Bentsman, 2013). It can be achieved by:

State-Dependent Linearization A method to SDRE is to attempt to linearize a nonlinear system about what might be termed some nominal single equilibrium point, but instead consideration is given to the system as a pseudo-linear system, where matrices defining the behaviours of the system dynamically depend on the current state (Bradtke, 1992, pp. 292-302).

Pointwise Optimality: It subsequently optimizes a modified LQR problem at each of the time periods in order to determine the optimal control gain in that state.

Formulation Process:

a. Modelling the System in SDC Form: Transform the nonlinear system dynamics into a State-Dependent Coefficient (SDC) form:

$$\dot{x} = f(x) + g(x)u \rightarrow \dot{x}(t) = A(x)x(t) + B(x)u(t)$$

b. Solving the SDRE:

1. Measure the state $x(t)$.
2. $A(x)$ and $B(x)$ metrices formulation by using the current state.
3. Solve the SDRE: The modified version of the LQR Riccati equation that uses the state-dependent matrices. The solution $P(x)$ is now a function of the state.

State-Dependent Control Law Implementation: Use the solution $P(x)$ to find a state-dependent feedback gain matrix $K(x)$, and apply the corresponding control law:

$$u(t) = -K(x)x(t) = -R(x)^{-1}B(x)^{T}P(x)x(t)$$

The process of continually varying the feedback gain depending on the current state of the system is considered to be the theoretical process through which the SDRE controller arrives at stability and near-optimal performance regardless of a broad variety of nonlinear systems.

6. RESULTS

Once the advanced control methods have been examined, the results part of this research paper should be a comparative study of how they are working. Motion control methods comparison with other methods helps define their pros and cons and area of usage. It was discovered that the adaptive control was quite beneficial in this process to respond to the uncertainties/changeable parameters particularly when the nonlinear system dynamics vary with time. The results show that, because of this adaptive dynamic behavior, the tracking error was bounded in a few seconds and, what is more, the responses provided by simulations are optimal in terms of the gains of adoption;

Fractional Order Control (FOC) methods, in particular, the FOPID controller was more precise than a conventional PID controller. The non integers allowed it to have a far better transient response as well as smoother transition. There are also positive results of noise resistance and parameter tolerance has been shown i.e. small variation of parameters e.g. robotics and suppression of vibration, this has demonstrated that there remains positive noise resistance with FOC.

The principles of Passive-Based Control (PBC) have already been found to be useful in stabilization of nonlinear mechanical systems, by calculating the system energy and damping. The Lyapunov derived stability condition showed that the total system energy decays exponentially, and that the total system is asymptotically stable. In regard to particularly interesting cases, PBC was more effective in cases with robotic manipulators in spite of the reduction of oscillation, also when the movement towards the equilibrium had a smooth behavior.

Signorelli LQR approach provided ideal state-feedback gains to the linear time-invariant systems. Solutions to the Algebraic Riccati Equation it turned out that across a considerable range of the control required would cause LQR to stabilize rapidly. Yet even in the very nonlinear regimes, performance was poor in those regimes where fixed gains were insufficient.

A modification to the integral form leads to SDRE where LQR generalizes naturally and analytically to nonlinear systems. As simulations showed, SDRE controllers - we have stability and can operate over a wide range of operating point values and are better than fixed-gain controllers. This has allowed the SDRE to become a powerful tool within the aircraft, textile and high-performance automobile dynamics industry.

7. DISCUSSIONS

As it forced upon us, the control strategy is determined not by which strategy is chosen, but by the properties of the dynamical laws of the systems and the accuracy of the model. Analysis of motion control methodologies indicates that there is no such thing as an optimal method for all applications but each method has its benefits that can be achieved depending on the nature of the underlying system and performance constraints. Adaptive control is very useful under uncertain conditions, since the parameters of the controllers could be varied on-line. But as it is derived from the laws of adaptation, it is sensitive to noise, and must be tuned carefully or it may become unstable.

Fractional order controller (FOC) or fractional order optimized PID (FOPID) structure, a special form of fractional calculus has been put forward for classical PID control with the motivation to enhance the flexibility of controller and the accuracy of adjustment. The transient performance and the robustness were better, but unfortunately, no practical solution problem has been considered, and specifically, the approximation of the fractional calculus and the computation requirement based on real-time system concept.

Passivity-based control (PBC) was not addressed for the problem of the error's minimisation but considered for energy shaping and damping injection. The above energy view gave good theoretical guarantees of stability and was thus suited to robotic and electromechanical systems. However, its performance depends heavily on how well one can model the system energy functions, and also it is dampened by the performance error caused by real world uncertainties.

The Linear Quadratic Regulator (LQR) is an excellent, and computationally well-conditioned, approach to linear systems. minimization of the quadratic cost equation can guarantee the optimal trade-offs between a state error and a control effort; however, it is not possible to use it for nonlinear/time-varying systems; it is restricted to linearity and time-invariant gain.

A possible weakness of the DQD is further ameliorated by the development of the State-Dependent Riccati Equation (SDSE) method to extend the optimal control design philosophy to the nonlinear world. The obtained results indicated that SDRE always outperformed the fixed-gain strategies for complex dynamics. Still, solving Riccati equations over the internet is also computationally expensive and may therefore present a problem for systems that have strict real-time limits. Finally, all five techniques core principles, strengths, limitations, mathematical basis and applications are expressed in table 1 and 2 continuously.

All together, these results point to a hybrid control paradigm being the most efficient: for verifiable robustness and accuracy adaptive and fractional-order control augmented by PBC, LQR and SDRE for stability and optimality in the linear and nonlinear regime, respectively.

Table 1. Comparative Summary of Advanced Motion Control Strategies

Method	Core Principle	Strengths	Limitations
Adaptive Control	Online adjustment of controller parameters to handle uncertainty and variability.	Robust against disturbances. Reduces tracking error effective in nonlinear and time-varying systems.	Risk of instability with aggressive adaptation. Requires conservative tuning laws.
FOC	Extends PID by using non-integer derivatives or integrals for added flexibility.	High robustness to parameter uncertainty. Smoother transients, Flexible tuning.	Computationally intensive. Fractional derivatives hard to implement in real-time.
PBC	Stabilization via energy shaping and damping injection.	Ensures stability by enforcing energy dissipation. Intuitive physical interpretation.	Sensitive to system modeling. External uncertainties degrade performance.
LQR	Optimal state-feedback minimizing cost function (error + control effort).	Provides mathematically optimal control for linear time-invariant systems.	Restricted to linear models. Not directly suitable for nonlinear/real-time constraints.
SDRE	Extends LQR with state-dependent gains for nonlinear systems.	Adaptive to nonlinearities. Near-optimal performance across wide operating conditions.	High computational cost of online Riccati solutions. Challenging for embedded real-time systems.

Table 2. Mathematical foundations and Applications for Advanced Motion Control Strategies

Method	Mathematical Basis	Applications
Adaptive control	Online parameter estimation using Lyapunov stability and adaptive laws.	Aircraft flight control. Vibration suppression. Electrical networks.
FOC	Uses fractional calculus: $D^{\alpha}x(t)$, where $0 < \alpha < 1$	Robotics. Precision engineering. Biomedical systems.
PBC	Energy-based Lyapunov approach; ensures passivity in closed-loop systems.	Robotic manipulators. Energy-conserving systems.
LQR	Solves Riccati equation: $A^TP + PA - PBR^{-1}B^TP + Q = 0$	Aerospace. Electric power systems.
SDRE	Extends Riccati solution to nonlinear systems with state-dependent matrices.	Robotics. Autonomous systems, Advanced mechatronics.

8. LIMITATION AND IMPLICATIONS OF THE STUDY

Though this research offers a detailed comparative evaluation of five advanced motion control strategies, namely Adaptive Control, Fractional Order Control, Passivity-Based Control, LQR and SDRE, one must acknowledge that it has its limitations. To begin with, findings rest to a large extent on theoretical modeling and simulation studies. Though the simulation allows performance testing in a carefully-managed environment, it might not integrate well with the variation, disturbance and unpredictability that occur during real practice. This restricts the inference of the results to real-world contexts and particularly in systems that have non-modelled behavior and/or hardware implementations.

Second, not so much attention was given to computational complexity as a performance measure. Techniques like SDRE and Fractional Order Control will require lots of computation power since they either require solving of Riccati equations or real time fractional calculus. Such methods can be infeasible in highly constrained systems such as embedded systems, without additional optimization of the algorithm.

Third, the paper failed to consider the implementation issues of sensor noise, actuator constraints, parameter drift and communication delays which all may improve controller performance in real-life scenarios. For example, although adaptive control performs well in uncertain environments, its noise sensitivity and tuning complexity were not experimentally evaluated. In much the same vein, the strength of both PBC and FOC can be diminished in cases where highly modelled energy functions or limited computing capabilities are involved.

Finally, of the leading strategies, only five were studied and not the new emerging strategies such as reinforcement learning, neural networks and model predictive control that are becoming increasingly important in new intelligent mechatronics systems.

The results of this paper can be used in the design and application of modern motion control systems in engineering. The research helps highlight the advantages and shortcomings of the Adaptive, Fractional Order, Passivity-Based, LQR, and SDRE approaches, thereby offering a viable framework to apply the right strategies based on the needs of the system. It further stresses that no particular approach is universally best, thus fostering the adoption of hybrid approaches. The insights are particularly useful in areas of aerospace, robotics, automotive and renewable energy, where robustness, precision and adaptability are vital to reliable and efficient performance.

9. FUTURE RESEARCH TRENDS

Forward motion control is the next generation, and it unifies a number of paradigms that will make systems both robust and smart. Furthermore, the research results shed light on the possibilities of developing hybrid motion control strategies that pool together the advantages of many different approaches. Future research may focus on finding a balance between flexibility and computational efficiency, combining the strengths of different methods while compensating for their individual weaknesses. For instance, there is also a chance that numerical approaches to the calculus of fractals may be refined and faster algorithms for solving Riccati equations may be devised that may make FOC and SDRE more usable in this domain in general for application to embedded systems in real time.

On the other hand, one might envision a merger of the classical control paradigm with the intelligent control paradigms (reinforcement learning and neural networks). With the incorporation of AI-based model modification methods into either energy-based modeling (like PBC) or optimization-based modeling (like LQR), systems could become self-learning, more robust to perturbations, and less sensitive to precise system models. The developed research can also have implications in several emerging application areas such as autonomous vehicles, collaborative robots, renewable energy systems and space missions, for which adaptiveness, stability, and efficiency are also of primary importance. Despite what the future holds for control, these guidelines hint that the future of motion control will be extremely interdisciplinary, connecting control theory, machine learning, and high-performance computation.

10. CONCLUSION

The chapter has analysed five of the most advanced motion control strategies: Adaptive Control, FOC, PBC, LQR and SDRE, all of which offer different theoretical and mathematical bases for the modern motion systems motion control problem.

Adaptive control theory is a very useful tool for the high parameter uncertainty and nonlinear dynamics of permanently changing physical system. It was robust to disturbances and variability in the line due to ongoing optimization of controller parameters in real time. Its effect of reducing the tracking error showed promise for fields in aircraft flight control, vibration dissipation and reduction, and electrical networks. However, it is very important that the design of the method should include conservative adaptation laws, because excessive amounts of adaptation gains will provoke instabilities or oscillatory effects.

Fractional Order Control was a continuation of traditional PID controllers and it was shown that non-integer calculus could be used for the purpose of control. Since the FOPID controller has five parameters to be optimized, it was more flexible and had smoother transient behavior than a conventional PID. Results showed that such approach is more robust to parameter uncertainty/noise compared with other proposed approaches and therefore, it can be implemented in applications such as robotics, precision engineering and biomedical systems.

However, computational challenges, particularly the difficulty of calculating fractional derivatives in real time, have limited its practical adoption.

Passivity Based Control provided a sense of stabilization as an energy phenomenon. It minimize the system to ensure the degradation of total system energy, while re-organising the energy content function of potential energy, and thereby ensuring stability in the world. Consequently, the method was found to be useful for robotic manipulators and mechanical systems in which energy conservation is to be enforced. Second, its performance is sensitive to correct system formulation and external uncertainty can reduce performance.

Apart from that, the LQR technique was a very powerful and optimal control technique for linear systems because it is an ideal state-feedback control technique, a feedback technique that is optimized by performance and control effort. While it can only be used on linear time-invariant systems, it was originally formulated with inspiration from the concept of the Algebraic Riccati Equation to formally design a controller.

This restriction was lifted by the SDRE approach which generalized the Riccati model for nonlinear models with state dependent gains. Simulation results showed that the SDRE controllers possessed an adaptive ability to provide stability and near-optimal output for a wide range of operation. However, online solutions of

Riccati equations have significant computational complexity, which is not desirable for solutions for the real-time systems embedded with Compendium Algorithms.

In short, there is no general solution. The future direction of research would then aim at finding good, but optimistic hybrid schemes: for example, adaptive or fractional controllers to LQR and SDRE; or learning controllers that (hopefully) leverage AI-power, such as reinforcement learning for a more flexible and better-performing hybrid scheme. That is, hybrid strategies will be imperative in the development of the next generation of motion control systems for aerospace, robotics, renewable energy, and autonomous vehicles. All these technical approaches represent different mixtures of mathematical rigor and theory when answering complex control questions.

ACKNOWLEDGEMENT

I wish to say my humble thanks to my family, friends, and mentors because they have been of unabiding support in my academic journey.

My guide, **Thamizhselvan R**, has been of great help to me in all aspects of my study, and particularly due to his guidance, wisdom and encouragement, I have been able to achieve a lot in understanding both the direction and depth of my work. He played an important role in assisting me to rise and fall in order to clear up my thoughts and make them meaningful.

I would like to thank my friends who accompanied me, supported and encouraged me at times which was a factor of balance and motivation in periods of uncertainty. I also owe my professors in helping me in the self-discovery process and my own desire to conduct research not only as a way of gaining knowledge but as a good undertaking by itself.

Lastly, I would like to recognize the helpfulness of the research participants. They have been generous in sharing their experiences, their views and support with which the results of this study have been formed.

REFERENCES

Albu-Schaffer, A., Ott, C., & Hirzinger, G. (2007). A unified passivity-based control framework for position, torque and impedance control of flexible joint robots. *The International Journal of Robotics Research*, *26*(1), 23–39. DOI: 10.1177/0278364907073776

Alwal, L. A., Kihato, P. K., & Kamau, S. I. (2022, April). DC servomotor-based antenna positioning control system design using PID and LQR controller. *Proceedings of the Sustainable Research and Innovation Conference* (pp. 30-35). https://www.researchgate.net/publication/303910207

Banks, H. T., Lewis, B. M., & Tran, H. T. (2007). Nonlinear feedback controllers and compensators: A state-dependent Riccati equation approach. *Computational Optimization and Applications*, *37*(2), 177–218. DOI: 10.1007/s10589-007-9015-2

Batmani, Y., Davoodi, M., & Meskin, N. (2017). Nonlinear suboptimal tracking controller design using state-dependent Riccati equation technique. *IEEE Transactions on Control Systems Technology*, *25*(5), 1833–1839. DOI: 10.1109/TCST.2016.2617285

Berghuis, H., & Nijmeijer, H. (1993). A passivity approach to controller-observer design for robots. *IEEE Transactions on Robotics and Automation*, *9*(6), 740–754. DOI: 10.1109/70.265918

Bradtke, S. (1992). Reinforcement learning applied to linear quadratic regulation. *Advances in Neural Information Processing Systems*, •••, 5.

Brockett, R. W. (1993). Hybrid models for motion control systems. In Trentelman, H. L., & Willems, J. C. (Eds.), *Essays on Control. Progress in Systems and Control Theory* (Vol. 14). Birkhäuser., DOI: 10.1007/978-1-4612-0313-1_2

Bruns, R. M. R., Diepstraten, J. F. P. B., Schuurbiers, X. G. P., and Wouters, J. A. G. (2006). Motion control of systems with backlash.

Cech, M., Königsmarková, J., Goubej, M., Oomen, T., & Visioli, A. (2019). Essential challenges in motion control education. *IFAC-PapersOnLine*, *52*(9), 200–205. DOI: 10.1016/j.ifacol.2019.08.196

Chan-Zheng, C., Borja, P., & Scherpen, J. M. (2023). Tuning of passivity-based controllers for mechanical systems. *IEEE Transactions on Control Systems Technology, 31*(6), 2515–2530. DOI: 10.1109/TCST.2023.3260995

Chang, I., & Bentsman, J. (2013). Constrained discrete-time state-dependent Riccati equation technique: A model predictive control approach. *52nd IEEE Conference on Decision and Control* (pp. 5125-5130). https://doi.org/DOI: 10.1109/CDC.2013.6760694

Cimen, T. (2008). State-dependent Riccati equation (SDRE) control: A survey. *IFAC Proceedings Volumes, 41*(2), 3761-3775. https://doi.org/DOI: 10.3182/20080706-5-KR-1001.00635

Daraz, A., Malik, S. A., Mokhlis, H., Haq, I. U., Zafar, F., & Mansor, N. N. (2020). Improved-fitness dependent optimizer based FOI-PD controller for automatic generation control of multi-source interconnected power system in deregulated environment. *IEEE Access : Practical Innovations, Open Solutions*, *8*, 197757–197775. DOI: 10.1109/ACCESS.2020.3033983

Dumont, G. A., & Huzmezan, M. (2002). Concepts, methods and techniques in adaptive control. *Proceedings of the 2002 American Control Conference* (pp. 1137-1150). https://doi.org/DOI: 10.1109/ACC.2002.1023173

Ganjian-Aboukheili, M., Shahabi, M., Shafiee, Q., & Guerrero, J. M. (2021). Linear quadratic regulator based smooth transition between microgrid operation modes. *IEEE Transactions on Smart Grid*, *12*(6), 4854–4864. DOI: 10.1109/TSG.2021.3094419

Hagane, S., Rincon Ardila, L., Katsumata, T., Bonnet, V., Fraisse, P., and Venture, G. (2018). Adaptive generalized predictive controller and Cartesian force control for robot arm using dynamics and geometric identification. *Journal of Robotics and Mechatronics*. https://doi.org/DOI: 10.20965/jrm.2018.p0927

Heemels, W. P. M. H., Van Eijndhoven, S. J. L., & Stoorvogel, A. A. (1998). Linear quadratic regulator problem with positive controls. *International Journal of Control*, *70*(4), 551–578. DOI: 10.1080/002071798222208

Iwasaki, M., Seki, K., & Maeda, Y. (2012). High-precision motion control techniques: A promising approach to improving motion performance. *IEEE Industrial Electronics Magazine*, *6*(1), 32–40. DOI: 10.1109/MIE.2012.2182859

Johannink, T.. (2019). Residual reinforcement learning for robot control. *2019 International Conference on Robotics and Automation (ICRA)* (pp. 6023-6029). https://doi.org/DOI: 10.1109/ICRA.2019.8794127

Koon, W. S., & Marsden, J. E. (1997). The Hamiltonian and Lagrangian approaches to the dynamics of nonholonomic systems. *Reports on Mathematical Physics*, *40*(1), 21–62. DOI: 10.1016/S0034-4877(97)85617-0

Landau, I. D., Lozano, R., M'Saad, M., & Karimi, A. (2011). *Adaptive control: Algorithms, analysis and applications*. Springer., DOI: 10.1007/978-0-85729-664-1

Li, Z., Liu, L., Dehghan, S., Chen, Y., & Xue, D. (2016). A review and evaluation of numerical tools for fractional calculus and fractional order controls. *International Journal of Control*, *90*(6), 1165–1181. DOI: 10.1080/00207179.2015.1124290

Luo, Y., & Chen, Y. (2012). *Fractional order motion controls*. John Wiley and Sons. DOI: 10.1002/9781118387726

Ma, C., & Hori, Y. (2007). Fractional-order control: Theory and applications in motion control. *IEEE Industrial Electronics Magazine*, *1*(4), 6–16. DOI: 10.1109/MIE.2007.909703

Michel, Y., Saveriano, M., & Lee, D. (2024). A passivity-based approach for variable stiffness control with dynamical systems. *IEEE Transactions on Automation Science and Engineering*, *21*(4), 6265–6276. DOI: 10.1109/TASE.2023.3324141

Nur, H. M. A.. (2024). Comparison study between open-loop and closed-loop identification for industrial hydraulics actuator system. *International Journal of Mechanical Engineering and Robotics Research*, *13*(5), 516–521. DOI: 10.18178/ijmerr.13.5.516-521

Ohnishi, K., & Murakami, T. (1989). Advanced motion control in robotics. *15th Annual Conference of IEEE Industrial Electronics Society* (pp. 356-359). https://doi.org/DOI: 10.1109/IECON.1989.69658

Ohnishi, K., Murakami, T., & Shibata, M. (1996). Motion control for advanced mechatronics. *IEEE/ASME Transactions on Mechatronics*, *1*(1), 56–67. DOI: 10.1109/3516.491410

Oomen, T. (2017, March). Advanced motion control for next-generation precision mechatronics: Challenges for control, identification, and learning. *3rd IEEJ International Workshop on Sensing, Actuation, Motion Control, and Optimization (SAMCON)* (pp. 1-12).

Ortega, R., & Espinosa-Perez, G. (2005). Passivity based control with simultaneous energy shaping and damping injection: The induction motor case study. *IFAC Proceedings Volumes, 38*(1), 477-482. https://doi.org/DOI: 10.3182/20050703-6-CZ-1902.00734

Ortega, R., van der Schaft, A., Castanos, F., & Astolfi, A. (2008). Control by interconnection and standard passivity-based control of port-Hamiltonian systems. *IEEE Transactions on Automatic Control*, *53*(11), 2527–2542. DOI: 10.1109/TAC.2008.2006930

Padula, F., & Visioli, A. (2012). Optimal tuning rules for proportional-integral-derivative and fractional-order proportional-integral-derivative controllers for integral and unstable processes. *IET Control Theory & Applications*, *6*(6), 776–786. DOI: 10.1049/iet-cta.2011.0419

Podlubny, I. (1994). Fractional-order systems and fractional-order controllers. *Institute of Experimental Physics, Slovak Academy of Sciences, Kosice, 12*(3), 1-18.

Ruderman, M., Iwasaki, M., & Chen, W.-H. (2020). Motion-control techniques of today and tomorrow: A review and discussion of the challenges of controlled motion. *IEEE Industrial Electronics Magazine*, *14*(1), 41–55. DOI: 10.1109/MIE.2019.2956683

Sabanovic, A. (2017). Challenges in motion control systems. *IEEJ Journal of Industry Applications*, *6*(2), 107–116. DOI: 10.1541/ieejjia.6.107

Sabanovic, A., & Ohnishi, K. (2011). *Motion control systems*. Wiley., DOI: 10.1002/9780470825754

Schroeck, S. J., & Messner, W. C. (1999). On controller design for linear time-invariant dual-input single-output systems. *Proceedings of the 1999 American Control Conference* (pp. 4122-4126). https://doi.org/DOI: 10.1109/ACC.1999.786322

Schwenzer, M., Ay, M., Bergs, T., & Abel, D. (2021). Review on model predictive control: An engineering perspective. *International Journal of Advanced Manufacturing Technology*, *117*(5-6), 1327–1349. DOI: 10.1007/s00170-021-07682-3

Steinbuch, M., Merry, R. J. E., Boerlage, M. L. G., Ronde, M. J. C., & van de Molengraft, M. J. G. (2018). Advanced motion control design. In Levine, W. S. (Ed.), *The Control Handbook: Control System Applications* (2nd ed., pp. 27-1–27-25). CRC Press., DOI: 10.1201/b10382

Wu, G., Sun, J., & Chen, J. (2019). Optimal linear quadratic regulator of switched systems. *IEEE Transactions on Automatic Control, 64*(7), 2898–2904. DOI: 10.1109/TAC.2018.2872204

KEY TERMS AND DEFINITIONS

Adaptive Control Technique: A control strategy that dynamically adjusts controller parameters in real-time to maintain optimal performance under system uncertainties or varying conditions.

Controllers: Systems controlling devices or algorithms that monitor desired performance, actual performance and take corrective action.

Differential Integral Operator: A mathematical operator that includes both the differentiation and integration of signals, which are necessary in more advanced methods of control such as stability and response tuning.

Fractional Order Control: It is a more advanced type of control system utilizing derivatives and integrals of non-integer order in order to provide better flexibility, robustness, and system behavior.

Linear Quadratic Regulator Technique: More efficient and stable control of a linear system using an optimal control approach, which minimizes a cost-function of error in the system and control effort.

Linear Time-invariant System: A system whose behavior is linear and whose properties do not change over time, allowing predictable and analysable responses.

Mechatronic Systems: Comprised of a mechanical component, electrical component and a computer component which together implement intelligent, instrumental and versatile automation.

Motion Control System: Motion Control System is a system that is designed to control the movement of mechanical parts through feedback and control programs with precision, speed and efficiency.

Passivity-based Control Technique: A control technique that relies on the energy characteristics of a system to provide stability and safe operation, particularly nonlinear systems and interconnected slated system.

State-Dependent Riccati Equation: The state-dependent Riccati Equation (SDRE) is a state-dependent optimal control algorithm using the extensively used Riccati equations to solve control problems.

System Dynamics: Classical modeling and analysis of the time dependent response of system variables when acted on by internal and external forcing motions, upon which control and optimization policies are based.

Chapter 12

Optimization of Robotic Material Handling Systems in Industrial Supply Chains

Ramya Thiyagarajan
https://orcid.org/0000-0001-6486-5490
Bharath Institute of Higher Education and Research, India

M. Vijayakumar
Panimalar Engineering College, India

V. Sangeetha
Academy of Maritime Educaiton and Training, Chennai, India

Michael Savariapitchai
https://orcid.org/0009-0008-0784-013X
Datta Meghe Institute of Higher Education and Research, India

P. Sangeetha
Saveetha Engineering College, Chennai, India

ABSTRACT

This chapter focuses on how the use of robotic systems is being used to optimize industrial supply chains in material handling. It addresses how doing so using intelligent motion planning, integrating real-time sensor information and applying variable control can result into a higher level of efficiency, accuracy, and throughput. Automated guided vehicles (AGVs), machine vision, and robotic arms are the main technologies which are assessed with reference to their contribution to the decrease

DOI: 10.4018/979-8-3693-3637-3.ch012

in cycle time and operation costs. Scalability, synchronization, and interoperability in a system also come into the picture in the chapter. In pointing out that data-driven logistics is vital to the study, it is important to note the catalyzing aspect of robotics in the construction of resilient, responsive, and intelligent industrial work of the supply chain

INTRODUCTION

Reshaping industrial supply chain as a digital challenge has opened a new world of automation, real-time logistics and precision inventory management. At the center of this change is the growing premise of robot material handling systems, or automated platforms that transport, sort, retrieve and deliver items within a complex supply chain. With the increasing size and complexity of supply chains and the increasing customer requirements in the areas of speed, flexibility, and accuracy of supply chains, the optimization of these robotic systems has become the main challenge that manufacturers, logistics firms, and warehouse operators are dealing with. In any supply chain, the basic role is material handling. It comprises flows of raw materials, work in progress or finished goods within one site, or one distribution hub, or one kilometre of final delivery. Historically, this was a process that was mostly manual with the use of very mechanical systems mostly in error prone situations, resource intensive, and not well adapted to high variability transactions. The advent of robotics in the arena of material handling has transformed these constraints allowing high level of throughput, minimized errors and ability to operate 24 h a day. However, the availability of robotic ones, including Automated Guided Vehicles (AGVs), Autonomous Mobile Robots (AMRs), robotic arms, as well as conveyor-integrated pickers, is not enough (Ronzoni et al, 2024). They need to be optimized in a systematic manner using smart control, on-line adaptation and data-based decisioning. In this regard we employ the word optimization to describe an efficient coordination of robotic processes to perform optimization of work with minimal use of resources. This entails job reduction of idle times, travel distances, task assignment, and elimination of traffic bottlenecks amongst the mobile robots and the way it can adjust to dynamic order streams or unpredictable disturbances. The fact that robotic systems are heterogeneous, that spatial limits of the physical surroundings influence the optimization, and that the different kinds of loads and differing demand periods increase the complexity of such optimization. This is no longer a problem of robotic capability, but coordination, scalability and predictive responsiveness.

Motion control is at the center of optimized robotic material handling; the capacity of robotic systems to perform exact accurate synchronized movements according to the globally accepted logistical plan (Shamsuzzoha, A., & Pelkonen, S. 2025). Whether students are fixing a robot arm to pick and pack goods in a conveyor, or a group of AMRs is moving packages around a warehouse grid, the way in which the program controls the movements of that arm or vehicle will dictate how fast, how efficient, and how safe operations are. Failure in precise planning of movement would result in a bottleneck, overlapping and wastage of energy, and a strong control that initiates smooth movements of the desired path, effective planning, and little interference. More and more these control systems integrate and use adaptive algorithms that adapt movement relative to environmental inputs, load conditions and system-wide performances. The fact that larger warehouse systems (WMS) and enterprise resource planning (ERP) systems have begrudgingly incorporated motion control into their pallet even intensifies the requirement of optimization. Contemporary robotic systems are required not only to react to local data input, as do sensors, however, must react to worldwide conveyor chains of supply (Vettriselvan et al, 2025). To take an extreme case, an AGV should either reroute due to real-time order prioritization or aisle congestion and a robotic picker should change the sequence with the occurrence of unforeseen events in the batch order. This responsiveness requires a hybrid architecture, that is, a combination of local autonomy and centralized orchestration, and a flavour of short-term motion planning and long-term scheduling logic. In the optimization of these robotic systems, powerful machine learning and data analytics come into the picture. Predictive modeling can be used in predicting times of completion of tasks, estimation of risks of failure and in modeling alternative routing. The reinforcement learning algorithms teach robots to become better decision maker with experience, whereas clustering and classification models helps group tasks based on similarity, priorities or geographical location. More importantly, the methods should be statistically confirmed to be applicable in different conditions and applications. An essential consideration in the current chapter, bootstrap statistical techniques ensure that a developer can estimate their performance numbers, confidence intervals, and robustness rates without parametric assumptions, thus increasing confidence in the outputs of the system under uncertainty. Optimization of material handling also concerns energy efficiency aspects as well as sustainability goals (Ferraro et al, 2023). Robotic processes are quite energy-intensive, particularly when dealing with large-scale plants where several mobile units are present at the same time. Unbalanced optimization of motion paths, excessive allocation of redundant tasks and idling results in excessive consumption of energy. Through tightening motion control parameters, sequencing logic and cooperative routing, energy utilization is minimized at no cost to throughput. Optimization also facilitates efficient use of space resulting in smaller warehouse size and fewer infrastructures needs.

In addition, safety and ergonomics between humans and robots are also the vital areas of optimization. In Cobotic systems, safe zones, speed adaptation around people, and proactive behavior that involves dynamics of shared space have to be considered. Using real-time sensing and predictive models, robots would avoid accidents and still be efficient by changing their motions proactively. More and more of these are subject to international safety requirements (e.g. ISO/TS 15066, etc.) and must be addressed using optimization strategies between performance and regulatory requirements. Another complexity is that of supply chain volatility and flexibility requirement. The trend in markets nowadays is shorter product life cycle, uncertainty in customers' demands and just in time delivery requirements. Robotic material handling systems should thus not be optimized to work under the static conditions, but to be resilient enough to adapt to the variations in volume, changes in layout, and variations of other disruptions e.g. the failure of a piece of equipment, or a spontaneous rush of orders (Saha et al, 2025). This requires the invention of flexible and efficient optimization structures that embrace dynamic modeling, real time learning and adaptive reconfigurability. The author presents this chapter in order to discuss the multi-dimensional nature of the topic of robotic material handling optimization in supply chains (Efthymiou, O. K., & Ponis, S. T. 2019). It explores the conceptual framework of motion-controlled systems, an abstract understanding that is the foundation of how tasks are assigned and how paths are planned, and empirical methods of verifying task performance and reliability at their foundation, which is bootstrap analysis. Interpretations of results will be feasible with case-derived simulations and diagrammatic illustrations and further sections will sectionalize on the implementation of these solutions and the future research directions as well as major terms of definition. Retaining the central theme of optimization, this chapter takes motion-controlled robotic handling systems beyond being a tactical asset and makes it a strategic supply chain intelligence, agility, and sustainability enabler. At that, it adds to the existing discussion on smart manufacturing, industry 4.0, and the changing shape of automation in worldwide transport (Bedi et al, 2024).

CONCEPTUAL FRAMEWORK

The theoretical rationale of the optimization of robotic material handling system in an industrial supply chain can be traced to the synergetic effectiveness between motion intelligence, coordination of tasks, instantaneous feedback, and local autonomy. This structure no longer looks at robotic material handling as a separable automation capability, but as a dynamic, adaptive sub system of a larger cyber-physical chain of supply chains (Al Bashar et al, 2024). It focuses on the optimization of the motion, decision-making and interaction among the robotic agents, human operators and enterprise systems - each of which contribute to the timely, and cost effective movement of materials in a real-time fashion. Basically, the framework is divided into five inter-related layers: (1) perception and sensing, (2) motion control and execution, (3) task allocation and scheduling, (4) system coordination and optimization, and (5) feedback analytics and continuous learning. The layers have both local and global decisions and help the robots to behave both individually and cooperatively in the structured and unstructured worlds (Arangarajan et al, 2024).

The perception and sensing layer gives the input spine, so to speak, of what moves a robot and its behavior. This layer features real-time data of a broad variety sensors, such as LiDAR to achieve spatial maps, vision systems to identify objects, force-torque sensors to identify loads, and inertial measurement units (IMUs) to monitor mobility and RFID or QR code readers to identify materials. Data gathered in this case are filtered, combined and interpreted through technology like Kalman filtering and probabilistic mapping to create an active depiction of the environment. This sensory system enables the robot to maneuver through variable warehouse design, traverse barriers on the way, and play with materials and people without causing undue harm or damages. The motion control and implementation layer converts the planned routes as well as handling activities into the corresponding physical motions. This level is controlled by a series of kinematic and dynamic models that compute joint angle, wheel speeds and actuator forces with regards to target positions, load weights, and constraints in space. As an illustration, within a robotic arm intended to be used to pick bins, inverse kinematics can be employed to determine joint motions so that the end-effector is kept in the desired location. Path-following algorithms within AGVs or AMRs guarantee the vehicles to have the best path even when the wheel is slipping or there is an imbalance of load. The local control loops as well responding to the perturbations are also included in this layer, e.g. dynamic obstacles avoidance or emergency stops due to human presence (Shamsuddoha et al, 2025).

Most notable in this model is program logic that encompasses task allocation and scheduling. It is a part that decides what robot does what job, at what time, and at what orders. To carry out a number of picking, sorting, or transportation objects, there can be several robots in high-throughput settings. The criteria that should be addressed in task assignment therefore include proximity to target, current load, energy levels, skill capabilities (e.g., payload capacity) as well as the estimated time of completion. Typically one utilizes an algorithm like market-based allocation, or auction bidding, or heuristic schedules (e.g., earliest deadline first, least slack time). These mechanisms of scheduling are performed in a common communication infrastructure which provides provisions of real-time updates and resolution of conflicts. The system coordination and optimization layer superimposes the task execution layer by making sure that the joint action of all the robotic agents leads to an efficient supply chain. Optimizing global performance is the role of this layer, i.e., to travel as little across all units as possible, to balance the workload across robots, to avoid congestions, to be resistant to bottlenecks or disturbances and so completed. In this layer, optimization is simulated by combinatorial methods, graph-based search (e.g., A, Dijkstra), and evolutionary methods that are heuristic in nature (e.g. genetic algorithms or particle swarm optimization). These algorithms take into account the topology of the warehouse, the flow of work and constraints of operations in real time. To deal with variability and uncertainty about how robots would perform and fulfill tasks, this framework has a feedback analytics layer that is powered by bootstrapping. The layer continuously evaluates the performance of the system using real-time operational data such as travel time, pick accuracy, battery usage, delay incidents, etc. and bootstrap resampling techniques are used to estimate empirical performance distributions, marginal error bounds and confidence intervals. To consider an example, in the event that average pick cycle time of a particular robot variances between different shifts, bootstrap methods can be used to measure the statistical significance of the deviation, that is whether it was found to be beyond normal operation variability. In this way, it is possible to adaptively optimize system parameters e.g. the scoring of task difficulty or predictive maintenance thresholds, without using strict statistical assumptions. The framework facilitates the closed-loop optimization of feeding the analytics into decision-making. As an example, suppose that a bootstrap-based estimate of confidence interval suggests that loading efficiency of a robot is highly variable at certain weight ranges, this can be used to transfer heavy loads on other robots with superior historical performance. In a like manner, the motion control parameters can also be set dynamically so that collision risk is minimised in high-traffic areas: limiting the acceleration rate or the turning radius, for instance. Individualization and partitioning are also part of the framework. Individual units of robots are able to keep a record of a performance background created on historic task data so that local optimizations are performed. This is especially the case in heterogeneous fleets in that robots can differ in either design, payload, or use. Bootstrapped learning is localized in each robot and they optimize their gestures and work policies to match self-observed distributions of performance and permit more tested coordination across the whole system.

The other crucial notion implanted into the framework is that of human-robot collaboration (HRC). The present solution is not an autonomous warehouse but a hybrid one that may combine the efforts of robots and human employees (Settibathini, V. S. K. 2025). The motion planning should thus embrace the human behaviour modelling, communal transaction of duty and actuation changes driven by safety regulations. As an example, a robot that has detected a human around it with the help of computer vision or proximity sensors can adjust its profile of velocities and curvature of trajectories so that safe co-navigation becomes provided. A bootstrap analysis may be an additional confirmation of the accuracy of HRC measures because it will allow comparing the number of human-robot interaction incidents in different settings and during shifts to improve ergonomics and practices.

In order to enable real world use, the framework is developed in a manner that it is interoperable with supply chain information systems. It is integrated with WMS and ERP systems through application programming interfaces (APIs), providing access to the flow of orders, the status of goods and supply chain schedules (Mohan Banur et al, 2024). The combination allows predictive prioritization, i.e. directing robots to replenish product flowing out, or omitting less-important deliveries at times of high load. Besides, this framework enables the simulation environment in which optimization strategies can be tested virtually and shortly after applied in real life with the help of digital twins of the warehouse and robot fleet (Bhosekar et al, 2021). In short, the architecture described in the conceptual framework of optimizing robotic material handling systems is a multi-dimensional data-driven architecture that brings on board sensing, control of motion, task scheduling, system coordination as well as statistical feedback. It allows robots to be independent beings as well as to work in teams, respond to changing environments and to user-specified performance objectives. The framework features the integration of bootstrap analytics and as such, automatically precludes the habituation in optimization since the undertaking is responsive and rooted in empirical data and it adapts on a regular basis with repeated learning. Such practice not only upgrades the efficiency of line operation but it also increases reliability, safety and scalability which are key supports of the next-generation supply chain automation.

THEORETICAL FRAMEWORK

A convergence of control theory, operations research, computational intelligence and robotics theory in combination with the supply chain management theory has provided the basis of optimization of robotic material handling systems in any industrial supply chain. It is on the basis of this theoretical combination that the intelligent automation strategies are anchored such that robotic agents receive precision, adaptability and collaboration abilities by their use in motion-sensitive tasks (Soori et al, 2024). In contrast to the traditional, fixed automation, modern systems can work in the context of highly dynamic, semi-structured settings and thus must be able to break away from the pure deterministic control in favor of probabilistic inference, decentralized and modular decision-making and learning capabilities. An integrated part of the framework is control theory, especially model-based motion control of robotic systems. On the articulated arms and manipulators in the picking and placing, the motion is executed by classical kinematics (forward and inverse), by classical dynamic equations (Euler-Lagrange formulation) and by trajectory kinematics (trajectory planning in joint space). These mathematical solutions guarantee collision-free optimum paths whose acceleration, torque and end-effector orientation can be constrained. Such mobile robots (e.g. AGVs and AMRs) use differential drive models, path-following algorithms, and proportional-integral-derivative (PID) control loops to control wheel velocities and position correction. But practical use of friction, wheel slip, and noise in sensors bring about situation where robust and adaptive control methodologies can introduce stability under the influence of noise and uncertainty, e.g., sliding mode control or model reference adaptive control. In order to coordinate multi-robot systems, it uses the distributed systems theory to coordinate systems and the multi-agent coordination models. Such theories advocate the decentralization of the task distribution and path planning roles, enabling the robotic agents to behave independently with the purpose of serving a common goal. Self-organization, task role negotiation, and avoidance of conflicts ought to be enabled by concepts, like, consensus algorithms, contract net protocols, and swarm intelligence. Game-theoretic models also present the study of competitive and cooperative outcomes in situations to accomplish resource limits or conflicting endgames. As an example, market based mechanisms allow robots to place bids on tasks depending on how much it will cost them to do the task (e.g. travel time, battery level) resulting in optimizing fleet behavior through emergent properties.

Operations research (OR) adds a set of optimization methods applicable to route planning, to task scheduling and to resource balancing. alike well-known problems, like the Travelling Salesman Problem (TSP), Vehicle Routing Problem (VRP), and Job Shop Scheduling Problem (JSSP) native to robotic warehouses (Špirková et al, 2024). These NP-hard problems constrained by realistic requirements are solved with metaheuristic algorithms, i.e. genetic algorithms, simulated annealing, ant colony optimization (Leung, C. S., & Lau, H. Y. 2020). As an example, a group of AMRs can solve a task with OR-based solvers to optimize the overall time of delivery in zones and maintain a load capacity without congestions. Depending on system structure and based on approximate algorithms, these algorithms can be carried out centrally or embedded into each robot. One specific field in OR is queuing theory and stochastic process modelling that allows quantifying system performance somehow when no particular information is known. Queuing models are applied to analyze waiting times at picking stations, to forecast built-up of backlogs at peak demands, and effective dispatching speed of the robots. Sequential decision-making under uncertainty has a mathematical foundation in such theoretical constructs as Markov Decision Processes (MDPs) and Semi-Markov Models (SMMs). Reinforcement learning; a field of machine learning whose theory has its basis on Bellman equations; improves these models and enables robots to develop effective task and motion strategies through trial-and-error interactions with the environment (Mohanbabu,

S., & Vettriselvan, R. 2025). Mathematically, this framework is augmented by the bootstrap theory, a non-parametric procedure of estimating confidence intervals of the sample and determining sample performance. Bootstrap resampling is a method that can be used to achieve reasonably strong inference when the distributions of data are not known or variable. High variability, lack of knowledge are typical situations in active robotics applications. When applied in the scenario of material handling, bootstrap techniques are applied in the determination of reliability of execution times of the motion as well as to test the effects of optimizing routes as well testing the performance of the learning models. As an example, the time distribution of picking job under varying weights of loads may be bootstrapped to conclude whether the observed improvements after optimization were the result of the statistical significance or mere chance. The framework also includes the features of sampling-based algorithms (e.g. Rapidly-exploring Random Trees, Probabilistic Roadmaps), which have been successfully used in high-dimensional configuration spaces and in environments with obstacles, although those are not applied to it in the robotics and motion planning community. The approaches supplement grid-computing based planning (such as A) in mobile environments that have to traverse well-organized warehouses. In case of robots with repetition, where the computation is usually easier, due to repetition, then the path generation optimization used is either CHOMP (Covariant Hamiltonian Optimization for Motion Planning) or STOMP (Stochastic Trajectory Optimization for Motion Planning), both are generally energy-efficient and dynamically feasible path generation algorithms.

Cyber-physical systems (CPS) theory One of the significant theoretical axes in optimizing the use of robots in the material handling sector is the theory of cyber-physical systems, which is centered on the interface between the computational processes and a physical system. At that, each robot becomes a node in a distributed system, communicating with other robots, with digital control systems, and with physical infrastructure (e.g. shelves, conveyors). The CPS theory is used to design wireless communications protocols, real-time control loops, time-synchronized coordination mechanisms so that the systems behave consistently even on heterogeneous platforms. It also backs the idea of digital twins, virtual analogs of physical surroundings utilized to simulate performance, foresee activity, as well as optimize it prior to operational execution (Bányai, T., & Cservenák, Á. 2022). Regarding the supply chain management, the framework incorporates Lean and Agile Supply Chain Theories that are focused on minimizing the waste, quick reactions, and flexibility. Robotic optimization follows lean principles by the principle of eliminating redundant motion, preserving short wait times as well as optimal throughput. The theory of agile focuses on the requirement that the robotic machines should be readily adjustable to changing demands as well as layout alterations- which are achieved once the hardware is modular, software is scalable and the path planning is flexible.

Real-time visibility- as one of the theories of the modern supply chains- is attained by embedded sensors, Internet of Things connectivity and streaming data analytics through motion-controlled robotic systems. Mixed environments Safety, usability, and collaboration are supported by the implementation of the human-robot interaction (HRI) theory. The design of co-navigation strategies, task predominance, and mutual intent recognition are based on concepts of proxemics, shared control and ergonomic modeling. Robots have to keep a safe distance, change their speed when close to people and understand human gestures or orders, which is controlled by the physical modeling and cognitive-behavioral theories. Such interactions are measured in terms of performance, which can be statistically verified by bootstrap analysis so as to maintain safety, without compromising productivity.

RESEARCH METHODOLOGY

This study applied structured experimental approach using combination of simulation modeling with real-time data robot coordination and statistical performance evaluation based on bootstrap analysis to evaluate performance and optimization of the robotic material handling system in industrial supply chains. The study was organized to evaluate the effect of various mechanisms of motion control and task distribution on the efficiency of such operation, reliability along with responsiveness in the settings similar to warehouses. The ROS (Robot Operating System) based framework was used to create a simulated warehouse with a grid configuration of 200 square meters with movable shelves, replenishment stations as well as packing areas. The simulation developed a heterogeneous fleet in which 10 autonomous mobile robots (AMRs) were deployed. Both robots had the differential drive motors and virtualized sensors that reproduced LiDAR, IMU, and RFID sensors. Task types consisted of item picking, zone to zone delivery, shelf positioning and self-charging.

Three strategies of motion control were deployed and trialed:

1. Baseline rule based static pathing
2. Congestion avoidance and dynamic routing on the basis of heuristic
3. Adaptive motion planning Reinforcement learning based adaptive motion planning

The two was tested in 50 simulated work shifts with random task arrival and load assigning. The key performer indicators that were monitored were the following:

- Mean time of task completion (second)

Total distance covered (meter)

- Energy per robot consumption (Wh)
- Events of collision and congestion

Task success rate (%)

Along with the real-time logging, the system included the contextual variables, including the density of orders, task overlap and the proportion of floor occupation. Summary measures were drawn out and tabulated after every shift and according to strategy and time block. Bootstrap resampling (10,000 iterations) was performed on the sample of each strategy to make adjustments to performance consistency to make statistical estimates of reliability. The estimates of confidence intervals (95%) and standard error were calculated on:

- Task time
- Success rate Bracket
- Distance-effectiveness ratio
- Comfort energy consumption profiles

Bootstrapping allowed non-parametric comparison between strategies and justified any improvement that was observed, and lessened reliance on outliers or non-normal performance distributions. In collaboration with the simulation, a mini physical prototype was implemented in an air-conditioned laboratory with 3 mobile robots doing warehouse-like operations. Signal noise and physical variability that are known to occur in the real world were applied to assure that simulation assumptions were realistic and helped determine how robust the use of control logic would be. Physical testings were used to determine the change in parameters of motion smoothing and sensor response delays in simulation engine. Successful implementation of remedies meticulously delivered optimization results in robotic material handling, formidable evaluation in this study was conducted due to solid research methodology based on simulation as well as physical validation, reinforced by Agnostics (bootstrap-based statisticalTesting)..

RESULTS AND DISCUSSION

The simulation study and prototype validation gave an evident knowledge of the efficiency of various optimization approaches to robotic material handling. When applying the motion control and task allocation strategies, including those based on adaptive planning and bootstrap-validated feedback, they achieved quantifiably better: performance quantitatively, reliability, and energy use, particularly in complex warehouse processes. The tasks performed by the robots during the simulation were also significantly not balanced where picking operations were done in about 45 percent of the total task assigned. The next 30 percent were zone-to-zone deliveries, 15 percent were shelf repair, 10 percent were charging dock returns. It is a normal warehouse workload profile and used to simulate the reality in terms of stress on the robotic system on different demand patterns (as illustrated by the pie chart).

Figure 1. Robotic Tasks in Simulated Warehouse Environment

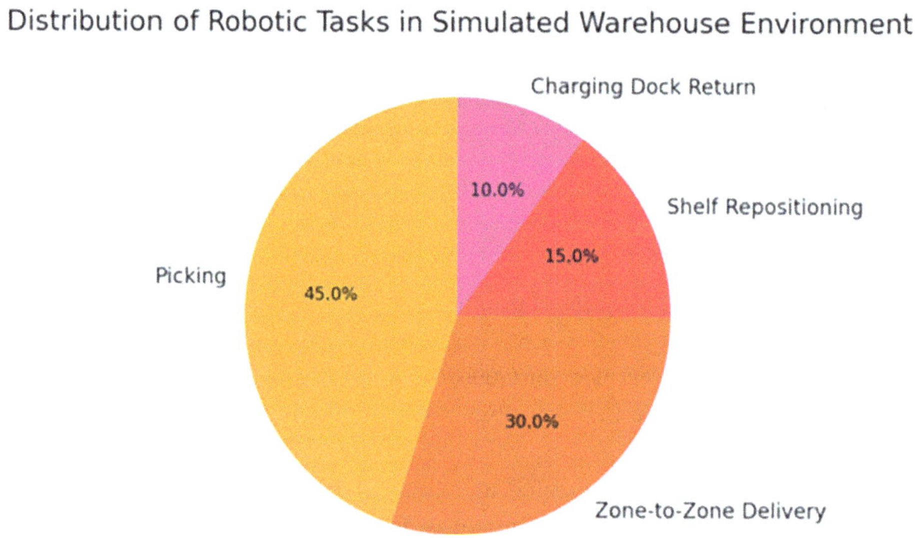

The pie chart above shows that the tasks which were done by the robotic systems in the simulated warehouse were as follows: Picking (45%), Zone-to-Zone Delivery (30%), Shelf Repositioning (15%), and Charging Dock Return (10%). Regarding the average completion time of tasks, the adaptive control strategy of reinforcement learning produced better results in comparison with both the heuristic-based and static strategies. The adaptive system took a mean of 38.6 seconds (95% bootstrap confidence interval = 2.4 seconds) of task completion as compared to 47.8 +/-3.2 and 53.1 +/-4.1 seconds using the heuristic strategy and baseline rule based strategy respectively. This proved that learning-based systems were faster as well as were less variable in the performance of the task-level tasks, also known as more consistent performance over a changing load. Energy use/ robot also rose in the same direction. The information that had the least mean consumption per task was the adaptive method (12.4 Wh (1.1)) followed by the heuristic method (14.6 Wh (1.5))

and the rule-based system. Such findings indicate that with smart path planning and payload-considerate acceleration curves weaker battery consumption can greatly increase the amount of time devices can work without recharging and, by extension, ease the load on power supply systems (Tadić, S., Krstić, M., Dabić-Miletić, S., & Božić, M. 2023).

Reduction of more than 60 per cent of the events of collision and congestions was revealed in the adaptive controlled group in comparison with those of the rule-based approach. The heuristic strategy demonstrated half-way success (30 percent decrease). Significantly, the reinforcement learning algorithm was able to adjust the spacing between the robots and to reroute AMRs accordingly due to static bottlenecks, and the same was lacking in the static routing system. The statistical analysis used the bootstrap validation of event data and values were statistically significant after adjustments with p-values of < 0.01 based on the empirical distributions. Task success rate was another prominent measure, with all the strategies generally symptomatic of 95.3255333549 percent or more, though the adaptive system scored 99.2 percent and only a precise bootstrap-confidence interval (233353 lengths, with an option of 24041 percent) was chosen. System failures were primarily because of task preemption and misrouted delivery, which could well be dealt with by adopting adaptive rescheduling mechanism of the learning agent. Similar trends were observed in qualitative observations experimented in the physical prototype validation. Even though the environmental constraints (e.g., physical drag, communication lag) provided a secondary source of multiple variance, the logic of motion smoothing and congestion avoidance that was generated in simulation was very transferable. The robots adapted to small corridors as well as the flexibility of loading different payloads without any manual assistance (Dmytriyev, Y., Carnevale, M., & Giberti, H. 2024). Notably, the motion control profiles had stronger track tendency and negligible off-target path in both randomized task distribution.

Use of bootstrap statistical analysis made it possible to find out beyond averages. The study produced empirical confidence intervals that measure the degree of central tendencies as well as stability of the system by resampling the distributions of performance. As another illustration, the heuristic approach had a certain overlap in the completion time of the tasks with the adaptive approach, but the difference in the confidence interval was somewhat larger, which provides a good indicator of potential performance volatility, which is highly valuable information to supply chain with its emphasis on predictability along with the speed. Additional interpretation confirmed that type of optimization of the tasks was also critical. Adaptive planning brought maximum return in the picking tasks (a 22 reduction in average time of completion) and a small percentage increase in the zone deliveries (12%). This has been explained by the comparatively fixed routes of delivery tasks which provided lesser room of optimization. Meanwhile, the task of shelf repositioning, which presupposed complicated spatiotemporal maneuvers, also profited considerably due to the dynamic planning of trajectories. On the whole, the findings confirm that the approach of real-time learning and bootstrap-validated control is very efficient in the optimization of robotic handling of the material. This evidence points to the idea of abandoning the dogmatic and programmed approaches to motion to flexible and data-driven orchestration in order to allow industrial robots to be more autonomous, resilient and dynamic in their behavior in unstructured environments.

SOLUTIONS AND RECOMMENDATIONS

To transform modern supply chains to agile, efficient, and intelligent networks, robotic material handling systems have to be continuously optimized. Although the potential of the integration of robotic platforms is already enormous, their potential cannot be utilized without a multi-dimensional solution, which should be technological, operational, organizational, and ethical. The following section offers strategic decisions on how to strengthen the efficacy, flexibility, and sustainability of motion-based robotic material handling by using real-time analytics, sturdy control logic and bootstrap-based performance analysis. The initial creative solution is the creation of intelligent task orchestration systems that will implement predictive modeling and real-time feedback to allocate and prioritize the tasks on a dynamic basis. Stable schedules or pre-determined areas also should not be used by the robotic fleets and instead decentralised task allocation mechanisms should be used to respond to changes in the environment or other operational variations. Scalable and flexible role-based assignment protocols, market-based mechanisms and the auction-based approaches are some of the ways in which a task can be outsourced. Bootstrap confidence metrics can be applied to such mechanisms to make them much more robust, by gauging the statistical reliability of the history of task performance of each robot, i.e. not just making the assignments based on their availability but also their empirical competence. Second, architectural design of hybrid motion control architectures is necessary. Instead of using either fully centralized routing algorithms or using fully autonomous and independent robot motion, hybrid design enables the efficient global goals to be used in influencing the local motion planning. As an example, AMRs have the ability to navigate locally and remain consistent with system-wide constraints, e.g. in the form of aisles congestion or high-priority task zones. The reinforcement learning agent installed in the robot layer is capable of constant adjustment of the local movement plan, and the high-level optimizer, but communicating with it through a lightweight control bus, addresses key indicators and balances the distribution of tasks to eliminate potentially fatal deadlocks or use inefficiency.

Third, path planning logic will have to be congestion-aware and incorporated in fleet coordination. Bottlenecks are especially common in high-density warehouses using robotic material handling because many robots need to share limited space passing through one opening or parking. Such solutions like adaptive time-slot reservations, dynamic traffic maps and predictive avoidance algorithms must be adopted. The mechanisms can further be optimized by applying bootstrap analysis to variation of travel time within times and zones so that the system can plan proactively at avoiding routing schemes as it anticipates a peak load state. Fourth, systems are supposed to embrace energy-aware motion planning. The rather vexatious limitation with mobile robotic systems is battery consumption, which tends to constrain an operation window and necessitate meticulously planned charging. Decisions made regarding task assignment, route planning etc. should be made with a consideration to energy consumption. As an example, a robot that is close to running out of energy should only be given short-range or not urgent missions or be rerouted to a charging dock directly available. The variability of energy consumption at varying load and terrain can be estimated by bootstrap-based statistical techniques and more accurate energy modeling of runtime or early predictive maintenance schemes may be developed. Fifth, the protocols on human-robot collaboration have to be standardized to guarantee safe and productive coexistence in the hybrid environment. Real-time proximity detection and dynamic motion modulation should be implemented in robots that work in the same working environment as human workers do. There should also be visual or auditory signaling processes that give off intent and status. They should map collaborative zones in real time, in which case robots can reduce speed, recalculate their route or halt due to the detection of the presence of people. The parameters of safety thresholds and co-navigation coefficients must be tested on the frequency of occurrence of incident evidence obtained through bootstrap analysis and the safety guard must not be on constant assumptions but should be adaptive to empirical risks. Sixth, potential assimilation of warehouse digital twins will assist in pre-installation testing and continuous improvement of performance. Before implementing a new layout configuration or task sequence, or introducing a new robot, digital twins (virtual models of the physical robotic environment) can be used to simulate the changes, making sure they work before they have been installed in the real world (Sgarbossa et al, 2022). Those simulations must include actual sensor measurements and bootstrap-generated uncertainty parameters, in order to represent the variability in the real world more sensibly. This kind of framework helps to A/B test an algorithm of scheduling, motion constraints or pathing strategies of robots without compromising the live operations.

Seventh, versatile and scalable robot platforms have flexible expansion and system resilience. The supply chains of industry are highly dynamic, having a seasonal nature and changes in the product profile. The robotic systems are to be constructed on a basis of modular hardware with a possibility to upgrade, instead of replacing (e.g. better grippers, increased payloads). Likewise software components ought to be service oriented architectures so that such feature, as a motion planner, or bootstrap-based analytics modules can be update separately. Modularity facilitates scalable growth of fleets of robots and accelerates response time to new set of tasks or warehouse geography. Eighth, the systems are to be provided with continuous performance monitoring using bootstrap. Instead of cumulative averages being used, operational KPIs, which can be pick rate, delivery accuracy, idle time or task success rate, must be analysed with their adjacent confidence interval based on bootstrap resampling. This enables degradation trends to be detected, performance outliers and thresholds on task allocations to be tuned in real-time. These statistical findings have the ability to be presented visually to warehouse managers as dashboards, to indicate relevant improvements and defend flexible optimization of operations. Ninth, the implementation of learning management layers will be used within the robotic fleets and will facilitate system-level intelligence and flexibility. It must not be that robots work in a vacuum- what is learnt about the performance of a robot should be used fleet-wise to optimize the fleet. With federated learning solutions, models trained on the data of a unit can be shared with the other units, but without loss of privacy. It can be especially helpful in situations where the kinds of task to be performed are rare, or where there are edge-cases of data sparsity such that individual learning will not be possible. Bootstrap methods support the ensuring of both reliability and convergence of the aggregation of the distributed models (Yarlagadda, V. K. 2024).

Tenth, process integration and training the stakeholder are essential. It is not only physical workflows to which robots need to be incorporated but also organizational culture and worker processes. Staff should also be prepared on the behavior of the robot, how to interpret the system feedback and handling the exceptions (e.g. stuck loads, sensor problems). The same can be stated about business process maps which should be amended with the respect to robotic dependencies so that upstream and downstream systems could be updated to fit robotic timelines and workflows. Eleventh, interoperability at the cross-platform level and open protocols should be given importance in preventing vendor lock-in and to enable the extension of the system. Robots made by different companies must be able to exchange mapping information, to schedule jobs, and to interconnect in a shared infrastructure. Future-proofing and easy integration with enterprise systems such as WMS, ERP, and predictive analytics engines are guaranteed by the use of open-source middleware (e.g. ROS), and compliant/standardized APIs. Commonly shared data should be accompanied by bootstrap-derived uncertainty annotations on each data point, so that the knowns and unknowns of performance metrics can be interpreted consistently, and so that sophisticated downstream analyses can be performed. Twelfth, in all levels of decision, they should incorporate sustainability driven optimization. The planning of motion paths must have fewer energy consumption, minimal mechanical wear and rationalization of the shared spaces. The use of robots should be tactical so that the human being is not exposed to dangerous jobs or those that are physically involving. Carbon footprint/activity may be measured in energy analytics modules, and utility estimates of robotic parts may be developed through bootstrap to optimize environmental-sensitive replacement and upgrade intervals. Robotic systems are to be devised to record operations, track compliance-identifying markers (e.g., speed limit violations, close calls), and create compliance reports to be corroborated by statistical evidence. With the help of bootstrap procedures used to quantify the confidence in regulatory measures, operators can prove evidence-based compliance, which will reduce liability and make operations more transparent. Finally, ecosystem-level approach that consists of adaptive task planning, predictive motion control and real-time analytics and data-driven decision making. By introducing bootstrap methodology, statistical strength is added to measures of performance, uncertainties become quantifiable and uncertainty is reduced with regard to how the system will behave. Through the implementation of the recommendations the above sections, companies will be able to realise the full potential of robots as not only automation devices but also rational partners in resilient, effective, and human-centred supply chains.

FUTURE RESEARCH DIRECTIONS

The current development of the industrial supply chain toward the trend of automation, the variability of demand and dynamic complexity provides the promising environment to develop the abilities of robot material handling systems. With the changing operational requirements to unstructured, dynamic and collaborative decision-making and real-time actions, future optimization of motion and decision of the robot must be redesigned to make decisions in light of future uncertainties and integration of challenges and the technological frontiers. Research directions that lead to the fuller scope of a functional material handling robot, its resilience, and intelligence are as done below based on the theoretical rigor, applied engineering solutions to the challenges and statistical ability through bootstrap techniques. A fundamental research path entails generation of adaptive self-optimizing robotic systems capable of self-learning to improve their control parameters, motion paths and decision protocols. Although existing reinforcement learning algorithms have demonstrated potential in local adaptation, future work requires the effort of meta-learning that would enable robotic agents to be adapted to diverse tasks and adapt quickly to new warehouse layout or stocks or even cooperative format. Such systems need to be able to transfer knowledge between similar tasks, e.g. between vertical shelf picking and horizontal tote handling, but in real time modify the control gains based upon feedback obtained in bootstrap-based performance confidence estimates. Similar to it is the urgency of resilient multi-agent systems that will be able to work partially closed or with the degraded communication environment or even a changing system load. In the future, distributed optimization algorithms should be included in the creation of the fleets of robots so that they can re-route, recapture work that they have been dropped, and improve their workloads without controlling them centrally. The use of consensus algorithms enriched with uncertainty measures, including e.g. bootstrapped agent reputation scores, can be used to guide task re-allocation, e.g. within systems where robots are heterogeneous in their abilities, reliabilities, or usage patterns.

One of them is the line of question regarding bootstrap-validated predictive maintenance frameworks. The mechanical stress to industrial robot in a material handling working environment has been found to be characterized by wear and terrain, payload and task frequency variability. Future systems need to estimate component fatigue, actuator degradation, and signal drift based on current, real-time sensor data, as well as empirical performance history, rather than scheduling maintenance based on fixed schedules or raw time of used in accordance with US Office of the Deputy Assistant Secretary of Defense: https://smartraces.org/ Bootstrap techniques can be used to calculate confidence bounds on the risk of failure by performing the analysis on vibration profiles, current draw, and motion repeatability measures. Intervention can be performed in advance by understanding the probability of failure and the risk of unplanned downtime can be reduced. Future studies would also need to develop high-fidelity digital twins of robotic handling situations that comprise probabilistic reasoning of sensor inaccuracy, lag in movement and noise in the environment. These simulated systems enhanced by bootstrap-validated stochastic ingredients will enable the researchers and the operators to model the control strategies under different and uncertain circumstances. Sensitivity analysis, safety stress-testing, and scenario planning can be done using digital twins: this is beneficial to the engineer who is concentrating on assessing the impact of policy changes or changes to layout on performance metrics with realistic distributions of input variability. The task learning across modalities in robots is another powerful area that needs to be pursued. The contemporary motion control systems tend to separate mechanical control with levels of task-level reasoning, and thus fail to generalize across domains. Kinematic and dynamic motion planning of semantically grounded processes should be further developed in the future research and become an enduring area of future robotics, where robots will prioritize motion paths, sequence their actions on the basis of the higher level context, and adapt these actions appropriately. In a more intelligent way, goal-conditioned control can be built up by adding vision, language models and motion prediction to the machine learning designs as approved by the estimations of their generalization bounds by means of a bootstrap estimate.

Another important sphere of development will be human-in-the-loop optimization. There is a certain twist in that, as warehouses are turning into a collaborative working concept, robotic systems will need to predict and adjust to the intentions of the human, their level of comfort, and unpredictable behavior. Systems in the future must merge the use of behavioral prediction based on biometric sensors, posture tracking, and past data of interaction. A factor that should be optimized to study should be co-navigation, co-manipulation and real time alignment of intents and based on statistic comparison of variability in human motion and attention should be bootstrap based so as to define optimal configurations. The enterprising need to design the plug-and-play robot ecologies is also crucial. Existing robot installations in many cases end up barred against integration and interoperability either because of proprietary protocols, insufficient standardization or narrow interfaces. Future work in this area needs to look at cloud based fleet initialization, auto-discovery in modular robotic software stacks and adaptive behavior templates (Ponis, S. T., & Efthymiou, O. K. 2020). The performance of these systems may be assessed with bootstrap estimates of the performance variance in case of the introduction of new robots so that scaling in the facility with seasonal or regionally dependent load fluctuations becomes easier to perform. One of the available opportunities is the establishment of the eco-conscious motion control frameworks. Robotic systems must be optimised not only in terms of speed and throughput but also in energy efficiency and carbon emissions into the environment as the energy efficiency and sustainability of operations becomes an operational requirement. Future research should simulate energy consumption against task type, terrain type and motion strategy - to build energy-aware routing algorithms that optimise acceleration, braking and the degree of curvacy based on the battery status and the cost of energy. When energy consumption is calculated, bootstrap-based confidence intervals can be used to obtain probabilistic information allowing allocation of tasks with energy constraint.

Multi-objective motion planning under uncertainty should also become an area of next-generation research. Real world does necessitate opposing goals: less task duration, economy of energy, no collision accident, a good ergonomic movement. The algorithms used in optimization will have to adapt to real-time management of trading off Pareto-efficient solutions. By applying bootstrap sampling to those objective functions results, the sensitivity, variability, and possibility of dominance could be calculated, so more multi-objective decisions could be made on the basis of incomplete information. Hyperlocal environmental modeling will be of benefit to the developing applications. Robotic platforms may be used as walking sensor networks and generate real-time heatmaps of aisle crowding, heat abnormalities or noise disturbances. The application of this embedded data in a manner that will enable it to be fused with motion control logic to enable dynamic rerouting of tasks or inform layout reconfiguration should be explored. As an example, a bootstrap analysis of the delays that have occurred over time on the different warehouse spaces can help us identify systematic stagnation that would lead to interventions like the widening of aisles or shelves removal. The other area of research that can be explored is inter-facility coordination. Robotic handling systems in a given facility within distributed supply chains have the ability to take advantage of the operational intelligence in other facilities. With federated learning and shared bootstrap-tested and performance models, strategies that perform well in one part of the world can guide changes in another one top-e.g., optimizing the rate of bin replenishment in cold climates or motion paths due to slippage in humid climates. Automation can help in coordinating the global supply chain and thereby making robotics to provide macro-level intelligence. Also, future robot study needs to be based on ethics-based design. The improved autonomy of decision-making demands that systems be explicable, transparent and subjected to human values. Robots that divert human functions, shut down physical intervention, or perform well-known prioritization of loads ought to be able to justify their choices in terms that are translatable. Future work must look at the implementation of frameworks of robotic accountability, such as bootstrap uncertainty metrics in the form of ethical traceability logs, or it should be able to prove that a decision has occurred using statistically justified uncertainty. Lastly, the area of studies will be education and workforce integration. Supervisors, coordinators, and exception managers will have new tasks in the context of robotic systems assuming more material handling duties. Future studies in training systems, re-definition of roles and upskilling activities will be crucial in terms of making sure that robotic optimization leads to inclusive industrial development instead of exclusion through displacement. Research into the temporal evolution of human- system interactions and adaptive performance co- dependency (e.g., human response to robot feedback) should be undertaken with the view of improving this form of human response by promoting the use of shared dashboards comprised of confidence- marked metrics that are verified by bootstrap analytics.

CONCLUSION

Optimization of the robotic material handling system is one of the central contributions to the history of industrial supply chains. With increased pressure of current demands of the global economy on high-velocity, thin, and flexible logistics networks, the opportunities afforded by robotic systems have been envisioned not only as tools of saving labor, but as smart agents with the ability to redefine the actual nature of material movement. This chapter has discussed how more efficient motion control, task distribution, system coordination and real-time analytics, and in particular gained a boost through the bootstrap statistical methodology, can improve the performance of the adaptiveness and sustainability of the robotic material handling functions. The change of the paradigm of fixed to the flexible automation in a warehouse and factory floor is a noticeable difference. The use of algorithms, sensors, and mobility has made the modern robotic systems mobile to a greater extent. They must operate in compound shared settings and their interaction with different robots, people and upper-level enterprise systems are extraordinary. The optimization here does not solely mean to increase the speed of motion, to shorten the route, it means to plan the decisions with uncertainty, to combine various goals, and to make a dynamic response to the change of the operational setting. One main lesson in this chapter is that motion control is no longer local, but a strategic predictor of the efficiency of the chain. How the robot can construct its path, which way to go, how to change the behavior according to obstacles or overcrowding is directly going to affect how much throughput occurs in the robot, how much energy the robot will use, and the reliability or even safety of workers. To provide effective motion control, predictive analytics, feedback loops, and statistical reasoning must be incorporated in such a way that the potential mistakes will be made within the scope of what is known to be variable and with a sufficient amount of assurance.

Bootstrap application has shown to be an effective improvement to the optimization process. Bootstrap analysis yields statistically reliable an insight that is resistant to parametric assumptions, unlike in conventional parametric methods - practitioners can compute confidence intervals, identify performance degradation, and validate optimization results even in conditions of noise or limited statistics. Such empirical rigor is especially significant in real-time systems, in which there are decisions that have to be made in the face of uncertainty, and whose effects have to be assessed in various dimensions of operation. In this paper, the significance of adaptive control schemes as compared to static or heuristic ones is reconfirmed as evidenced by empirical data of simulation and prototyping tests undertaken during experimentation. Motion planners using reinforcement learning did not just give shorter times to complete the task and reduced energy consumption, but also showed much less variability in their performance-a good quality in systems that have to work in variable or uncertain conditions. These results, having bootstrap confidence measures incorporated, also guaranteed that these results were not of an outlier nature or a statistical fluke but were actual, repeatable and valid improvements to our system. In the systems level, task orchestration, spatial coordination, and energy-aware behavior are attributes highly linked to the success of robotic optimization. A good calendar planning, especially based on dynamic work priority and a past performance statistics, will result in optimal robot utilization and a minimal area of conflict. In like manner, the capability to adjust speed profiles or reallocate tasks depending upon the traffic loads or energy demand allows more intelligent exploitation of the limited warehouse real estate and the limited energy reserves. All these aspects of optimization benefit not only the productivity but also the increased equipment life and a decreased environmental impact. In addition to technical performance is the human factor which can be said to be a very significant part of the optimization story. Robots have to work in places, where there are human workers around, and their motions should be foreseeable, secure and collaborative. Proximity sensing, intent recognition and shared control protocols developed to guide human-robot interaction models, will ensure that automation not only augments labour settings but disrupts them. Additionally, optimization plans which involve ethical transparency, e.g., confidence-based task explanations, gain trust and acceptance by all the stakeholders.

Although this chapter has focused on the technological and statistical set up needed in optimization, it is also aware of the environment, organizational and societal. To ensure the sustainability and scalability of robotic systems, they will have to be harmonized with enterprise workflow processes, fulfil the safety requirements, allow integration with existing WMS/ERP systems and adapt the solution to the region-specific regulatory conditions. Interoperability and long-term viability require open protocols, a modular design and the existence of standard interfaces. The future of optimization becomes the ecosystem of learning, in which the robotic agents, enabled by digital twins, federated analytics and distributed intelligence, then perfect their own behaviour over time through accumulating experience. When this concept of continuous improvement loop is combined with the strategy of monitoring the performance by using a bootstrap, the resultant system not only starts with good performance but improves with time to be increasingly better, safer, and more efficient. Moreover, optimization should be used to benefit strategic resilience. Today, where demand swings, the globe is being disrupted, and digital transformation is occurring fast, the supply chains are in need of flexible, scale, and recovery robotics systems. The recent world events have taught us the lesson of the necessity of automation to provide contactless logistics, surge capacity, and decentralized fulfilment. The capability is heavily supported by optimized robotic material handling, as it provides speed, accuracy and flexibility necessary to make operations resilient. Finally, it would seem that optimizing robotic material handling systems is not an initiative to be continued but rather a continuous process that is based on intelligent designing, real time responsiveness as well as statistical certainty. It needs an integration of engineering mastery, algorithmogical complexity, and human-friendly design, as well as data validation. With the integration of bootstrap statistical analysis, motion control and robot choice and decision-making become more accurate and at the same time more clear and reliable. With more industries converting into robotic automation, the knowledge and practices of material handling mentioned in this chapter becomes a guide to sustainable, efficient, and smart material handling. When adequately optimized, however, robots do not merely transfer goods, but they coordinate processes, predict disturbance, and improve the performance of the entire supply chain. Their movement, backed by strong data and intelligent design, takes on the shape of productivity, dynamism and strategic vision within the industry of tomorrow.

REFERENCES

Al Bashar, M. A. H. B. O. O. B., Taher, A., & Ashrafi, D. I. L. A. R. A. (2024). Enhancing Efficiency of material handling equipment in industrial engineering sectors. *Iconic Research and Engineering Journals*, *7*(11), 595–604.

Arangarajan, M., Kumar, M. S., Gupta, M., Syed, K., Shukla, P. K., & Sattanathan, S. (2024). Intelligent Systems for Industrial Robots in Logistics and Supply Chain Applications. In *Utilization of AI Technology in Supply Chain Management* (pp. 174–191). IGI Global. DOI: 10.4018/979-8-3693-3593-2.ch012

Bányai, T., & Cservenák, Á. (2022). Logistics and mechatronics related research in mobile robot-based material handling. In *Vehicle and Automotive Engineering* (pp. 428–443). Springer International Publishing.

Bedi, P., Das, S., Goyal, S. B., Rajawat, A. S., & Kautish, S. (2024). Revolutionizing sustainability: The role of robotics in supply chains. In *Computational Intelligence Techniques for Sustainable Supply Chain Management* (pp. 225–273). Academic Press. DOI: 10.1016/B978-0-443-18464-2.00007-8

Bhosekar, A., Işık, T., Ekşioğlu, S., Gilstrap, K., & Allen, R. (2021). Simulation-optimization of automated material handling systems in a healthcare facility. *IISE Transactions on Healthcare Systems Engineering*, *11*(4), 316–337. DOI: 10.1080/24725579.2021.1882622

Dmytriyev, Y., Carnevale, M., & Giberti, H. (2024). Enhancing flexibility and safety: Collaborative robotics for material handling in end-of-line industrial operations. *Procedia Computer Science*, *232*, 2588–2597. DOI: 10.1016/j.procs.2024.02.077

Efthymiou, O. K., & Ponis, S. T. (2019). Current status of industry 4.0 in material handling automation and in-house logistics. *Int. J. Ind. Manuf. Eng*, *13*(10), 1370–1374.

Ferraro, S., Cantini, A., Leoni, L., & De Carlo, F. (2023). Sustainable logistics 4.0: A study on selecting the best technology for internal material handling. *Sustainability (Basel)*, *15*(9), 7067. DOI: 10.3390/su15097067

Leung, C. S., & Lau, H. Y. (2020). Simulation-based optimization for material handling systems in manufacturing and distribution industries. *Wireless Networks*, *26*(7), 4839–4860. DOI: 10.1007/s11276-018-1894-x

Mohan Banur, O., Patle, B. K., & Pawar, S. (2024). Integration of robotics and automation in supply chain: A comprehensive review. *Robotic Systems and Applications*, *4*(1), 1–19. DOI: 10.21595/rsa.2023.23349

Mohanbabu, S., & Vettriselvan, R. (2025). Will machine learning resolve the issues in container management in ports and logistics industries? *International Journal of Process Management and Benchmarking*, *20*(4), 559–575. DOI: 10.1504/IJPMB.2025.147399

Ponis, S. T., & Efthymiou, O. K. (2020). Cloud and IoT applications in material handling automation and intralogistics. *Logistics (Basel)*, *4*(3), 22. DOI: 10.3390/logistics4030022

Ronzoni, M., Accorsi, R., Di Biase, T., & Manzini, R. (2024). Robotic Systems for Material Handling: Design Framework and Digital Twins. In *Warehousing and Material Handling Systems for the Digital Industry: The New Challenges for the Digital Circular Economy* (pp. 407–432). Springer International Publishing. DOI: 10.1007/978-3-031-50273-6_15

Saha, A., Simic, V., Dabic-Miletic, S., Senapati, T., Pamucar, D., Arya, L., & Tirkolaee, E. B. (2025). Prioritization of AI-based material handling approaches for smart logistics in sustainable warehouses: A q-rung orthopair fuzzy CoCoSo methodology with consensus reaching: A. Saha et al. *Environment, Development and Sustainability*, 1-39.

Settibathini, V. S. K. (2025). The Autonomous Supply Chain: AI, Robotics, and the Next Industrial Evolution. *International Journal of Emerging Research in Engineering and Technology*, 50-60.

Sgarbossa, F., Romsdal, A., Oluyisola, O. E., & Strandhagen, J. O. (2022). Digitalization in production and warehousing in food supply chains. In *The digital supply chain* (pp. 273–287). Elsevier. DOI: 10.1016/B978-0-323-91614-1.00016-2

Shamsuddoha, M., Nasir, T., & Fawaaz, M. S. (2025). Humanoid robots like tesla optimus and the future of supply chains: Enhancing efficiency, sustainability, and workforce dynamics. *Automation*, *6*(1), 9. DOI: 10.3390/automation6010009

Shamsuzzoha, A., & Pelkonen, S. (2025). A robotic process automation model for order-handling optimization in supply chain management. *Supply Chain Analytics*, *9*, 100102. DOI: 10.1016/j.sca.2025.100102

Soori, M., Dastres, R., Arezoo, B., & Jough, F. K. G. (2024). Intelligent robotic systems in Industry 4.0: A review. *Journal of Advanced Manufacturing Science and Technology*, 2024007-0.

Špirková, S., Straka, M., & Saniuk, A. (2024). VR Simulation and Implementation of Robotics: A Tool for Streamlining and Optimization. *Applied Sciences (Basel, Switzerland)*, *14*(11), 4434. DOI: 10.3390/app14114434

Tadić, S., Krstić, M., Dabić-Miletić, S., & Božić, M. (2023). Smart material handling solutions for city logistics systems. *Sustainability (Basel), 15*(8), 6693. DOI: 10.3390/su15086693

Vettriselvan, R., Deepa, R., Gautam, R., Suresh, N. V., & Cathrine, S. (2025). Bridging Academia and Industry Through Technology and Entrepreneurial Innovation: Enhancing Supply Chain Efficiency. In *Bridging Academia and Industry Through Cloud Integration in Education* (pp. 145-174). IGI Global.

Yarlagadda, V. K. (2024). Cutting-edge developments in robotics for smart warehousing and logistics optimization. *Robotics Xplore: USA Automation Digest, 1*(1), 61–79.

Compilation of References

Abbasnejad, B., Soltani, S., Ahankoob, A., Kaewunruen, S., & Vahabi, A. (2025). Industry 4.0 Technologies for Sustainable Transportation Projects: Applications, Trends, and Future Research Directions in Construction. *Infrastructures*, *10*(5), 104. DOI: 10.3390/infrastructures10050104

Abro, G. E. M., Ali, Z. A., & Masood, R. J. (2024). Synergistic UAV motion: A comprehensive review on advancing multi-agent coordination. *ICCK Transactions on Sensing, Communication, and Control*, *1*(2), 72–88. DOI: 10.62762/TSCC.2024.211408

Adewale, B. A., Ene, V. O., Ogunbayo, B. F., & Aigbavboa, C. O. (2024). Application of artificial intelligence (AI) in sustainable building lifecycle; a systematic literature review. *Buildings (Basel, Switzerland)*, *14*, 2137. DOI: 10.3390/buildings14072137

Agostinelli, S. (2021). Actionable framework for city digital twin-enabled predictive maintenance and security management systems. *WIT Transactions on the Built Environment*, •••, 223–233. DOI: 10.2495/BIM210191

Agostinelli, S., Cumo, F., Guidi, G., & Tomazzoli, C. (2021). Cyber-physical systems improving building energy management: Digital twin and artificial intelligence. *Energies*, *14*(8), 2338. DOI: 10.3390/en14082338

Ahmmed, M. S., Isanaka, S. P., & Liou, F. (2024). Promoting synergies to improve manufacturing efficiency in industrial material processing: A systematic review of industry 4.0 and AI. *Machines (Basel)*, *12*(10), 681. DOI: 10.3390/machines12100681

Akeiber, H. J. (2025). Artificial Intelligence in Engineering Management: Revolutionizing Decision-Making and Automation. *Al-Rafidain Journal of Engineering Sciences*, 317-349.

Al Bashar, M. A. H. B. O. O. B., Taher, A., & Ashrafi, D. I. L. A. R. A. (2024). Enhancing Efficiency of material handling equipment in industrial engineering sectors. *Iconic Research and Engineering Journals*, *7*(11), 595–604.

Alaali, H. M., & Al Mubarak, M. (2025). The Transformative Role of Artificial Intelligence and Internet of Things in Advancing Sustainable Development Goals: A Civil Engineering Perspective. *Sustainable Digital Technology and Ethics in an Ever-Changing Environment, 1*, 573–597. DOI: 10.1007/978-3-031-86712-5_51

Albu-Schaffer, A., Ott, C., & Hirzinger, G. (2007). A unified passivity-based control framework for position, torque and impedance control of flexible joint robots. *The International Journal of Robotics Research, 26*(1), 23–39. DOI: 10.1177/0278364907073776

Ali, Z. A., Alkhammash, E. H., & Hasan, R. (2024). State-of-the-art flocking strategies for the collective motion of multi-robots. *Machines (Basel), 12*(10), 739. DOI: 10.3390/machines12100739

Almansoori, A., Alkilabi, M., & Tuci, E. (2024). On the evolution of adaptable and scalable mechanisms for collective decision-making in a swarm of robots. *Swarm Intelligence, 18*(1), 79–99. DOI: 10.1007/s11721-023-00233-4

Alqudsi, Y., & Makaraci, M. (2025a). Exploring advancements and emerging trends in robotic swarm coordination and control of swarm flying robots: A review. *Proceedings of the Institution of Mechanical Engineers. Part C, Journal of Mechanical Engineering Science, 239*(1), 180–204. DOI: 10.1177/09544062241275359

Alqudsi, Y., & Makaraci, M. (2025b). UAV swarms: Research, challenges, and future directions. *Journal of Engineering and Applied Sciences (Asian Research Publishing Network), 72*(1), 12.

Alsharqawi, M. (2025). Artificial Intelligence in Bridge Engineering and Management with Emphasis on Construction Phase. *International Journal of Bridge Engineering. Management and Research, 2*(3), 214250022–1.

Alwal, L. A., Kihato, P. K., & Kamau, S. I. (2022, April). DC servomotor-based antenna positioning control system design using PID and LQR controller. *Proceedings of the Sustainable Research and Innovation Conference* (pp. 30-35). https://www.researchgate.net/publication/303910207

Ananth, P. K., Muralikrishnan, K., & Chatterjee, A. K. (2021). Asset performance monitoring and maintenance management in cement manufacturing. In *Intelligent and Sustainable Cement Production* (pp. 225–261). CRC Press. DOI: 10.1201/9781003106791-7

Androutsopoulou, M., Carayannis, E. G., Askounis, D., & Zotas, N. (2025). Towards AI-Enabled Cyber-Physical Infrastructures—Challenges, Opportunities, and Implications for a Data-Driven eGovernment Theory, Policy, and Practice. *Journal of the Knowledge Economy*, •••, 1–38. DOI: 10.1007/s13132-025-02726-5

Arangarajan, M., Kumar, M. S., Gupta, M., Syed, K., Shukla, P. K., & Sattanathan, S. (2024). Intelligent Systems for Industrial Robots in Logistics and Supply Chain Applications. In *Utilization of AI Technology in Supply Chain Management* (pp. 174–191). IGI Global. DOI: 10.4018/979-8-3693-3593-2.ch012

Arputharaj, J. V., & Pal, S. K. (2024). Transforming Industry 5.0: Real time monitoring and decision making with IIOT. In *Sustainability in Industry 5.0* (pp. 76-106). CRC Press

Aspalli, M. S., & Patil, S. J. (2017, December). Study of AI and PI controller using SVPWM technique for induction motor speed control. In *2017 International Conference on Electrical, Electronics, Communication, Computer, and Optimization Techniques (ICEECCOT)* (pp. 1-6). IEEE. DOI: 10.1109/ICEECCOT.2017.8284589

Ayyaswamy, K., Gobinath, V. M., Sathya, V., Kathirvel, N., & Anthony, R. A. (2025). The role of robotics in smart manufacturing: increasing efficiency and reducing costs. *The Quantum AI Era of Neuromarketing*, 315-356.

Bai, Z., Kong, B., Chen, J., Miao, L., & Zhang, Z. (2023). Environment design and construction of real-time dispatching and intelligent decision making simulation system for integrated coal supply chain of large energy group. In *International Conference on Intelligent Systems, Communications, and Computer Networks (ISCCN 2023)* (Vol. 12702, pp. 101-111). SPIE DOI: 10.1117/12.2679959

Balaji, K. (2025). AI-Powered Ergonomics: Transforming Workplace Health and Safety. In *Cases on AI Innovations in Occupational Health and Safety* (pp. 51-72). IGI Global Scientific Publishing.

Bals, L., Kirchoff, J. F., & Foerstl, K. (2016). Exploring the reshoring and insourcing decision making process: Toward an agenda for future research. *Operations Management Research : Advancing Practice Through Research*, *9*(3), 102–116. DOI: 10.1007/s12063-016-0113-0

Banks, H. T., Lewis, B. M., & Tran, H. T. (2007). Nonlinear feedback controllers and compensators: A state-dependent Riccati equation approach. *Computational Optimization and Applications*, *37*(2), 177–218. DOI: 10.1007/s10589-007-9015-2

Bányai, T., & Cservenák, Á. (2022). Logistics and mechatronics related research in mobile robot-based material handling. In *Vehicle and Automotive Engineering* (pp. 428–443). Springer International Publishing.

Batmani, Y., Davoodi, M., & Meskin, N. (2017). Nonlinear suboptimal tracking controller design using state-dependent Riccati equation technique. *IEEE Transactions on Control Systems Technology*, *25*(5), 1833–1839. DOI: 10.1109/TCST.2016.2617285

Bedi, P., Das, S., Goyal, S. B., Rajawat, A. S., & Kautish, S. (2024). Revolutionizing sustainability: The role of robotics in supply chains. In *Computational Intelligence Techniques for Sustainable Supply Chain Management* (pp. 225–273). Academic Press. DOI: 10.1016/B978-0-443-18464-2.00007-8

Berghuis, H., & Nijmeijer, H. (1993). A passivity approach to controller-observer design for robots. *IEEE Transactions on Robotics and Automation*, *9*(6), 740–754. DOI: 10.1109/70.265918

Berglund, E. Z., Monroe, J. G., Ahmed, I., Noghabaei, M., Do, J., Pesantez, J. E., & Levis, J. (2020). Smart infrastructure: A vision for the role of the civil engineering profession in smart cities. *Journal of Infrastructure Systems*, *26*(2), 03120001. DOI: 10.1061/(ASCE)IS.1943-555X.0000549

Bhagat, D., & Dorle, S. (2024). The Power of Intelligent. *Hyperautomation in Business and Society, 27*.

Bhambri, P., Rani, S., & Khang, A. (2024). AI-driven Digital Twin and Resource Optimization in Industry 4.0 Ecosystem. *Intelligent Techniques for Predictive Data Analytics*, 47-69.

Bhanji, S., Shotz, H., Tadanki, S., Miloudi, Y., & Warren, P. (2021). Advanced enterprise asset management systems: Improve predictive maintenance and asset performance by leveraging Industry 4.0 and the Internet of Things (IoT). In *ASME/IEEE Joint Rail Conference* (Vol. 84775, p. V001T12A002). American Society of Mechanical Engineers Borgi, T., Hidri, A., Neef, B., & Naceur, M. S. (2017). Data analytics for predictive maintenance of industrial robots. In *2017 International Conference on Advanced Systems and Electric Technologies (IC_ASET)* (pp. 412-417). IEEE. DOI: 10.1115/JRC2021-58346

Bhimana, S., & Ravindran, S. (2024). Swarm intelligence: Applications and implementations in autonomous systems. In *Collective Intelligence* (pp. 21–57). CRC Press. DOI: 10.1201/9781032690711-2

Bhosekar, A., Işık, T., Ekşioğlu, S., Gilstrap, K., & Allen, R. (2021). Simulation-optimization of automated material handling systems in a healthcare facility. *IISE Transactions on Healthcare Systems Engineering*, *11*(4), 316–337. DOI: 10.1080/24725579.2021.1882622

Borges, G. D., Reis, A. M., Ariente Neto, R., Mattos, D. L. D., Cardoso, A., Gonçalves, H., & Arezes, P. (2021). Decision-making framework for implementing safer human-robot collaboration workstations: System dynamics modeling. *Safety (Basel, Switzerland)*, *7*(4), 75. DOI: 10.3390/safety7040075

Bossens, D. M., Ramchurn, S., & Tarapore, D. (2022). Resilient robot teams: A review integrating decentralised control, change-detection, and learning. *Current Robotics Reports*, *3*(3), 85–95. DOI: 10.1007/s43154-022-00079-4

Boulaala, M., Elmessaoudi, D., Buj-Corral, I., El Mesbahi, J., Ezbakhe, O., Astito, A., El Mrabet, M., & El Mesbahi, A. (2020). Towards design of mechanical part and electronic control of multi-material/multicolor fused deposition modeling 3D printing. *International Journal of Advanced Manufacturing Technology*, *110*(1-2), 45–55. DOI: 10.1007/s00170-020-05847-0

Bradtke, S. (1992). Reinforcement learning applied to linear quadratic regulation. *Advances in Neural Information Processing Systems*, •••, 5.

Brockett, R. W. (1993). Hybrid models for motion control systems. In Trentelman, H. L., & Willems, J. C. (Eds.), *Essays on Control. Progress in Systems and Control Theory* (Vol. 14). Birkhäuser., DOI: 10.1007/978-1-4612-0313-1_2

Brucherseifer, E., Winter, H., Mentges, A., Mühlhäuser, M., & Hellmann, M. (2021). Digital Twin conceptual framework for improving critical infrastructure resilience. *Automatisierungstechnik*, *69*(12), 1062–1080. DOI: 10.1515/auto-2021-0104

Bruns, R. M. R., Diepstraten, J. F. P. B., Schuurbiers, X. G. P., and Wouters, J. A. G. (2006). Motion control of systems with backlash.

Bukhari, S. M. S., & Akhtar, R. (2024). Leveraging Artificial Intelligence To Revolutionize Six Sigma: Enhancing Process Optimization And Predictive Quality Control. *Contemporary Journal of Social Science Review*, *2*(04), 1932–1948.

Cech, M., Königsmarková, J., Goubej, M., Oomen, T., & Visioli, A. (2019). Essential challenges in motion control education. *IFAC-PapersOnLine*, *52*(9), 200–205. DOI: 10.1016/j.ifacol.2019.08.196

Chang, I., & Bentsman, J. (2013). Constrained discrete-time state-dependent Riccati equation technique: A model predictive control approach. *52nd IEEE Conference on Decision and Control* (pp. 5125-5130). https://doi.org/DOI: 10.1109/CDC.2013.6760694

Chan-Zheng, C., Borja, P., & Scherpen, J. M. (2023). Tuning of passivity-based controllers for mechanical systems. *IEEE Transactions on Control Systems Technology, 31*(6), 2515–2530. DOI: 10.1109/TCST.2023.3260995

Cimen, T. (2008). State-dependent Riccati equation (SDRE) control: A survey. *IFAC Proceedings Volumes, 41*(2), 3761-3775. https://doi.org/DOI: 10.3182/20080706-5-KR-1001.00635

Coito, T., Firme, B., Martins, M. S., Vieira, S. M., Figueiredo, J., & Sousa, J. M. (2021). Intelligent sensors for real-Time decision-making. *Automation, 2*(2), 62–82. DOI: 10.3390/automation2020004

Daraz, A., Malik, S. A., Mokhlis, H., Haq, I. U., Zafar, F., & Mansor, N. N. (2020). Improved-fitness dependent optimizer based FOI-PD controller for automatic generation control of multi-source interconnected power system in deregulated environment. *IEEE Access : Practical Innovations, Open Solutions, 8*, 197757–197775. DOI: 10.1109/ACCESS.2020.3033983

Das, S. (2024). Bio-Inspired Communication Strategies in Swarm Robotics. In *Bio-inspired Swarm Robotics and Control: Algorithms, Mechanisms, and Strategies* (pp. 77-100). IGI Global. DOI: 10.4018/979-8-3693-1277-3.ch006

Derradji, R., & Hamzi, R. (2024). The convergence of process-risk management and decision-making: An integrated vision. *International Journal of Business Continuity and Risk Management, 14*(3), 305–326. DOI: 10.1504/IJBCRM.2024.141606

Devi, K. V. R., Smitha, B. S., Lakhanpal, S., Kalra, R., Sethi, V. A., & Thajil, S. K. (2024). A review: Swarm robotics: Cooperative control in multi-agent systems. In *E3S Web of Conferences* (Vol. 505, p. 03013). EDP Sciences. DOI: 10.1051/e3sconf/202450503013

Dias, L. S., & Ierapetritou, M. G. (2017). From process control to supply chain management: An overview of integrated decision making strategies. *Computers & Chemical Engineering, 106*, 826–835. DOI: 10.1016/j.compchemeng.2017.02.006

Dias, P. G. F., Silva, M. C., Rocha Filho, G. P., Vargas, P. A., Cota, L. P., & Pessin, G. (2021). Swarm robotics: A perspective on the latest reviewed concepts and applications. *Sensors (Basel), 21*(6), 2062. DOI: 10.3390/s21062062 PMID: 33804187

Dimitrakieva, S., Marinova-Stoyanova, M., & Gramchev, B. (2025). Integrating Artificial Intelligence into the Management Decision-Making Process in the New Era of Industry 4.0. In *environment. Technology. Resources. Proceedings of the International Scientific and Practical Conference* (Vol. 4, pp. 91-98)

Dmytriyev, Y., Carnevale, M., & Giberti, H. (2024). Enhancing flexibility and safety: Collaborative robotics for material handling in end-of-line industrial operations. *Procedia Computer Science*, *232*, 2588–2597. DOI: 10.1016/j.procs.2024.02.077

Dobaj, J., Riel, A., Macher, G., & Egretzberger, M. (2023). Towards devops for cyber-physical systems (cpss): Resilient self-adaptive software for sustainable human-centric smart cps facilitated by digital twins. *Machines (Basel)*, *11*(10), 973. DOI: 10.3390/machines11100973

Dumont, G. A., & Huzmezan, M. (2002). Concepts, methods and techniques in adaptive control. *Proceedings of the 2002 American Control Conference* (pp. 1137-1150). https://doi.org/DOI: 10.1109/ACC.2002.1023173

Duraimutharasan, N., Deepan, A., Swadhi, R., Velmurugan, P. R., & Varshney, K. R. (2025). Enhancing Control Engineering Through Human-Machine Collaboration: AI for Improved Efficiency and Decision-Making. In *Harnessing AI for Control Engineering* (pp. 155-176). IGI Global Scientific Publishing

Duraimutharasan, N., Deepan, A., Swadhi, R., Velmurugan, P. R., & Varshney, K. R. (2025). Enhancing Control Engineering Through Human-Machine Collaboration: AI for Improved Efficiency and Decision-Making. In Harnessing AI for Control Engineering (pp. 155-176). IGI Global Scientific Publishing.

Duraimutharasan, N., Deepan, A., Swadhi, R., Velmurugan, P. R., & Varshney, K. R. (2025). Enhancing control engineering through human-machine collaboration: AI for improved efficiency and decision-making. In *Harnessing AI for Control Engineering* (pp. 155–176). IGI Global. DOI: 10.4018/979-8-3693-7812-0.ch008

Dutta, U., & Baruah, S. (2025). AI-driven 3D printing: Unlocking new frontiers in healthcare. In *AI-Powered Systems for Healthcare Diagnostics and Treatment* (pp. 233–266). IGI Global. DOI: 10.4018/979-8-3693-9735-0.ch010

Efthymiou, O. K., & Ponis, S. T. (2019). Current status of industry 4.0 in material handling automation and in-house logistics. *Int. J. Ind. Manuf. Eng*, *13*(10), 1370–1374.

Epiphaniou, G., Hammoudeh, M., Yuan, H., Maple, C., & Ani, U. (2023). Digital twins in cyber effects modelling of IoT/CPS points of low resilience. *Simulation Modelling Practice and Theory*, *125*, 102744. DOI: 10.1016/j.simpat.2023.102744

Evans, O., & Patel, M. (2024). Robotic Swarm Intelligence: Coordination and Collaboration in Multi-Robot Systems. *International Journal on Mechanical Engineering and Robotics*, *13*(1), 23–30.

Fatorachian, H., & Smith, C. (2022). Impact of CPS on enhancing supply chain resilience, with a focus on solutions to pandemic challenges. In *Cyber-physical systems* (pp. 109–125). CRC Press.

Ferraro, S., Cantini, A., Leoni, L., & De Carlo, F. (2023). Sustainable logistics 4.0: A study on selecting the best technology for internal material handling. *Sustainability (Basel)*, *15*(9), 7067. DOI: 10.3390/su15097067

Ferrell, O. C., Gresham, L. G., & Fraedrich, J. (1989). A synthesis of ethical decision models for marketing. *Journal of Macromarketing*, *9*(2), 55–64. DOI: 10.1177/027614678900900207

Flammini, F. (2021). Digital twins as run-time predictive models for the resilience of cyber-physical systems: A conceptual framework. *Philosophical Transactions. Series A, Mathematical, Physical, and Engineering Sciences*, *379*(2207), 20200369. DOI: 10.1098/rsta.2020.0369 PMID: 34398658

Fu, G., Gu, T., Gao, H., & Lu, C. (2020). A postprocessing and path optimization based on nonlinear error for multijoint industrial robot-based 3D printing. *International Journal of Advanced Robotic Systems*, *17*(5), 1729881420952249. DOI: 10.1177/1729881420952249

Gadekallu, T. R., Dev, K., Khowaja, S. A., Wang, W., Feng, H., Fang, K., & Wang, W. (2025). Framework, Standards, Applications and Best practices of Responsible AI: A Comprehensive Survey. *arXiv preprint arXiv:2504.13979*.

Ganjian-Aboukheili, M., Shahabi, M., Shafiee, Q., & Guerrero, J. M. (2021). Linear quadratic regulator based smooth transition between microgrid operation modes. *IEEE Transactions on Smart Grid*, *12*(6), 4854–4864. DOI: 10.1109/TSG.2021.3094419

Gao, Y., Wang, S., Dragicevic, T., Wheeler, P., & Zanchetta, P. (2023). Artificial intelligence techniques for enhancing the performance of controllers in power converter-based systems—An overview. *IEEE Open Journal of Industry Applications*, *4*, 366–375. DOI: 10.1109/OJIA.2023.3338534

Gbadamosi, A. Q., Oyedele, L. O., Delgado, J. M. D., Kusimo, H., Akanbi, L., Olawale, O., & Muhammed-Yakubu, N. (2021). IoT for predictive assets monitoring and maintenance: An implementation strategy for the UK rail industry. *Automation in Construction*, *122*, 103486. DOI: 10.1016/j.autcon.2020.103486

George, A. S. (2024). AI-enabled intelligent manufacturing: A path to increased productivity, quality, and insights. *Partners Universal Innovative Research Publication, 2*(4), 50–63.

Ghazaly, N. M., & Chaudhari, P. B. (2024). AI-Based Adaptive Control Systems for Power Electronics. *Research Journal of Computer Systems and Engineering, 5*(1), 34–44.

Goldsby, T. J., Hoang, T. T., Stank, T. P., & Bell, J. E. (2023). A Modernized Framework for Transportation Decision-Making in a Hyper-Integrated Global Supply Chain Environment. *Transportation Journal, 62*(1), 16–42. DOI: 10.5325/transportationj.62.1.0016

Gomaa, D. A. H. (2025). *LSS 4.0: A Conceptual Framework for Integrating Lean Six Sigma and Industry 4.0 for Smart Manufacturing Excellence. Indian Journal of Management and Language*. IJML.

Govindaraj, M., Shakila, P., & Lawrence, J. (2025). AI-Driven Cybersecurity for Industrial Automation: Resilient Solutions for Industry 4.0. In *AI-Enhanced Cybersecurity for Industrial Automation* (pp. 83-102). IGI Global Scientific Publishing.

Gros, I. C., Lü, X., Oprea, C., Lu, T., & Pintilie, L. (2023, August). Artificial intelligence (AI)-based optimization of power electronic converters for improved power system stability and performance. In 2023 IEEE 14th International Symposium on Diagnostics for Electrical Machines, Power Electronics and Drives (SDEMPED) (pp. 204-210). IEEE.

Groshev, M., Guimaraes, C., Martín-Pérez, J., & de la Oliva, A. (2021). Toward intelligent cyber-physical systems: Digital twin meets artificial intelligence. *IEEE Communications Magazine, 59*(8), 14–20. DOI: 10.1109/MCOM.001.2001237

Gunasekara, L., El-Haber, N., Nagpal, S., Moraliyage, H., Issadeen, Z., Manic, M., & De Silva, D. (2025). A Systematic Review of Responsible Artificial Intelligence Principles and Practice. *Applied System Innovation, 8*(4), 97. DOI: 10.3390/asi8040097

Habib, A., Habib, M., Bashir, B., & Bachir, H. (2025). Exploring the Sustainability Benefits of Digital Twin Technology in Achieving Resilient Smart Cities During Strong Earthquake Events. *Arabian Journal for Science and Engineering, 50*(20), 1–15. DOI: 10.1007/s13369-025-10017-z

Habib, M. K. (Ed.). (2023). *Global perspectives on robotics and autonomous systems: Development and applications*. IGI Global. DOI: 10.4018/978-1-6684-7791-5

Hagane, S., Rincon Ardila, L., Katsumata, T., Bonnet, V., Fraisse, P., and Venture, G. (2018). Adaptive generalized predictive controller and Cartesian force control for robot arm using dynamics and geometric identification. *Journal of Robotics and Mechatronics*. https://doi.org/DOI: 10.20965/jrm.2018.p0927

Hakimi, O., Liu, H., Abudayyeh, O., Houshyar, A., Almatared, M., & Alhawiti, A. (2023). Data fusion for smart civil infrastructure management: A conceptual digital twin framework. *Buildings (Basel, Switzerland)*, *13*(11), 2725. DOI: 10.3390/buildings13112725

Handfield, R., Sroufe, R., & Walton, S. (2005). Integrating environmental management and supply chain strategies. *Business Strategy and the Environment*, *14*(1), 1–19. DOI: 10.1002/bse.422

Hang, P., Lv, C., Huang, C., Cai, J., Hu, Z., & Xing, Y. (2020). An integrated framework of decision making and motion planning for autonomous vehicles considering social behaviors. *IEEE Transactions on Vehicular Technology*, *69*(12), 14458–14469. DOI: 10.1109/TVT.2020.3040398

Hassan, A., Dutta, P. K., Gupta, S., Mattar, E., & Singh, S. (Eds.). (2024). *Human-centered approaches in Industry 5.0: Human-machine interaction, virtual reality training, and customer sentiment analysis*. IGI Global. DOI: 10.4018/979-8-3693-2647-3

Heemels, W. P. M. H., Van Eijndhoven, S. J. L., & Stoorvogel, A. A. (1998). Linear quadratic regulator problem with positive controls. *International Journal of Control*, *70*(4), 551–578. DOI: 10.1080/002071798222208

Houston, C., Gooberman-Hill, S., Mathie, R., Kennedy, A., Li, Y., & Baiz, P. (2017). Case study for the return on investment of internet of things using agent-based modelling and data science. *Systems*, *5*(1), 4. DOI: 10.3390/systems5010004

Hozdić, E., Kendić, S., & Jurković, Z. (2022). Complex adaptive manufacturing system concept as a cyber-physical production system: solutions to the Covid-19 pandemic challenges. In *Cyber-Physical Systems* (pp. 129–156). CRC Press.

Huang, Y., & Fu, J. (2019). Review on application of artificial intelligence in civil engineering. *Computer Modeling in Engineering & Sciences*, *121*(3), 845–875. DOI: 10.32604/cmes.2019.07653

Huang, Z., Gong, J., Xiao, X., Gao, Y., Xia, Y., Wheeler, P., & Ji, B. (2025). Artificial Intelligence and Digital Twin Technologies for Power Converter Control in Transportation Applications: A Review. *IET Power Electronics*, *18*(1), e70013. DOI: 10.1049/pel2.70013

Hu, Y., Wu, L., Li, N., & Zhao, T. (2024). Multi-Agent Decision-Making in Construction Engineering and Management: A Systematic Review. *Sustainability (Basel)*, *16*(16), 7132. DOI: 10.3390/su16167132

Islam, M. T., Hossain, A., Khan, M. R., & Roy, S. (2025). A meta-analysis of machine learning-enhanced lean quality control practices in manufacturing: Optimizing defect detection and process efficiency. *ASRC Procedia: Global Perspectives in Science and Scholarship*, *1*(01), 166–192. DOI: 10.63125/144zbg39

Islam, M. T., Sepanloo, K., Woo, S., Woo, S. H., & Son, Y. J. (2025). A review of the industry 4.0 to 5.0 transition: Exploring the intersection, challenges, and opportunities of technology and human–machine collaboration. *Machines (Basel)*, *13*(4), 267. DOI: 10.3390/machines13040267

Ivanova, S., Kuznetsov, A., Zverev, R., & Rada, A. (2023). Artificial intelligence methods for the construction and management of buildings. *Sensors (Basel)*, *23*(21), 8740. DOI: 10.3390/s23218740 PMID: 37960440

Ivanov, D., & Dolgui, A. (2021). A digital supply chain twin for managing the disruption risks and resilience in the era of Industry 4.0. *Production Planning and Control*, *32*(9), 775–788. DOI: 10.1080/09537287.2020.1768450

Ivanov, D., & Petrova, A. (2024). The Future of Swarm Robotics: Advancing Efficient Task Execution with Coordinated Multi-Agent Systems. *Journal of Computational Innovation*, *4*(1).

Iwasaki, M., Seki, K., & Maeda, Y. (2012). High-precision motion control techniques: A promising approach to improving motion performance. *IEEE Industrial Electronics Magazine*, *6*(1), 32–40. DOI: 10.1109/MIE.2012.2182859

Jakhar, S. K., & Barua, M. K. (2014). An integrated model of supply chain performance evaluation and decision-making using structural equation modelling and fuzzy AHP. *Production Planning and Control*, *25*(11), 938–957. DOI: 10.1080/09537287.2013.782616

Jia, Y. (2025). Collective Systems, Collective Intelligence, and Collective Strategies. In *Distributed Coordination Control System: Application Scenarios and Countermeasures* (pp. 9–14). Springer Nature Singapore. DOI: 10.1007/978-981-96-0365-7_2

Jin, A. S., Hogewood, L., Fries, S., Lambert, J. H., Fiondella, L., Strelzoff, A., & Linkov, I. (2022). Resilience of cyber-physical systems: Role of AI, digital twins, and edge computing. *IEEE Engineering Management Review*, *50*(2), 195–203. DOI: 10.1109/EMR.2022.3172649

Johannink, T.. (2019). Residual reinforcement learning for robot control. *2019 International Conference on Robotics and Automation (ICRA)* (pp. 6023-6029). https://doi.org/DOI: 10.1109/ICRA.2019.8794127

Jung, H., & Oh, S. (2021). Data-driven optimization of integrated control framework for flexible motion control system. *IEEE Transactions on Industrial Informatics*, *18*(7), 4762–4772. DOI: 10.1109/TII.2021.3100872

Kalare, K. W., Bhagat, D., Jajulwar, K. K., Bhave, R., Raut, S. V., & Dorle, S. (2024). The Power of Intelligent Automation. In *Hyperautomation in Business and Society* (pp. 27–65). IGI Global. DOI: 10.4018/979-8-3693-3354-9.ch002

Kaur, M. J., Riaz, S., & Mushtaq, A. (2019). Cyber-physical cloud computing systems and internet of everything. In *Principles of Internet of Things (IoT) Ecosystem: Insight Paradigm* (pp. 201–227). Springer International Publishing.

Kazeem, K. O., Olawumi, T. O., & Osunsanmi, T. (2023). Roles of artificial intelligence and machine learning in enhancing construction processes and sustainable communities. *Buildings (Basel, Switzerland)*, *13*(8), 2061. DOI: 10.3390/buildings13082061

Kazim, R. M., Wang, G., Ming, Z., Allen, J. K., & Mistree, F. (2025). Design and Deployment of Swarm Engineering Systems: Insights, Challenges, and Innovations. *IEEE Access : Practical Innovations, Open Solutions*, *13*, 139345–139376. DOI: 10.1109/ACCESS.2025.3592950

Keung, K. L., Lee, C. K., Ji, P., & Ng, K. K. (2020). Cloud-based cyber-physical robotic mobile fulfillment systems: A case study of collision avoidance. *IEEE Access : Practical Innovations, Open Solutions*, *8*, 89318–89336. DOI: 10.1109/ACCESS.2020.2992475

Kezron, I. E. (2025). Novel cybersecurity framework for AI-driven drone integration by critical SMEs in economically distressed US rural communities: Advancing secure precision operations in high-risk environments. *Well Testing Journal*, *34*(S3), 1–44.

Khalid, M. (2025). Cyber-Physical Systems and their Role in Industry 4.0: Architecture, Applications, and Future Prospects. *International journal of applied biology and forensic*, *9*(1), 46-59.

Khalifa, R. B., Nhassa, G., & Snoussi, A. (2025). Techniques and methods for optimizing 3D printing parameters. In *Modeling, Analysis, and Control of 3D Printing Processes* (pp. 329–354). IGI Global.

Khan, M. W., Saad, S., Ammad, S., Rasheed, K., & Jamal, Q. (2024). Smart infrastructure and AI. In *AI in Material Science* (pp. 193–215). CRC Press. DOI: 10.1201/9781003438489-9

Khan, M. Z. (2024). *Introduction To Artificial Intelligence And Human-centric Computing*. OrangeBooks Publication.

Khan, S. A., Ansari, J. A., Chandio, R. H., Munir, H. M., Alharbi, M., & Alkuhayli, A. (2022). AI based controller optimization for VSC-MTDC grids. *Frontiers in Energy Research*, *10*, 1008099. DOI: 10.3389/fenrg.2022.1008099

Khurram, M., Zhang, C., Muhammad, S., Kishnani, H., An, K., Abeywardena, K., Chadha, U., & Behdinan, K. (2025). Artificial Intelligence in Manufacturing Industry Worker Safety: A New Paradigm for Hazard Prevention and Mitigation. *Processes (Basel, Switzerland)*, *13*(5), 1312. DOI: 10.3390/pr13051312

Koon, W. S., & Marsden, J. E. (1997). The Hamiltonian and Lagrangian approaches to the dynamics of nonholonomic systems. *Reports on Mathematical Physics*, *40*(1), 21–62. DOI: 10.1016/S0034-4877(97)85617-0

Körwer, N., Bischoff, M., Laumen, M., & De Doncker, R. W. (2024, September). AI-Based Motion Control with System Dynamics Flexibility. In 2024 Energy Conversion Congress & Expo Europe (ECCE Europe) (pp. 1-7). IEEE.

Kumar, V., & Das, S. (2024). Harmony in Motion: Exploring Bio-Inspired Swarm Robotics and Control Strategies for Mechanical Systems. In *Bio-inspired Swarm Robotics and Control: Algorithms, Mechanisms, and Strategies* (pp. 46-66). IGI Global.

Kyzyrkanov, A., Atanov, S., Aljawarneh, S., Tursynova, N., Otarbay, Z., & Saltanat, A. (2025). Decentralized Coordination of Intelligent Robot Swarms. In *2025 IEEE 5th International Conference on Smart Information Systems and Technologies (SIST)* (pp. 1-6). IEEE. DOI: 10.1109/SIST61657.2025.11139307

Lalitha, S. D., Danamaraju, C., Raj, M. P., Kumar, D., & Sethuraman, R. (2024, April). AI-driven control strategies for power electronics converters. In *2024 Ninth International Conference on Science Technology Engineering and Mathematics (ICONSTEM)* (pp. 1-6). IEEE. DOI: 10.1109/ICONSTEM60960.2024.10568766

Landau, I. D., Lozano, R., M'Saad, M., & Karimi, A. (2011). *Adaptive control: Algorithms, analysis and applications*. Springer., DOI: 10.1007/978-0-85729-664-1

Lazaroiu, G., Androniceanu, A., Grecu, I., Grecu, G., & Neguri ă, O. (2022). Artificial intelligence-based decision-making algorithms, Internet of Things sensing networks, and sustainable cyber-physical management systems in big data-driven cognitive manufacturing. *Oeconomia Copernicana, 13*(4), 1047–1080. DOI: 10.24136/oc.2022.030

Lee, J., Kao, H. A., Ardakani, H. D., & Siegel, D. (2015). Intelligent factory agents with predictive analytics for asset management. In *Industrial Agents* (pp. 341–360). Morgan Kaufmann. DOI: 10.1016/B978-0-12-800341-1.00019-X

Lee, J., Ni, J., Singh, J., Jiang, B., Azamfar, M., & Feng, J. (2020). Intelligent maintenance systems and predictive manufacturing. *Journal of Manufacturing Science and Engineering, 142*(11), 110805. DOI: 10.1115/1.4047856

Leung, C. S., & Lau, H. Y. (2020). Simulation-based optimization for material handling systems in manufacturing and distribution industries. *Wireless Networks, 26*(7), 4839–4860. DOI: 10.1007/s11276-018-1894-x

Li, H., Parikh, D., He, Q., Qian, B., Li, Z., Fang, D., & Hampapur, A. (2014). Improving rail network velocity: A machine learning approach to predictive maintenance. *Transportation Research Part C, Emerging Technologies, 45*, 17–26. DOI: 10.1016/j.trc.2014.04.013

Liu, B., Qian, B., Liang, Y., Dai, P., Li, R., & Wei, Q. (2025). Fast and accurate auto-disturbances-rejection temperature control system based on particle swarm optimized fuzzy control: Applied for particle three-dimensional (3D) printing prosthetic orthotic plate. *Additive Manufacturing, 101*, 104704. DOI: 10.1016/j.addma.2025.104704

Liu, K., Meng, Q., Kong, Q., & Zhang, X. (2022). Review on the developments of structure, construction automation, and monitoring of intelligent construction. *Buildings (Basel, Switzerland), 12*(11), 1890. DOI: 10.3390/buildings12111890

Li, X., Zhang, X., Lin, F., & Blaabjerg, F. (2021). Artificial-intelligence-based design for circuit parameters of power converters. *IEEE Transactions on Industrial Electronics, 69*(11), 11144–11155. DOI: 10.1109/TIE.2021.3088377

Li, Z., Liu, L., Dehghan, S., Chen, Y., & Xue, D. (2016). A review and evaluation of numerical tools for fractional calculus and fractional order controls. *International Journal of Control, 90*(6), 1165–1181. DOI: 10.1080/00207179.2015.1124290

Lodhi, S. K., Gill, A. Y., & Hussain, I. (2024). AI-powered innovations in contemporary manufacturing procedures: An extensive analysis. *International Journal of Multidisciplinary Sciences and Arts, 3*(4), 15–25. DOI: 10.47709/ijmdsa.v3i4.4616

Longa, M. E., Tsourdos, A., & Inalhan, G. (2022). Human–machine network through bio-inspired decentralized swarm intelligence and heterogeneous teaming in SAR operations. *Journal of Intelligent & Robotic Systems*, *105*(4), 88. DOI: 10.1007/s10846-022-01690-5

Louis, J., & Dunston, P. S. (2018). Integrating IoT into operational workflows for real-time and automated decision-making in repetitive construction operations. *Automation in Construction*, *94*, 317–327. DOI: 10.1016/j.autcon.2018.07.005

Luo, Y., & Chen, Y. (2012). *Fractional order motion controls*. John Wiley and Sons. DOI: 10.1002/9781118387726

Ma, C., & Hori, Y. (2007). Fractional-order control: Theory and applications in motion control. *IEEE Industrial Electronics Magazine*, *1*(4), 6–16. DOI: 10.1109/MIE.2007.909703

MacDonald, E., & Wicker, R. (2016). Multiprocess 3D printing for increasing component functionality. *Science*, *353*(6307), aaf2093. DOI: 10.1126/science.aaf2093 PMID: 27708075

Mahajan, P. (2025). What is Ethical: AIHED Driving Humans or Human-Driven AIHED? A Conceptual Framework enabling the Ethos of AI-driven Higher education. *arXiv preprint arXiv:2503.04751.*

Malatji, M. (2025). Augmented Intelligence Framework for Human–Artificial Intelligence Teaming in Cybersecurity. *Human-Centric Intelligent Systems*, 1-30.

Maniraj, M., & Kumar, S. R. (2025). From theory to practice: Control system challenges in additive manufacturing. In *Modeling, Analysis, and Control of 3D Printing Processes* (pp. 237–262). IGI Global.

Manzoor, B., Othman, I., Durdyev, S., Ismail, S., & Wahab, M. H. (2021). Influence of artificial intelligence in civil engineering toward sustainable development—A systematic literature review. *Applied System Innovation*, *4*(3), 52. DOI: 10.3390/asi4030052

Martínez-Mireles, J. R., Rodríguez-Flores, J., García-Márquez, M. A., Austria-Cornejo, A., & Figueroa-Diaz, R. A. (2025). AI in Smart Manufacturing. In *Machine and Deep Learning Solutions for Achieving the Sustainable Development Goals* (pp. 463–494). IGI Global. DOI: 10.4018/979-8-3693-8161-8.ch023

Martin-Gomez, A., Avila-Gutierrez, M. J., & Aguayo-Gonzalez, F. (2021). Holonic reengineering to foster sustainable cyber-physical systems design in cognitive manufacturing. *Applied Sciences (Basel, Switzerland)*, *11*(7), 2941. DOI: 10.3390/app11072941

Martini, B., Bellisario, D., & Coletti, P. (2024). Human-centered and sustainable artificial intelligence in industry 5.0: Challenges and perspectives. *Sustainability (Basel)*, *16*(13), 5448. DOI: 10.3390/su16135448

Meena, G., Vettriselvan, R., Rajesh, D., & Velmurugan, P. R. (2025). Diversity and Inclusion: Harnessing the Power of Inclusivity for Business Success. In *Security and Strategy Models for Key-Solving Institutional Frameworks* (pp. 203-234). IGI Global Scientific Publishing.

Mehra, I. (2024). AI-Driven Autonomous Vehicles: Safety, Ethics, and Regulatory Challenges. *Journal of Science. Technology and Engineering Research*, *1*(2), 18–31.

Merchán-Cruz, E. A., Gabelaia, I., Savrasovs, M., Hansen, M. F., Soe, S., Rodriguez-Cañizo, R. G., & Aragón-Camarasa, G. (2025). Trust by Design: An Ethical Framework for Collaborative Intelligence Systems in Industry 5.0. *Electronics (Basel)*, *14*(10), 1952. DOI: 10.3390/electronics14101952

Meshram, V. S., Becchi, L., Alfonso, C. G., Paolucci, L., Grasso, F., & Reatti, A. (2024, September). Advancements in Artificial Intelligence for Design, Control, and Maintenance of Power Converters: A Comprehensive Review. In 2024 IEEE 8th Forum on Research and Technologies for Society and Industry Innovation (RTSI) (pp. 506-511). IEEE.

Michel, Y., Saveriano, M., & Lee, D. (2024). A passivity-based approach for variable stiffness control with dynamical systems. *IEEE Transactions on Automation Science and Engineering*, *21*(4), 6265–6276. DOI: 10.1109/TASE.2023.3324141

Mills, C. (2022). The design and implementation of a low-cost high-precision 3D-printed translation stage using an optimized PID controller for optical applications. *Unpublished manuscript.*

Mintoo, A. A., Saimon, A. S. M., Bakhsh, M. M., & Akter, M. (2022). National Resilience Through Ai-Driven Data Analytics And Cybersecurity For Real-Time Crisis Response And Infrastructure Protection. *American Journal of Scholarly Research and Innovation*, *1*(01), 137–169. DOI: 10.63125/sdz8km60

Mohammed, M., Mohammed, A., Ihmedee, F., Adam, T., Betar, B., & Gopinath, S. (2025). Emerging artificial intelligence methods in civil engineering: A Comprehensive Review. *Al-Rafidain Journal of Engineering Sciences*, 280-293.

Mohan Banur, O., Patle, B. K., & Pawar, S. (2024). Integration of robotics and automation in supply chain: A comprehensive review. *Robotic Systems and Applications*, *4*(1), 1–19. DOI: 10.21595/rsa.2023.23349

Mohanbabu, S., & Vettriselvan, R. (2025). Will machine learning resolve the issues in container management in ports and logistics industries? *International Journal of Process Management and Benchmarking*, *20*(4), 559–575. DOI: 10.1504/IJPMB.2025.147399

Molęda, M., Małysiak-Mrozek, B., Ding, W., Sunderam, V., & Mrozek, D. (2023). From corrective to predictive maintenance A review of maintenance approaches for the power industry. *Sensors (Basel)*, *23*(13), 5970. DOI: 10.3390/s23135970 PMID: 37447820

Mon, B. F., Hayajneh, M., Ali, N. A., Ullah, F., Ullah, H., & Alkobaisi, S. (2025). Digital Twins in the IIoT: Current Practices and Future Directions Toward Industry 5.0. *Computers, Materials & Continua*, *83*(3), 3675–3712. DOI: 10.32604/cmc.2025.061411

Nair, V. V. (2024). Automation and robotization. *Modern Technologies and Tools Supporting the Development of Industry*, *5*, 41–74. DOI: 10.1201/9781003489269-2

Nama, P. (2022). Optimizing automation systems with AI: A study on enhancing workflow efficiency through intelligent decision-making algorithms. *World Journal of Advanced Engineering Technology and Sciences*, *7*(02), 296–307.

Napoleone, A., Negri, E., Macchi, M., & Pozzetti, A. (2023). How the technologies underlying cyber-physical systems support the reconfigurability capability in manufacturing: A literature review. *International Journal of Production Research*, *61*(9), 3122–3144. DOI: 10.1080/00207543.2022.2074323

Naranjo, I., Torre, R. D. L., & Boza, A. (2025). Sustainable Decision-Making in SMEs of the Textile Industry: Integrating Methodologies, Practices, and Barriers. *Business Strategy and the Environment*, *34*(6), 6385–6441. DOI: 10.1002/bse.4295

Nguyen, L. V. (2024). Swarm intelligence-based multi-robotics: A comprehensive review. *AppliedMath*, *4*(4), 1192–1210. DOI: 10.3390/appliedmath4040064

Noor, A. M., Zakaria, Z., Mazalan, M., Mahmud, M. F., Norali, A. N., & Zaidi, A. F. A. (2025). Smart Maintenance System (SMAT): Predictive Maintenance of Electrical Motor Applications. *Malaysian Journal of Science and Advanced Technology*, 107-115.

Nur, H. M. A.. (2024). Comparison study between open-loop and closed-loop identification for industrial hydraulics actuator system. *International Journal of Mechanical Engineering and Robotics Research*, *13*(5), 516–521. DOI: 10.18178/ijmerr.13.5.516-521

Ohnishi, K., & Murakami, T. (1989). Advanced motion control in robotics. *15th Annual Conference of IEEE Industrial Electronics Society* (pp. 356-359). https://doi.org/DOI: 10.1109/IECON.1989.69658

Ohnishi, K., Murakami, T., & Shibata, M. (1996). Motion control for advanced mechatronics. *IEEE/ASME Transactions on Mechatronics*, *1*(1), 56–67. DOI: 10.1109/3516.491410

Olaseni, I. O.Iyiola Oladehinde Olaseni. (2020). Digital Twin and BIM synergy for predictive maintenance in smart building engineering systems development. *World J Adv Res Rev*, *8*(2), 406–421. DOI: 10.30574/wjarr.2020.8.2.0409

Oomen, T. (2017, March). Advanced motion control for next-generation precision mechatronics: Challenges for control, identification, and learning. *3rd IEEJ International Workshop on Sensing, Actuation, Motion Control, and Optimization (SAMCON)* (pp. 1-12).

Ortega, R., & Espinosa-Perez, G. (2005). Passivity based control with simultaneous energy shaping and damping injection: The induction motor case study. *IFAC Proceedings Volumes, 38*(1), 477-482. https://doi.org/DOI: 10.3182/20050703-6-CZ-1902.00734

Ortega, R., van der Schaft, A., Castanos, F., & Astolfi, A. (2008). Control by interconnection and standard passivity-based control of port-Hamiltonian systems. *IEEE Transactions on Automatic Control*, *53*(11), 2527–2542. DOI: 10.1109/TAC.2008.2006930

Padula, F., & Visioli, A. (2012). Optimal tuning rules for proportional-integral-derivative and fractional-order proportional-integral-derivative controllers for integral and unstable processes. *IET Control Theory & Applications*, *6*(6), 776–786. DOI: 10.1049/iet-cta.2011.0419

Pai, R., Chattopadhyay, G., & Karmakar, G. (2023). Maintenance and asset management practices of industrial assets: Importance of tribological practices and digital tools. *International Journal of Process Management and Benchmarking*, *13*(2), 233–256. DOI: 10.1504/IJPMB.2023.128508

Panetto, H., Iung, B., Ivanov, D., Weichhart, G., & Wang, X. (2019). Challenges for the cyber-physical manufacturing enterprises of the future. *Annual Reviews in Control*, *47*, 200–213. DOI: 10.1016/j.arcontrol.2019.02.002

Papamichail, K. N., & Robertson, I. (2005). Integrating decision making and regulation in the management control process. *Omega*, *33*(4), 319–332. DOI: 10.1016/j.omega.2004.05.002

Parnianifard, A., Jearavongtakul, S., Sasithong, P., Sinpan, N., Poomrittigul, S., Bajpai, A., & Wuttisittikulkij, L. (2022). Digital-twins towards cyber-physical systems: A brief survey. *Engineering Journal (New York)*, *26*(9), 47–61.

Patel, N., Chauhan, H., & Joshi, H. (2025). Orchestrating Intelligence: Governance and Leadership Frameworks for Human-AGI Collaboration in Quantum Systems. In *Interplay of Artificial General Intelligence with Quantum Computing: Towards Sustainability* (pp. 255–272). Springer Nature Switzerland. DOI: 10.1007/978-3-031-87931-9_18

Pech, M., Vrchota, J., & Bednář, J. (2021). Predictive maintenance and intelligent sensors in smart factory. *Sensors (Basel)*, *21*(4), 1470. DOI: 10.3390/s21041470 PMID: 33672479

Pena, W., Nadkarni, A., Galvan, G., Yin, X., Zhang, H., & Fung, J. (2024, October). Integrating IoT and Predictive Analytics for Remote Asset Health Monitoring. In *2024 Global Reliability and Prognostics and Health Management Conference (PHM-Beijing)* (pp. 1-7). IEEE. DOI: 10.1109/PHM-Beijing63284.2024.10874715

Podlubny, I. (1994). Fractional-order systems and fractional-order controllers. *Institute of Experimental Physics, Slovak Academy of Sciences, Kosice, 12*(3), 1-18.

Ponis, S. T., & Efthymiou, O. K. (2020). Cloud and IoT applications in material handling automation and intralogistics. *Logistics (Basel)*, *4*(3), 22. DOI: 10.3390/logistics4030022

Qudus, L. (2025). Resilient systems: Building secure cyber-physical infrastructure for critical industries against emerging threats. *Int J Res Publ Rev*, *6*(1), 3330–3346. DOI: 10.55248/gengpi.6.0125.0514

Radanliev, P., De Roure, D., Nicolescu, R., Huth, M., & Santos, O. (2022). Digital twins: Artificial intelligence and the IoT cyber-physical systems in Industry 4.0. *International Journal of Intelligent Robotics and Applications*, *6*(1), 171–185. DOI: 10.1007/s41315-021-00180-5

Rafsanjani, H. N., & Nabizadeh, A. H. (2023). Towards human-centered artificial intelligence (AI) in architecture, engineering, and construction (AEC) industry. *Computers in Human Behavior Reports*, *11*, 100319. DOI: 10.1016/j.chbr.2023.100319

Rajan, P. (2024). Integrating iot analytics into marketing decision making: A smart data-driven approach. *International Journal of Data Informatics and Intelligent Computing*, *3*(1), 12–22. DOI: 10.59461/ijdiic.v3i1.92

Rajput, S., & Yadav, S. (2025). Cyber-physical fusion architecture: To mitigate risk for resilient supply chain. In *Risk, Reliability and Resilience in Operations Management* (pp. 291-312). Elsevier.

Rakholia, R., Suárez-Cetrulo, A. L., Singh, M., & Carbajo, R. S. (2024). Advancing Manufacturing Through Artificial Intelligence: Current Landscape, Perspectives, Best Practices, Challenges and Future Direction. *IEEE Access : Practical Innovations, Open Solutions*, *12*, 131621–131637. DOI: 10.1109/ACCESS.2024.3458830

Rana, M. M., & Ibrahim, U. M. (2024). Exploring the Role of Reinforcement Learning in Area of Swarm Robotic. *European Journal of Electrical Engineering and Computer Science*, *8*(3), 15–24. DOI: 10.24018/ejece.2024.8.3.619

Ranawaka, A., Alahakoon, D., Sun, Y., & Hewapathirana, K. (2024). Leveraging the Synergy of Digital Twins and Artificial Intelligence for Sustainable Power Grids: A Scoping Review. *Energies*, *17*(21), 5342. DOI: 10.3390/en17215342

Rane, N. (2023). Integrating leading-edge artificial intelligence (AI), internet of things (IOT), and big data technologies for smart and sustainable architecture, engineering and construction (AEC) industry: Challenges and future directions. *Engineering and Construction (AEC) Industry: Challenges and Future Directions (September 24, 2023).*

Raol, J. R., & Ayyagari, R. (2019). *Control systems: classical, modern, and AI-based approaches.* CRC Press. DOI: 10.1201/9781351170802

Razali, M. N., Othman, S. H., Jamaludin, A. F., Maimun, N. H. A., Jalil, R. A., Adnan, Y. M., & Zulkarnain, S. H. (2021). Big data analytics for preventive maintenance management. *Planning Malaysia, 19.*

Renwick, J. T., & Babson, P. E. (2007). Vibration analysis---A proven technique as a predictive maintenance Tool. *IEEE Transactions on Industry Applications*, *IA-21*(2), 324–332. DOI: 10.1109/TIA.1985.349652

Revathi, S., Babu, M., Rajkumar, N., Meti, V. K. V., Kandavalli, S. R., & Boopathi, S. (2024). Unleashing the future potential of 4D printing: Exploring applications in wearable technology, robotics, energy, transportation, and fashion. In Hassan, A., Dutta, P. K., Gupta, S., Mattar, E., & Singh, S. (Eds.), *Human-centered approaches in Industry 5.0* (pp. 131–153). IGI Global. DOI: 10.4018/979-8-3693-2647-3.ch006

Rojas, L., Peña, Á., & Garcia, J. (2025). AI-driven predictive maintenance in mining: A systematic literature review on fault detection, digital twins, and intelligent asset management. *Applied Sciences (Basel, Switzerland)*, *15*(6), 3337. DOI: 10.3390/app15063337

Ronzoni, M., Accorsi, R., Di Biase, T., & Manzini, R. (2024). Robotic Systems for Material Handling: Design Framework and Digital Twins. In *Warehousing and Material Handling Systems for the Digital Industry: The New Challenges for the Digital Circular Economy* (pp. 407–432). Springer International Publishing. DOI: 10.1007/978-3-031-50273-6_15

Ruderman, M., Iwasaki, M., & Chen, W.-H. (2020). Motion-control techniques of today and tomorrow: A review and discussion of the challenges of controlled motion. *IEEE Industrial Electronics Magazine*, *14*(1), 41–55. DOI: 10.1109/MIE.2019.2956683

Sabanovic, A. (2017). Challenges in motion control systems. *IEEJ Journal of Industry Applications*, *6*(2), 107–116. DOI: 10.1541/ieejjia.6.107

Sabanovic, A., & Ohnishi, K. (2011). *Motion control systems*. Wiley., DOI: 10.1002/9780470825754

Sadasivan, M., Vasumathi, M. T., & Jayanthi, M. (2024). Artificial intelligence for smart manufacturing using digital twin technology. In *Artificial Intelligence based Solutions for Industrial Applications* (pp. 176–194). CRC Press. DOI: 10.1201/9781003534761-7

Saha, A., Simic, V., Dabic-Miletic, S., Senapati, T., Pamucar, D., Arya, L., & Tirkolaee, E. B. (2025). Prioritization of AI-based material handling approaches for smart logistics in sustainable warehouses: A q-rung orthopair fuzzy CoCoSo methodology with consensus reaching: A. Saha et al. *Environment, Development and Sustainability*, 1-39.

Sajadieh, S. M. M., & Noh, S. D. (2025). From Simulation to Autonomy: Reviews of the Integration of Artificial Intelligence and Digital Twins. *International Journal of Precision Engineering and Manufacturing-Green Technology*, 1-32.

Sajadieh, S. M. M., & Noh, S. D. (2025). A Review of Digital Twin Integration in Circular Manufacturing for Sustainable Industry Transition. *Sustainability (Basel)*, *17*(16), 7316. DOI: 10.3390/su17167316

Saleh, S., & Ahmed, M. (2024). A Cyclic Framework for Ethical Implications of Artificial Intelligence in Autonomous Vehicles. *International Journal of Advanced Computer Science and Applications*, *15*(3). Advance online publication. DOI: 10.14569/IJACSA.2024.0150335

Samuel, P., Saini, A., Poongodi, T., & Nancy, P. (2023). Artificial intelligence–driven digital twins in Industry 4.0. In *Digital Twin for Smart Manufacturing* (pp. 59–88). Academic Press. DOI: 10.1016/B978-0-323-99205-3.00002-X

Sanyal, S., Sharma, P., & Dudani, C. (2024). *A complex adaptive system framework to regulate artificial intelligence* (No. 26). Working paper.

Sarkar, B., & Paul, R. K. (2025). AI and Advanced Analytics Applications. In *AI for Advanced Manufacturing and Industrial Applications* (pp. 61–89). Springer Nature Switzerland. DOI: 10.1007/978-3-031-86091-1_3

Sarker, I. H. (2024). CyberAI: a comprehensive summary of AI variants, explainable and responsible AI for cybersecurity. In *AI-driven cybersecurity and threat intelligence: cyber automation, intelligent decision-making and explainability* (pp. 173–200). Springer Nature Switzerland. DOI: 10.1007/978-3-031-54497-2_10

Schmitt, R., Permin, E., Kerkhoff, J., Plutz, M., & Böckmann, M. G. (2016). Enhancing resiliency in production facilities through cyber physical systems. In *Industrial Internet of Things: Cybermanufacturing Systems* (pp. 287–313). Springer International Publishing.

Schroeck, S. J., & Messner, W. C. (1999). On controller design for linear time-invariant dual-input single-output systems. *Proceedings of the 1999 American Control Conference* (pp. 4122-4126). https://doi.org/DOI: 10.1109/ACC.1999.786322

Schwenzer, M., Ay, M., Bergs, T., & Abel, D. (2021). Review on model predictive control: An engineering perspective. *International Journal of Advanced Manufacturing Technology*, *117*(5-6), 1327–1349. DOI: 10.1007/s00170-021-07682-3

Segovia-Ferreira, M., Rubio-Hernan, J., Cavalli, A., & Garcia-Alfaro, J. (2024). A survey on cyber-resilience approaches for cyber-physical systems. *ACM Computing Surveys*, *56*(8), 1–37. DOI: 10.1145/3652953

Seifi, N., Ghoodjani, E., Majd, S. S., Maleki, A., & Khamoushi, S. (2025). Evaluation and prioritization of artificial intelligence integrated block chain factors in healthcare supply chain: A hybrid Decision Making Approach. *Computer and Decision Making: An International Journal*, *2*, 374–405. DOI: 10.59543/comdem.v2i.11029

Senthil, T. S., Puviyarasan, M., Babu, S. R., Surakasi, R., & Sampath, B. (2023). Industrial robot-integrated fused deposition modelling for the 3D printing process. In *Development, Properties, and Industrial Applications of 3D Printed Polymer Composites* (pp. 188–210). IGI Global.

Settibathini, V. S. K. (2025). The Autonomous Supply Chain: AI, Robotics, and the Next Industrial Evolution. *International Journal of Emerging Research in Engineering and Technology*, 50-60.

Sevcik, M. J., Bjerke, G., Wilson, F., Kline, D. J., Morales, R. C., Fletcher, H. E., & Eliasson, V. (2024). Extrusion parameter control optimization for DIW 3D printing using image analysis techniques. *Progress in Additive Manufacturing*, *9*(2), 517–528. DOI: 10.1007/s40964-023-00470-3

Sgarbossa, F., Romsdal, A., Oluyisola, O. E., & Strandhagen, J. O. (2022). Digitalization in production and warehousing in food supply chains. In *The digital supply chain* (pp. 273–287). Elsevier. DOI: 10.1016/B978-0-323-91614-1.00016-2

Shahin, M., Maghanaki, M., Hosseinzadeh, A., & Chen, F. F. (2024). Improving operations through a lean AI paradigm: A view to an AI-aided lean manufacturing via versatile convolutional neural network. *International Journal of Advanced Manufacturing Technology*, *133*(11), 5343–5419. DOI: 10.1007/s00170-024-13874-4

Shamsuddoha, M., Nasir, T., & Fawaaz, M. S. (2025). Humanoid robots like tesla optimus and the future of supply chains: Enhancing efficiency, sustainability, and workforce dynamics. *Automation*, *6*(1), 9. DOI: 10.3390/automation6010009

Shamsuzzoha, A., & Pelkonen, S. (2025). A robotic process automation model for order-handling optimization in supply chain management. *Supply Chain Analytics*, *9*, 100102. DOI: 10.1016/j.sca.2025.100102

Shandilya, S. K., Datta, A., Kartik, Y., & Nagar, A. (2024). Role of artificial intelligence and machine learning. In *Digital Resilience: Navigating Disruption and Safeguarding Data Privacy* (pp. 313–399). Springer Nature Switzerland. DOI: 10.1007/978-3-031-53290-0_6

Singh, P. K., & Kumar, A. (2023). Cyber Physical Systems in Supply Chain Management. In *Cyber Physical Systems* (pp. 85–110). Chapman and Hall/CRC.

Soori, M., Dastres, R., Arezoo, B., & Jough, F. K. G. (2024). Intelligent robotic systems in Industry 4.0: A review. *Journal of Advanced Manufacturing Science and Technology*, 2024007-0.

Sowan, B. I., Zighan, S., & Altarawneh, A. (2025). Supply Chain 4.0: Driving Operational Efficiency, Resilience, and Cybersecurity Using Advanced Technologies. In *2025 1st International Conference on Computational Intelligence Approaches and Applications (ICCIAA)* (pp. 1-8). IEEE.

Špirková, S., Straka, M., & Saniuk, A. (2024). VR Simulation and Implementation of Robotics: A Tool for Streamlining and Optimization. *Applied Sciences (Basel, Switzerland)*, *14*(11), 4434. DOI: 10.3390/app14114434

Stank, T. P., & Goldsby, T. J. (2000). A framework for transportation decision making in an integrated supply chain. *Supply Chain Management, 5*(2), 71–78. DOI: 10.1108/13598540010319984

Steinbuch, M., Merry, R. J. E., Boerlage, M. L. G., Ronde, M. J. C., & van de Molengraft, M. J. G. (2018). Advanced motion control design. In Levine, W. S. (Ed.), *The Control Handbook: Control System Applications* (2nd ed., pp. 27-1–27-25). CRC Press., DOI: 10.1201/b10382

Suárez-Riveros, E., Mejia-Mantilla, Á., Jaimes-Suárez, S., & Jimenez, J. F. (2021). Cyber-physical systems in logistics and supply chain. In *International Workshop on Service Orientation in Holonic and Multi-Agent Manufacturing* (pp. 260-271). Cham: Springer International Publishing. DOI: 10.1007/978-3-030-80906-5_18

Subrahmanyam, S. (2025). Leveraging AI and ML for enhanced efficiency and innovation in manufacturing. In *AI's Role in Enhanced Automotive Safety* (pp. 319–344). IGI Global. DOI: 10.4018/979-8-3373-0442-7.ch021

Suhasini, S., Divya, J. C., Sankara Mahalingam, M., Ramshankar, N., & Pandiaraj, A. (2024). A Technological Paradigm Shift for Enhancing Equipment Maintenance and Asset Management Through Augmented Reality. In *Navigating the Augmented and Virtual Frontiers in Engineering* (pp. 186–198). IGI Global.

Sun, Z., Jayasinghe, S., Sidiq, A., Shahrivar, F., Mahmoodian, M., & Setunge, S. (2024). Approach towards the development of Digital Twin for structural health monitoring of civil infrastructure: A comprehensive review. *Sensors (Basel), 25*(1), 59. DOI: 10.3390/s25010059 PMID: 39796850

Suthar, A., Kolhe, K., Gutte, V., & Patil, D. (2024). Predictive maintenance and real time monitoring using iot and cloud computing. In *2024 5th International Conference on Image Processing and Capsule Networks (ICIPCN)* (pp. 814-820). IEEE. DOI: 10.1109/ICIPCN63822.2024.00141

Tadić, S., Krstić, M., Dabić-Miletić, S., & Božić, M. (2023). Smart material handling solutions for city logistics systems. *Sustainability (Basel), 15*(8), 6693. DOI: 10.3390/su15086693

Tafida, A., Alaloul, W. S., Zawawi, N. A. B. W., Musarat, M. A., & Abubakar, A. S. (2024). A review of eco-friendly road infrastructure innovations for sustainable transportation. *Infrastructures, 9*(12), 216. DOI: 10.3390/infrastructures9120216

Tahaei, M., Constantinides, M., Quercia, D., & Muller, M. (2023). A systematic literature review of human-centered, ethical, and responsible AI. *arXiv preprint arXiv:2302.05284*.

Tan, K. K., Lee, T. H., & Huang, S. (2007). *Precision motion control: Design and implementation*. Springer.

Tan, Q., Tong, Y., Wu, S., & Li, D. (2020). Towards a next-generation production system for industrial robots: A CPS-based hybrid architecture for smart assembly shop floors with closed-loop dynamic cyber physical interactions. *Frontiers of Mechanical Engineering*, *15*(1), 1–11. DOI: 10.1007/s11465-019-0563-9

Tariq, U., & Ahmed, I. (2025). Reasoning About Responsibility in Autonomous Systems: Navigating the Challenges and Charting Future Directions. *Ubiquitous Technology Journal*, *1*(2), 46–60. DOI: 10.71346/utj.v1i2.21

Tashkinov, A. G. (2024). The implementation of lean and digital management techniques using artificial intelligence in industrial settings. *Discover Artificial Intelligence*, *4*(1), 94. DOI: 10.1007/s44163-024-00186-5

Teoh, Y. K., Gill, S. S., & Parlikad, A. K. (2021). IoT and fog-computing-based predictive maintenance model for effective asset management in Industry 4.0 using machine learning. *IEEE Internet of Things Journal*, *10*(3), 2087–2094. DOI: 10.1109/JIOT.2021.3050441

Thakar, S., Srinivasan, S., Al-Hussaini, S., Bhatt, P. M., Rajendran, P., Jung Yoon, Y., & Gupta, S. K. (2023). A survey of wheeled mobile manipulation: A decision-making perspective. *Journal of Mechanisms and Robotics*, *15*(2), 020801. DOI: 10.1115/1.4054611

Theresa, S., Nour, M., & Turnbull, B. (2025). Responsible Resilience in Cyber–Physical–Social Systems: A New Paradigm for Emergent Cyber Risk Modeling. *Future Internet*, *17*(7), 282. DOI: 10.3390/fi17070282

Tonelli, F., Demartini, M., Pacella, M., & Lala, R. (2021). Cyber-physical systems (CPS) in supply chain management: From foundations to practical implementation. *Procedia CIRP*, *99*, 598–603. DOI: 10.1016/j.procir.2021.03.080

Uddin, M. Z. (2025). Trustworthy AI and Explainability. In *Trustworthy Multimodal Intelligent Systems for Independent Living* (pp. 111–137). Springer Nature Switzerland. DOI: 10.1007/978-3-031-97359-8_4

Urrea, C. (2025). Hybrid Fault-Tolerant Control in Cooperative Robotics: Advances in Resilience and Scalability. [). MDPI.]. *Actuators*, *14*(4), 177. DOI: 10.3390/act14040177 van der Veen, G., Langelaar, M., van der Meulen, S., Laro, D., Aangenent, W., & van Keulen, F. (2017). Integrating topology optimization in precision motion system design for optimal closed-loop control performance. *Mechatronics*, *47*, 1–13. DOI: 10.1016/j.mechatronics.2017.06.003

Venice, J. A., Vettriselvan, R., Jain, S., Madusudanan, K., & Aarthy, C. C. J. (2025). Performance Evaluation and Metrics in Blockchain Powered AI/ML: Data Analytics for Cognitive Internet of Things (CIoT). In *Transforming Education With AI-Powered Personalized Learning* (pp. 143-178). IGI Global Scientific Publishing

Vettriselvan, R., Deepa, R., Gautam, R., Suresh, N. V., & Cathrine, S. (2025). Bridging Academia and Industry Through Technology and Entrepreneurial Innovation: Enhancing Supply Chain Efficiency. In *Bridging Academia and Industry Through Cloud Integration in Education* (pp. 145-174). IGI Global.

Vettriselvan, R., Velmurugan, P. R., Suresh, N. V., & Catherine, S. (2025). Strategies, Best Practices, and Pitfalls in the Era of Digital Transformation: Mastering Mobile Marketing. In *Impact of Digital Transformation on Business Growth and Performance* (pp. 67-98). IGI Global Scientific Publishing.

Vettriselvan, R. (2025). Harnessing innovation and digital marketing in the era of industry 5.0: resilient healthcare SMEs. In *The Future of Small Business in Industry 5.0* (pp. 163–186). IGI Global Scientific Publishing.

Vidyalakshmi, G., Gopikrishnan, S., Boulila, W., Koubaa, A., & Srivastava, G. (2025). Digital Twins and Cyber-Physical Systems: A New Frontier in Computer Modeling. *Computer Modeling in Engineering & Sciences*, *143*(1), 51–113. DOI: 10.32604/cmes.2025.057788

Vijayan, D. S., Sivasuriyan, A., Devarajan, P., Krejsa, M., Chalecki, M., Żółtowski, M., & Koda, E. (2023). Development of intelligent technologies in SHM on the innovative diagnosis in civil engineering—A comprehensive review. *Buildings (Basel, Switzerland)*, *13*(8), 1903. DOI: 10.3390/buildings13081903

Wang, D., & Zhang, J. (2022). Two improved scanning path planning algorithms and a 3D printing control system with circular motion controller. *Rapid Prototyping Journal*, *28*(4), 695–703. DOI: 10.1108/RPJ-08-2020-0190

Wang, L., & Wang, X. V. (2018). *Cloud-based cyber-physical systems in manufacturing*. Springer. DOI: 10.1007/978-3-319-67693-7

Wang, Y., Zhao, C., Liu, W., & Li, S. (2025). Path planning and real-time optimization of robotic arm 3D printing. *International Journal of Modeling, Simulation and Scientific Computing*, *16*(3), 2541013. DOI: 10.1142/S1793962325410132

Werbińska-Wojciechowska, S., Giel, R., & Winiarska, K. (2024). Digital twin approach for operation and maintenance of transportation system—Systematic review. *Sensors (Basel)*, *24*(18), 6069. DOI: 10.3390/s24186069 PMID: 39338814

Wu, G., Sun, J., & Chen, J. (2019). Optimal linear quadratic regulator of switched systems. *IEEE Transactions on Automatic Control*, *64*(7), 2898–2904. DOI: 10.1109/TAC.2018.2872204

Xie, B., Gao, H., & Zhang, T. (2023). A review of the control system of a 3D printer. *Recent Patents on Engineering*, *17*(1), 25–41.

Xu, H., & Rong, X. (2024). Design and application research on control system of 3D printer based on multi-DOF robotic arm. *Advances in Computer and Communication*, *5*(6).

Xu, H., Berres, A., Yoginath, S. B., Sorensen, H., Nugent, P. J., Severino, J., & Sanyal, J. (2023). Smart mobility in the cloud: Enabling real-time situational awareness and cyber-physical control through a digital twin for traffic. *IEEE Transactions on Intelligent Transportation Systems*, *24*(3), 3145–3156. DOI: 10.1109/TITS.2022.3226746

Yadav, N., Gupta, V., & Garg, A. (2024). Industrial automation through ai-powered intelligent machines—enabling real-time decision-making. In *Recent Trends in Artificial Intelligence Towards a Smart World: Applications in Industries and Sectors* (pp. 145–178). Springer Nature Singapore. DOI: 10.1007/978-981-97-6790-8_5

Yan, Y., & Kunhui, Y. (2024). Novel cyber-physical architecture for optimal operation of renewable-based smart city considering false data injection attacks: Digital twin technologies for smart city infrastructure management. *Sustainable Energy Technologies and Assessments*, *65*, 103733. DOI: 10.1016/j.seta.2024.103733

Yarlagadda, V. K. (2024). Cutting-edge developments in robotics for smart warehousing and logistics optimization. *Robotics Xplore: USA Automation Digest*, *1*(1), 61–79.

Yildirim, M., Gebraeel, N. Z., & Sun, X. A. (2017). Integrated predictive analytics and optimization for opportunistic maintenance and operations in wind farms. *IEEE Transactions on Power Systems*, *32*(6), 4319–4328. DOI: 10.1109/TPWRS.2017.2666722

Yildirim, M., Gebraeel, N. Z., & Sun, X. A. (2019). Leveraging predictive analytics to control and coordinate operations, asset loading, and maintenance. *IEEE Transactions on Power Systems*, *34*(6), 4279–4290. DOI: 10.1109/TPWRS.2019.2922380

Yu, D., & He, Z. (2022). Digital twin-driven intelligence disaster prevention and mitigation for infrastructure: Advances, challenges, and opportunities. *Natural Hazards*, *112*(1), 1–36. DOI: 10.1007/s11069-021-05190-x PMID: 35125651

Yu, D., Li, J., Wang, Z., & Li, X. (2023). An overview of swarm coordinated control. *IEEE Transactions on Artificial Intelligence*, *5*(5), 1918–1938. DOI: 10.1109/TAI.2023.3314581

Zhang, K., Shi, Y., Karnouskos, S., Sauter, T., Fang, H., & Colombo, A. W. (2022). Advancements in industrial cyber-physical systems: An overview and perspectives. *IEEE Transactions on Industrial Informatics*, *19*(1), 716–729. DOI: 10.1109/TII.2022.3199481

Zhang, L., Pan, Y., Wu, X., & Skibniewski, M. J. (2021). *Artificial intelligence in construction engineering and management* (Vol. 163). Springer. DOI: 10.1007/978-981-16-2842-9

Zhang, Y., & Abellera, J. U. (2025). Autonomous mobile robotics in smart warehousing: A cyber-physical systems approach to inventory management. *Future Technology*, *4*(4), 59–71. DOI: 10.55670/fpll.futech.4.4.6

Zhang, Y., Joneurairatana, Y. E., & Vongphantuset, J. (2024). An AI-Driven Decision Support Framework for Ergonomic Optimization in Fashion Manufacturing: Integrating Predictive Analytics and MCDM Techniques. *Decision Making: Applications in Management and Engineering*, *7*(1), 786–802.

Zhao, Z., Zhang, M., Wu, W., Huang, G. Q., & Wang, L. (2024). Spatial-temporal traceability for cyber-physical industry 4.0 systems. *Journal of Manufacturing Systems*, *74*, 16–29. DOI: 10.1016/j.jmsy.2024.02.017

Zhou, R., Jiang, J., Li, P., Lu, S., Li, B., & Du, C. (2025). Optimized compound control strategy for high-speed 3D printing: Enhancing surface quality and printing precision. *Journal of the Brazilian Society of Mechanical Sciences and Engineering*, *47*(8), 47. DOI: 10.1007/s40430-025-05684-w

About the Contributors

R. Vettriselvan, an Associate Professor at AMET University, specializes in HRM and Marketing. He is acting as a mentor at Saraswathi Institute of Medical Sciences, Hapur. He served as a Review Board member of the National Council for Higher Education, Malawi; Head of the Department of Commerce and Management Studies at St. Eugene University, Zambia; and Director of Research and Publication at St. John the Baptist University, Malawi. He has published 30 books and 100+ articles in peer-reviewed journals and edited book chapters. Under his guidance, two have completed and seven are pursuing their PhD.

N. V. Suresh is a distinguished academician from Chennai, Tamil Nadu. He completed his B.A. from D. G. Vaishnav College (Madras University, 2012), followed by an MBA and PGDM from Bharathiar University, and a Ph.D. in Management from the University of Madras. With over a decade of academic experience, he has served at IIKM, Remo International College, and AMET University, and currently holds the position of Associate Professor and Vice-Principal at ASET College of Science and Technology. He has published 30 international papers, authored two textbooks, guided Ph.D. scholars, completed major funded projects, and received the 2023 Lions International Best Faculty Award.

***

A. Deepan is an accomplished academic and management professional with over eight years of teaching experience across India, Central Africa, and Central Asia. Currently serving as Assistant Professor at Sambhram University, Uzbekistan, he has previously held faculty roles at Ananda College, India, and DMI St. John the Baptist University, Malawi. His expertise spans business, marketing, finance, and strategic management, supported by multiple research publications and a patent in financial data management. With prior industry leadership roles in insurance, batteries, and trading, he blends academic rigor with practical insight, fostering student success, innovation, and institutional growth.

Priscila A, Saveetha School of Management, Saveetha Institute of Medical and Technical Sciences, Saveetha University, Chennai, Tamil Nadu

Sadeeshkumar Arumugam is an Assistant Professor at St. Joseph's College of Engineering, OMR, Chennai, India, where he has been serving since 2014 in the Departments of Electronics & Instrumentation Engineering and Electrical & Electronics Engineering. He is currently pursuing his Doctorate at Anna University, Chennai, specializing in Power Electronics and Renewable Energy Systems. He holds a Postgraduate degree in Power Electronics and Drives from SSN College of Engineering (2014) and a Bachelor's degree in Electronics and Instrumentation Engineering from St. Peter's Engineering College (2011). His teaching and research expertise spans power electronic converters, inverters, and renewable energy systems, with multiple publications in these areas. He has also made significant contributions to biomedical instrumentation, the Internet of Things (IoT), and embedded systems design, where he has guided student teams that have twice won the Indian National Academy of Engineering (INAE) awards. Several of these works have been extended into patent applications with potential industrial impact. Mr. Sadeeshkumar's scholarly contributions include works published in Scopus-indexed journals and conferences, such as: Advanced Deep Learning Framework for Early Breast Cancer Detection with Graph Convolutional Networks (ICICACS 2025) Advanced Wireless Charging Solutions for Military Armored Vehicles (ICICCS 2025) A Study on the Inspection of Fundus Retinal Picture (Springer AISC, 2020) Wireless Power Transmission for Medical and Other Applications (IJITEE, 2019) Examination of Glioblastoma Images by Thresholding Using Heuristic Approach (ICBSII, 2018) Optimal Multilevel Image Thresholding to Improve the Visibility of Plasmodium sp. in Blood Smear Images (Springer, 2016) Segmentation of Noise Stained Gray Scale Images with Otsu and Firefly Algorithm (IJST, 2016) He is a Lifetime Member of the Indian Society for Technical Education (ISTE) and a Life Member of the International Association of Engineers (IAENG), Hong Kong, reflecting his commitment to professional growth, technical education, and global collaboration. His recent research interests include robotics, manipulators, and their practical deployments, aligning with the evolving needs of industry and interdisciplinary innovation.

Duggirala Aravind is currently associated with the Department of Aviation, bearing the registration number 249207990030, at ASET Institute of Science and Technology, Chennai, India. He previously completed his Bachelor's degree in

Electrical and Electronics Engineering at MITS University. His research interests include aircraft maintenance engineering, aviation safety management, renewable energy technologies in air transport, and sustainable energy systems. (aravindaset.edu@gmail.com).

Rajani Balakrishnan is a Senior Lecturer II at INTI International University, Nilai, specializing in Information Technology and Business Information Systems. She holds a PhD in IT from Infrastructure University Kuala Lumpur and has over 15 years of teaching, research, and academic leadership experience across Malaysian higher education institutions. Her research focuses on AI adoption, green technologies, HRM practices, telehealth, and sustainable digital transformation, with numerous publications in Scopus-indexed journals. Dr. Rajani is also active in supervising postgraduate research, presenting at international conferences, and contributing to community service initiatives, reflecting her commitment to scholarship and societal impact.

K. Prakash is an Assistant Professor (Selection Grade) in the College of Management, SRM Institute of Science and Technology, Ramapuram Campus. He holds a Ph.D. in Management from Bharathiar University and is UGC-NET certified in Management. With academic qualifications spanning management, psychology, and corporate secretaryship, his teaching focuses on Integrated Marketing Communication, Consumer Behavior, and Legal and Business Environment. His research interests lie in marketing communication, rural marketing strategies, and consumer perception. He has published extensively in reputed journals and edited volumes and has presented papers at national and international conferences across India.

Anitha Rexalin Devaraj is an Associate Professor in the Department of Physics at AMET University, Chennai, and currently serves as the Director of the Institution's Innovation Council (IIC). She holds M.Sc., M.Phil., and Ph.D. degrees in Physics, with research specializations in Quantum Computations, Density Functional Theory (DFT) analysis, and Vibrational Spectroscopy. With over eight years of teaching and research experience, she handles core physics and electronics courses for engineering and nautical science programs. Her research contributions include studies on molecular structure investigations, heterocyclic compounds, and nanocomposites, published in reputed journals indexed in ResearchGate and Google Scholar. She has participated in numerous conferences, is a member of professional scientific bodies, and actively promotes innovation and entrepreneurship through her leadership in AMET's Innovation Council.

E P John (Retd), BE, MBA, PhD, Assistant Professor (Sr G), Department of Management Studies, SRM Valliammai Engineering College, Kattankulathur – 603203, Tamilnadu, is a passionate teacher, Tamilnadu SET and UGC NET qualified, a fellow of IEI (India), a PoSH consultant, with two decades of Industrial experience (Indian Army), certified in HR analytics, Design Thinking, UHV and related management and technical areas. He always incorporates technology in his teaching and research methodologies, with subject expertise in Quantitative techniques, Operations Research, Bio-statistics, Management concepts & principles, Research methodology, Strategic Management, Human resource management, and general management. Has quality publications in Scopus, ABDC, and UGC care on his account alongside authoring textbooks in management and he improves every day towards a balanced human being in service of the student community and higher education.

J. Herbert Sinduja, M.E., Ph.D., is an Assistant Professor (OG) in Civil Engineering with over 11 years of teaching experience. A specialist in recycled aggregate concrete, optimization techniques, microstructural characterization, and statistical analysis, she has published seven journal papers with an H-index of 4. She received the prestigious Anna Centenary Research Fellowship (2021–2023) and has presented her research internationally, including at the MDPI Conference on Advanced Materials in Barcelona. Actively engaged in professional development, she has completed several NPTEL and Coursera certifications and is a member of the American Society of Civil Engineers.

Sridevi J is an Associate Professor in the Department of Commerce with over 26 years of teaching experience at both undergraduate and postgraduate levels. She holds multiple qualifications, including MBA, M.Com., M.Phil. (Commerce), and Ph.D. (Management), and has successfully completed four NPTEL certification courses. Her areas of expertise span commerce and management, with a strong focus on integrating interdisciplinary perspectives into business education. Dr. Sridevi has contributed significantly to curriculum design, academic regulations, and quality enhancement initiatives in higher education. She has published numerous research papers and book chapters in reputed journals and edited volumes, including nine publications indexed in Scopus. In addition to teaching, she actively mentors students, coordinates academic programs, and plays a key role in institutional development activities.

K. Selvi serves as Associate Professor in MBA Department at Saveetha Engineering College, bringing 24 years of academic excellence to her role. With a Ph.D. in Finance from Sri Venkateswara University and multiple master's degrees including MBA (Gold Medalist), M.Com, M.Phil, and M.F.M, she has established herself as a distinguished educator and researcher. Dr. Selvi is a recognized research supervisor under University of Madras with 3 completed Ph.D. scholars and one in progress. Her research portfolio includes 70 publications in reputed journals, including 12 Scopus-indexed papers focusing on finance, management, and AI applications. Dr. Selvi has secured 5 patents, authored 3 books, and received numerous awards including "Senior Woman Educator and Scholar Award" and "Best Professor Award - Pride of Education" (2024). As NBA Coordinator, she successfully led the MBA program to achieve 3-year accreditation (2024-2027) while also serving as Chairperson for the Women Empowerment Cell, Served as Director for Business Incubation Centre.

M. Durgarani MBA.,M.Phil., Ph.D., working as an Associate Professor in the School of Business and MAnagement, Christ University, Bangalore.

Manimaran M is a dynamic HR professional with experience in recruitment, teaching, and organizational development. He has worked as an HR Recruiter and now working as an Assistant Professor at Jaya College of Arts and Science, Thiruninravur, Chennai, mentoring students in HRM and business ethics. He holds an MBA degree in Anna University and a Bachelor's degree in Hotel Management in University of Madras. His achievements include a publication on adaptive leadership in Future-Forward Commerce and Management. He is passionate about aligning HR strategies with business goals to drive growth and foster positive work culture.

M.Nasrin Sulthana, Assistant professor SG School of Management Hindustan Institute of Technology and Science Padur, Chennai

Vijayakumar. M is an academician, trainer, and researcher with extensive experience in teaching, training, and student development. He is currently serving as a Professor in the Department of Management Studies at Panimalar Engineering College, Chennai, affiliated with Anna University. With strong expertise in communication, corporate engagement, and talent development, he has dedicated his career to preparing students for professional success. Over the years, he has coordinated numerous campus recruitment drives, collaborated with leading organizations, and mentored students to enhance their employability, confidence, and career readiness. His teaching and research interests span across communication skills, behavioral sciences, employability studies, and management training. He has authored and presented several research papers on contemporary issues including skill versus degree in employability, communication apprehension in campus placements, and the role of mind set and motivation in student success. As a trainer and mentor, Dr. Vijayakumar. M has designed impactful programs on soft skills, leadership, personality development, and corporate preparedness, which have benefitted thousands of engineering and management students. He is particularly recognized for his ability to connect with students, simplify complex concepts, and inspire them to pursue their goals with clarity and confidence. In addition to his academic pursuits, he is a passionate writer and communicator, deeply committed to exploring the evolving relationship between education, employability, and human behavior. His work continues to focus on empowering young minds with the skills and mind set needed to thrive in a competitive global environment.

P.Praba Devi is an Associate Professor in the Sona School of Management. She has completed her Doctoral research with 23+ years of experience in academic and industry. She teaches marketing courses for the students and her area of interest is Service Quality, marketing and consumer behavior. She is an approved supervisor by the Anna University to guide research scholars pursuing Doctoral Program. Besides attending several conferences, seminars and training programs she has also organized national seminar, faculty development programs and workshops. She has published articles in several Scopus/ WoS indexed journals, refereed journals and journals with high impact factor. She has also reviewed the books published by leading publishers. She has been a member of organising committee for few International conference. She is a member of the team preparing the Human Development Report and the State Balanced Growth Fund report for the State Planning Commission.

R.Akila working as Professor and Head, Department of Management Studies, Jeppiaar Engineering College Chennai, Tamil Nadu, India

R.Divya Ranjani Assistant Professor Department of Management Studies Saveetha Engineering College Thandalam

Gokilavani R, Background: Area of interest align with Finance, Marketing and. General management. With 15+ years of teaching experience, international exposure, research, training & consulting experience, contributed effectively for the development of the student's community and institution at large. Published more than 25 articles in reputed Journals including SCOPUS, WOS and UGC listed Category. Received Best Faculty award in my previous three institutions for Self and Students Development Various positions held in my previous experience includes Head-Student Development, Career Counselling Associate, Curriculum Development committee member, Research coordinator, Lead-Trainer and Student Association, Club and accreditation co ordinator -NBA,NAAC etc, Corportae Consultancy and Training Experience: Extended consultancy service on Bespoke finishing program in Sales Management for ITC's South India Sales trainees during 2019 . Training

R. Deepika working as Assistant Professor in Rajalakshmi School of Business, Chennaii, Tamil Nadu

Delecta Rajendren, Assistant professor, EDP in Talent Management at XLRI Jamshedpur, an experienced researcher, and first-time initiator in her university to assess and map the Cross-cultural competence of the middle-level professionals, working in the footwear industry in India and China. Extensively traveled in China and examined the cross-cultural behavior, and working environment, co-established the presence, while presenting, and attended International conferences across the continents. A Sr. HR professional with over 9+years of working experience, with an array to manage the various roles, handling the HR specimen roles, to giving an exclusivity, as starting the 'Campus Placement' drive, during the academic head for the MBA department in the J.M.C.E institution. Extensively researched over the years in Cross-Cultural competence, published papers, presented at various International and National conferences, and co-authored the book on Managerial Behavior and Effectiveness. An experienced soft-skill trainer, established a reputation in conducting the mock-drills sessions, for the upcoming professionals; to handling the grievance addressal systems in the manufacturing unit

M. Daniel Rajkumar is an Assistant Professor in the Directorate of Online and Distance Education at SRM Institute of Science and Technology, Chennai, with more than twenty years of academic experience. He holds a Ph.D. in Finance. Dr. M. Daniel Rajkumar's research interests span financial markets, banking, and strategic management, and he has published extensively in Scopus and ABDC indexed journals. His recent article on contemporary retail services appeared in the Northern Economic Review. He serves as Editor of the peer-reviewed ASET Journal of Management Science and has guided over 600 MBA projects. Recognized with the 2024 Inspirational Mentor Award, he actively contributes to curriculum development, academic audits, and professional mentoring.

Nithya.R - Accomplished Professor, PhD. M.Phil., M.Com, with over 16 years of Experience in Education and Research in the areas of accounting, finance and entrepreneurship. Team member, Quality Assurance department working for OAAA accreditation.

Palanivel Rathinasabapathi Velmurugan is associated with BSBI, Berlin as Lecturer in Human Resource Management & Finance Stream. Prior to joining BSBI, he held the designation "Head" Faculty of Management in Sharda University, Uzbekistan, Central Asia and "Dean of Arts" assignments in DMI St John the Baptist University, Malawi, Central Africa. Also, he has associated earlier with the DMI St Eugene University in Zambia, MAM Business School and various Engineering

Colleges in India. He has done his Ph.D. in Management from Mannonmanium Sundaranar University in India. Besides this, he holds an MBA Degree from SRM University in India and a BE Chemical Engineering degree from Annamalai University in India. Also, he completed six NPTEL Certificate Programs conducted by various IITs in India. He brings with him 16 years of rich working experience in teaching and research. Furthermore, he has a keen interest in the Case Method of teaching and research. He is an academic out of passion. Not only that, but he also takes keen interest in guiding Ph.D. students. He has supervised three Ph.D. research scholars under his supervision at DMI St Eugene University in Zambia.

Catherine.S, is currently working as an Associate professor in Saveetha Engineering College, Thandalam, Chennai. She has specialized in Marketing and Finance, and has been in to academics for over a decade. She did her Ph. D in Management from Sathyabama Institute of Science and Technology, Chennai. Published a Patent Office Journal No. 16/2023 in titled as "instigating machine learning methodology to review monetary advisory records in financial sector", in Application No.202341021851 A, 21/04/2023. She had published over 30 research papers in the journals that are indexed in Scopus, Web of Science, ABDC and UGC Care list, and published book chapters in the IGI Global publisher and national publisher. She was acted as convenor for International Conference and acted as session chair for conferences.

Muthumani S, Professor and Head, Department of Management Studies, Jerusalem College of Engineering, Chennai India

Michael Savariapitchai, Associate Professor, Datta Meghe Institute of Higher Education and Research, Wardah, Maharastra have educational and organizational leader offering 15 years of career excellence in university academic and administrative roles. Demonstrates in-depth knowledge of community involvement methods. Esteemed educator and communicator proficient in management Studies instruction and team-building exercises across Students and staff.

Muthuraj Subramani is working as an Assistant Professor in Department of Information Technology, Sanjivani College of Engineering, Kopargaon, Maharastra. He received his Bachelor`s Degree in Information Technology, Master's Degree in Computer Science and Engineering from Anna University, Chennai. He has 15 years of Teaching, Industry & Research experience. His major areas of interest include Artificial Intelligence and Machine Learning, Data Analytics, Digital Twin, . He holds membership in various professional bodies like ISTE,CSI, . He has published papers in referred National and International Conferences and Journals. He has published 2 patents and holds a role of Reviewer & Editorial board member in various International Journals & Conferences. He has delivered more than 30 guest lectures. . He has participated in more than 50 National FDP`s/ Seminars / Workshops and also organized programmes sponsored by agencies like DST, DRDO, CSIR and TNSCST.

V. Sangeetha Albin, an accomplished academic and administrator, holds a Ph.D. in Management from the University of Madras, an M.B.A. in Shipping and Logistics, and a B.E. in Electronics and Electrical Engineering. Formerly South India Manager at Kuehne-Nagel (Germany), she has served as Director of Administration and Joint Registrar. Currently, she is Additional Registrar (DGS Programs), overseeing Admissions, Placements, Finance, and Purchasing.

Selvalakshmi V is an academician, educator, researcher, POSH Trainer and mentor currently serving Professor, Saveetha School of Management, Saveetha Institute of Medical and Technical Sciences,SIMATS, Chennai. With more than two decades of rich experience, her experience and interest likes in the fields of research, incubation, entrepreneurship and have published more than 35 research papers and 5 books. She has served as one of the Expert Committee member for course development at NAHEP – IPD, Dapoli and have reviewed a full book under Taylor and Francis Publication.

V.P.Rameshkumar is Associate Professor and Public Liaison Officer of Sona College of Technology. He has 11 years of experience in industry and academia. His teaches Logistics and Supply chain management, Business Process Reengineering, World Class Manufacturing, Service Operations Management and Enterprise Resource Planning to MBA graduates. For engineering graduates, he has taught Principles of Management and Total quality management. Being a part of Industry Institute Interface (3i) organized various FDPs, MDPs, Workshops, Trainings, Seminars and Conferences. Belonging to the team of Placements, assisted in Campus Recruitment, Students Training & Development. Expertise in providing consultation to industry clients in the areas of Market Research and Surveys.

Savitha Venkatesan received the B.E. degree in Electrical and Electronics Engineering from Madras University, Chennai, India, in 2003, and the M.E. degree in Power Systems Engineering from the College of Engineering, Guindy, Anna University, Chennai, India, in 2007. She completed her Ph.D. degree in 2025 from Anna University, Chennai, India. She has over 12 years of teaching experience. Her research interests include power electronic converters, renewable energy sources, and control systems.

Index

Symbols

A

B

C

D

N

O

P

R

S

T

W

www.ingramcontent.com/pod-product-compliance
Lightning Source LLC
LaVergne TN
LVHW080904300926
849162LV00003B/552